# MECHANICS OF MATERIALS

# MECHANICS OF MATERIALS

**David Roylance**

JOHN WILEY & SONS, INC.
New York    Chichester    Brisbane    Toronto    Singapore

ACQUISITIONS EDITOR   Charity Robey
MARKETING MANAGER   Debra Riegert
SENIOR PRODUCTION EDITOR   Cathy Ronda
PRODUCTION ASSISTANT   Raymond Alvarez
DESIGNER   David Levy
MANUFACTURING MANAGER   Dorothy Sinclair
PHOTO RESEARCHER   Mary Ann Price
ILLUSTRATION COORDINATOR   Jamie Perea

This book was set in Times Roman by Publication Services and
printed and bound by Hamilton Printing. The cover was printed by Phoenix Color.

Recognizing the importance of preserving what has been written, it is a
policy of John Wiley & Sons, Inc. to have books of enduring value published
in the United States printed on acid-free paper, and we exert our best
efforts to that end.

The paper on this book was manufactured by a mill whose forest management programs include
sustained yield harvesting of its timberlands. Sustained yield harvesting principles ensure that
the number of trees cut each year does not exceed the amount of new growth.

ISBN 0-471-59399-0

Printed in the United States of America

10  9  8  7  6  5  4  3  2  1

# PREFACE

This text is intended to meet the needs of undergraduate courses in mechanics of materials, required of students in many engineering curricula. It is written primarily for use as an introduction to the area, but, depending on which topics are emphasized, it can be used in more advanced levels as well.

Mechanics of materials has been taught pretty much the same way for many years, using topics and methods described beautifully over 50 years ago in classic texts by Stephen Timoshenko. There are now many texts based on his approach, which emphasizes deriving formulas and working problems involving stresses and deformations in simple structures built with linear elastic materials. This text retains many of these traditional topics, which are fundamental to engineering practice. However, this text is a definite departure from the Timoshenko approach and should be considered by instructors willing to consider something a little different.

I have included a number of topics not found in most mechanics of materials texts, such as time-dependent effects and fracture phenomena. Such coverage is needed because today's engineers use many methods and materials not treated by the traditional texts. Beyond these "modern" topics, university instruction is most effective and interesting when it includes at least some coverage of "context" material. Several topics in this text, such as finite element methodology and statistical aspects of materials properties, may be scheduled as separate subjects later in the engineering curriculum. I feel that an introductory overview of these topics is valuable both for students who do not study these subjects later and for those who do. It makes the subjects easier to understand if they are taken later, and it helps put the traditional, introductory material in its proper context even for those who do not study the more advanced subjects.

In what is probably the most striking departure from traditional texts, I have sought to relate the mechanics of materials to the chemistry and microstructure of modern engineered materials, building on materials topics in chemistry and other subjects in the engineering science core curriculum. An appreciation of the relations between a material's processing, microstructure, properties, and eventual performance is extremely useful in mechanical design. Intuition is the designer's most important tool, and an important objective in teaching mechanics of materials is to help the student develop a sense of how the materials actually respond to mechanical loading. Coverage should encompass the materials in use today, to include those with appreciable time dependency and anisotropy in their mechanical response, in addition to the linear elastic materials usually studied. Formal concepts in stress and strain are important, but so is an appreciation of the molecular and microscopic response of materials to these stresses. This is my natural bias as a professor of materials engineering, but I am convinced that this mindset is useful to all engineers working with materials.

Today's design engineers use computer software for much of their work, and this text takes advantage of symbolic manipulation and spreadsheet methods in addition to programmed code. (Several examples employ the MAPLE™ package, which is available at steep student discounts on many campuses.) The computer allows the student to sidestep many tedious manipulations that have only marginal educational value and to see more easily the underlying mechanics. The programmed codes for stress transformations and finite element calculations included in the text diskette fall short of commercial standards in graphics-based user friendliness, but will allow the student to begin to acquire expertise in numerical modeling.

I have tried to describe the various physical concepts simply and at some length initially, and then gradually generalize the treatment both geometrically and mathematically. The issue of mathematical level and notation is an important one, as the more concise methods found in most of the professional literature tend to be daunting to the beginning student. This is true even of students with good mathematical skills, since the problem in most engineering subjects is not so much carrying out the various mathematical operations—such as algebraic manipulations and differentiation—as grasping mathematics as a *descriptive language* for engineering problems.

However, it is vital that students become comfortable with this language, just as medical students must grasp the jargon of their field to become effective physicians. I have made greater use of vector and matrix notation than in traditional texts, out of the belief that the extra abstraction is more than compensated by the value of immersion in the proper technical language. Further, the geometrical aspects of mechanics problems can lead to lengthy, tedious expressions that tend to distract the reader when only scalar algebraic methods are used, in spite of their being more direct.

The mathematical level of the text is not overly demanding. Most of the topics and problems require only algebraic manipulation and numerical evaluation, although a few topics—such as polymer viscoelasticity and fracture theory—employ mathematical techniques that are not needed in traditional mechanics of materials texts. Laplace transforms and convolution integrals are examples. However, these techniques are included in most first subjects in differential equations, which the student is assumed to have taken before or perhaps concurrently with this subject. These are not difficult techniques, especially if used in conjunction with symbolic manipulation software, and neither students nor instructors should shy away from them.

The text begins with a survey of the response of materials to simple tension. Here the geometry is simple, so that the basics of stress and strain and materials response can be described with scalar expressions. This permits a broad range of response to be covered—including energy-controlled elasticity of metals and ceramics, entropic elasticity of rubber, viscoelasticity of leathery polymers, and anisotropy of composite materials—leading to a walk along the stress-strain curve to preview the ways materials respond to increasingly higher stresses. This sets the stage for coverage in subsequent chapters of the structural types used in most mechanical designs: trusses, pressure vessels, torsion rods, beams, plates, and laminates. As the dimensionality increases, appropriate mathematical expressions and notational forms are introduced and used to preserve the basic simplicity of the concepts. After the mathematical foundations of solid mechanics have been established through this step-by-step process, an overview of closed-form, experimental, and finite element stress analysis can be presented in a natural and succinct way.

The final two chapters of the book describe the response of materials to stresses beyond the elastic limit. Yield and plastic flow are very materials dependent, and the engineer has access to a wide variety of materials processing methods to tailor and optimize the yielding process. This is one of the common grounds of mechanical and materials engineers, and is of great practical as well as theoretical value. The final chapter treats modern methods of designing for fracture and fatigue, including the elements of statistical variability, fracture mechanics, and time-dependent damage accumulation.

# CONTENTS

# LIST OF SYMBOLS

| | |
|---|---|
| $A$ | Area; free energy; Madelung constant |
| $\mathbf{A}$ | Transformation matrix |
| $\mathscr{A}$ | Plate extensional stiffness |
| $a$ | Length; transformation matrix element; crack length |
| $a_T$ | Time-temperature shifting factor |
| $B$ | Design allowable for strength |
| $\mathbf{B}$ | Matrix of derivatives of interpolation functions |
| $\mathscr{B}$ | Plate coupling stiffness |
| $b$ | Width; thickness |
| $\overline{b}$ | Burgers' vector |
| $C$ | Stress optical coefficient; compliance |
| $\mathscr{C}$ | Viscoelastic compliance operator |
| $c$ | Numerical constant; length; speed of light, wavespeed |
| C.V. | Coefficient of variation |
| $\mathbf{D}$ | Stiffness matrix; flexural rigidity of plate |
| $\mathscr{D}$ | Plate bending stiffness |
| $d$ | Diameter; distance; grain size |
| $E$ | Modulus of elasticity; electric field |
| $E^*$ | Activation energy |
| $\mathscr{E}$ | Viscoelastic stiffness operator |
| $e$ | Electronic charge |
| $e_{ij}$ | Deviatoric strain |
| $F$ | Force |
| $f_s$ | Form factor for shear |
| $G$ | Shear modulus |
| $\mathscr{G}$ | Viscoelastic shear stiffness operator, strain energy release rate |
| $\mathscr{G}_c$ | Critical strain energy release rate |
| $g$ | Acceleration of gravity |
| GF | Gage factor for strain gages |
| $H$ | Brinell hardness |

| | |
|---|---|
| $h$ | Depth of beam |
| $I$ | Moment of inertia; stress invariant |
| $\mathbf{I}$ | Identity matrix |
| $J$ | Polar moment of inertia |
| $K$ | Bulk modulus; global stiffness matrix; stress intensity factor |
| $\mathcal{K}$ | Viscoelastic bulk stiffness operator |
| $k$ | Spring stiffness; element stiffness; shear yield stress; Boltzmann's constant |
| $L$ | Length, beam span |
| $\mathbf{L}$ | Matrix of differential operators |
| $\mathcal{L}$ | Laplace transform |
| $M$ | Bending moment |
| $N$ | Crosslink or segment density; moiré fringe number; interpolation function; cycles to failure |
| $\mathbf{N}$ | Traction per unit width on plate |
| $N_A$ | Avogadro's number |
| $\mathcal{N}$ | Viscoelastic Poisson operator |
| $n$ | Refractive index; number of fatigue cycles |
| $\hat{\mathbf{n}}$ | Unit normal vector |
| $P$ | Concentrated force |
| $P_f$ | Fracture load; probability of failure |
| $P_s$ | Probability of survival |
| $p$ | Pressure; moiré gridline spacing |
| $Q$ | Force resultant; first moment of area |
| $q$ | Distributed load |
| $R$ | Radius; reaction force; strain or stress rate; gas constant; electrical resistance |
| $\mathbf{R}$ | Reuter's matrix |
| $r$ | Radius; area reduction ratio |
| $S$ | Entropy; moiré fringe spacing; total surface energy; alternating stress |
| $\mathbf{S}$ | Compliance matrix |
| $s$ | Laplace variable; standard deviation |
| SCF | Stress concentration factor |
| $T$ | Temperature; tensile force; stress vector; torque |
| $\mathbf{T}$ | Traction vector |
| $T_g$ | Glass transition temperature |
| $t$ | Time; thickness |
| $t_f$ | Time to failure |
| $U$ | Strain energy |
| $U^*$ | Strain energy per unit volume |
| $\mathbf{u}$ | Displacement vector |
| $\tilde{u}$ | Approximate displacement function |
| $V$ | Shearing force; volume; voltage |
| $V^*$ | Activation volume |
| $v$ | Velocity |

| | |
|---|---|
| $W$ | Weight; work |
| $u, v, w$ | Components of displacement |
| $x, y, z$ | Rectangular coordinates |
| $X$ | Standard normal variable |
| $\alpha, \beta$ | Curvilinear coordinates |
| $\alpha_L$ | Coefficient of linear thermal expansion |
| $\gamma$ | Shear strain; surface energy per unit area; weight density |
| $\Delta V$ | Change in volume |
| $\delta$ | Deflection, Dirac function |
| $\delta_{ij}$ | Kronecker delta |
| $\epsilon$ | Normal strain |
| $\boldsymbol{\epsilon}$ | Strain pseudovector |
| $\epsilon_{ij}$ | Strain tensor |
| $\boldsymbol{\epsilon}_T$ | Thermal strain vector |
| $\eta$ | Viscosity |
| $\theta$ | Angle; angle of twist per unit length |
| $\boldsymbol{\kappa}$ | Curvature |
| $\lambda$ | Extension ratio, wavelength |
| $\nu$ | Poisson's ratio |
| $\rho$ | Mass density; electrical resistivity; radius of curvature |
| $\Sigma_{ij}$ | Distortional stress |
| $\sigma$ | Normal stress |
| $\boldsymbol{\sigma}$ | Stress pseudovector |
| $\sigma_{ij}$ | Stress tensor |
| $\sigma_e$ | Endurance limit |
| $\sigma_f$ | Failure stress (ultimate tensile strength) |
| $\sigma_m$ | Mean stress |
| $\sigma_t$ | True stress |
| $\sigma_Y$ | Yield stress |
| $\tau$ | Shear stress; relaxation time |
| $\phi$ | Airy stress function |
| $\xi$ | Dummy length or time variable |
| $\Omega$ | Configurational probability |
| $\omega$ | Angular frequency |
| $\nabla$ | Gradient operator |

# TENSILE RESPONSE OF MATERIALS

## 1.1 INTRODUCTION

The progress of human development is often reckoned in terms of the materials used by society, with the Stone Age in prehistory and progressing through bronze, iron, and successively more sophisticated materials. Early people used materials largely as they were produced by nature, but today we have come to rely increasingly on *engineered* materials: materials we have optimized for various applications by careful selection and processing. For centuries the relations between processing and properties were developed and used empirically, but this approach becomes cumbersome and expensive as requirements grow increasingly complex. As can be depicted by the tetrahedron shown in Fig. 1.1, modern materials science and engineering deals with a general rational approach to materials optimization: A material's *processing* is chosen so as to develop a desired *microstructure;* the microstructure controls *properties;* and finally, the material's properties are important in dictating the *performance* of the structure in which the material is used.

Most engineering designs involve selection and manipulation of materials, and many of these designs are largely or partly *mechanical:* They require that the resulting structure support applied loads without fracture or excessive deformation. This requires that the designer be able to determine the magnitude and direction of internal forces that may cause rupture or slippage of molecular bonds and to provide enough material of suitable strength to ensure that these events do not occur. The techniques required to accomplish these tasks constitute the major part of the engineering science base in mechanical design.

As we study mechanics of materials, we must keep in mind that real designs must satisfy a number of criteria in addition to mechanical reliability. Cost is almost always very important: Material costs money, and the designer must use only enough material to satisfy the strength requirements. Other important criteria might include thermal and electrical requirements; toxicity and flammability of the material, which could endanger both the user and manufacturing personnel; and the environmental impact associated with both use and eventual disposal. At its core, however, the design problem for load-bearing structures involves ensuring the mechanical integrity of the material, and this aspect of the design process is the major goal of this text.

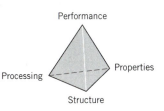

**FIGURE 1.1** Modern materials science and engineering.

Analysis or design problems in the mechanics of materials generally involve two major areas:

- Determination of the internal forces set up within the material by loads or displacements imposed on it; this is a largely mathematical undertaking, termed *stress analysis*. These internal forces are often independent of the choice of material used in the structure, and it is often possible to carry out this analysis without much specific knowledge of the material itself. This is the *mechanics* in "mechanics of materials."
- Understanding the material's response to these internal forces. The material may stretch or distort, this deformation may be reversible or permanent, or the material may fracture in any of several ways. This part of the problem is most certainly materials-specific; it is the *materials* in "mechanics of materials."

This chapter will outline some of the basic concepts underlying both of these aspects. This will be done by describing the internal force distribution set up by a simple tensile load and how materials can respond to these forces. Subsequent chapters will extend these concepts to geometrically more complicated situations and gradually introduce the mathematical language used by the literature of the field to describe them.

## 1.2 TENSILE STRENGTH AND TENSILE STRESS

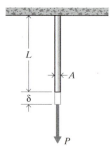

**FIGURE 1.2**
The tension test.

Perhaps the most natural test of a material's mechanical properties is the *tension test,* in which a strip or cylinder of the material, having length $L$ and cross-sectional area $A$, is anchored at one end and subjected to an axial load $P$, acting along the specimen's long axis, at the other (see Fig. 1.2). As the load is increased gradually, the axial deflection $\delta$ of the loaded end will increase also. Eventually the test specimen breaks or does something else catastrophic, often fracturing suddenly into two or more pieces. (Materials can fail mechanically in many different ways; for instance, recall how blackboard chalk, a piece of fresh wood, and bouncing putty break.) As engineers we naturally want to understand such matters as how $\delta$ is related to $P$ and what ultimate fracture load we might expect in a specimen of different size than the original one. As materials technologists we wish to understand how these relationships are influenced by the constitution and microstructure of the material.

One of the pivotal historical developments in our understanding of material mechanical properties was the realization that the strength of a uniaxially loaded specimen is related to the magnitude of its *cross-sectional area*. This notion is reasonable when we consider the strength to arise from the number of chemical bonds connecting one cross section with the one adjacent to it, as depicted in Fig. 1.3, where each bond is visualized as a spring with a certain stiffness and strength. Obviously, the number of such bonds will increase proportionally with the section's area. The axial strength of a piece of blackboard chalk will therefore increase as the *square* of its diameter. In contrast, increasing the *length* of the chalk will not make it stronger (in fact it will likely become weaker, because the longer specimen will be statistically more likely to contain a strength-reducing flaw).

Galileo (1564–1642)[1] is said to have used this observation to note that giants, should they exist, would be very fragile creatures. Their strength would be greater

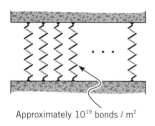

Approximately $10^{19}$ bonds / m$^2$

**FIGURE 1.3** Interplanar bonds.

---

[1]Galileo, *Two New Sciences,* English translation by H. Crew and A. de Salvio, The Macmillan Co., New York, 1933. Also see S.P. Timoshenko, *History of Strength of Materials,* McGraw-Hill, New York, 1953.

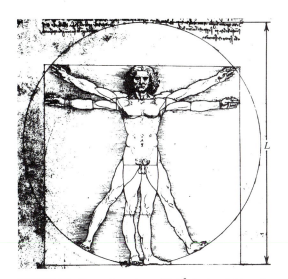

**FIGURE 1.4** Strength scales with $L^2$, but weight scales with $L^3$. (Bettman Archive.)

than ours, because the cross-sectional areas of their skeletal and muscular systems would be larger by a factor related to the square of their height (denoted $L$ in Fig. 1.4). On the other hand, their *weight*, and thus the loads they must sustain, would increase as their volume—that is, by the *cube* of their height. A simple fall would probably do them great damage. Conversely, the "proportionate" strength of the famous arachnid mentioned weekly in the *Spider-Man* comic strip is mostly just this same size effect. There's nothing magical about the muscular strength of insects, but the ratio of $L^2$ to $L^3$ works in their favor when strength per body weight is reckoned (see Fig. 1.5). This cautions us that simple scaling of a previously proven design is not a safe design procedure. A jumbo jet is not just a small plane scaled up; if that were done, the load-bearing components would be too small in cross-sectional area to support the much greater loads they would be called upon to resist.

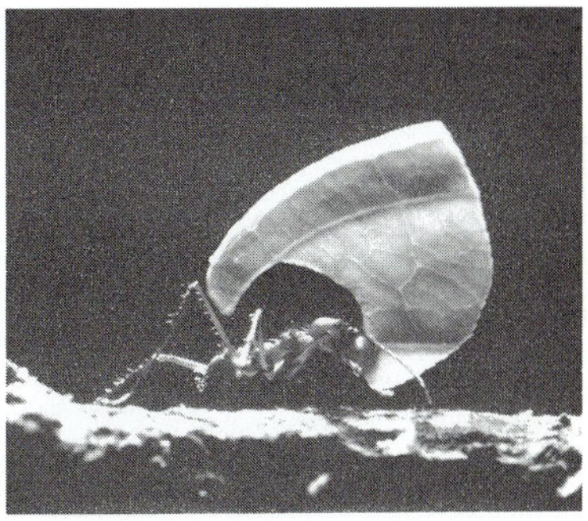

When reporting the strength of materials loaded in tension, it is customary to account for this effect of area by dividing the breaking load by the cross-sectional area:

$$\sigma_f = \frac{P_f}{A_0} \tag{1.1}$$

where $\sigma_f$ is the *ultimate tensile stress,* often abbreviated as UTS; $P_f$ is the load at fracture; and $A_0$ is the original cross-sectional area. (Some materials exhibit substantial reductions in cross-sectional area as they are stretched, and using the original rather than final area gives the so-called *engineering* strength.) The units of stress are obviously load per unit area: $N/m^2$ (also called pascals, or Pa) in the SI system, and $lb/in^2$ (or psi) in units still used commonly in the United States.

**FIGURE 1.5** Small animals can have very high strength-to-weight ratios, as this ant makes clear. (Paul McCormick/The Image Bank.)

**EXAMPLE 1.1**

In many design problems the loads to be applied to the structure are known at the outset, and we wish to compute how much material will be needed to support them. As a very simple case, consider using a steel rod, circular in cross-sectional shape

**FIGURE 1.6** Steel rod supporting a 10,000 lb weight.

as shown in Fig. 1.6, to support a load of 10,000 lb. What should the rod diameter be?

Directly from Eq. 1.1, the area $A_0$ that will be just on the verge of fracture at a given load $P_f$ is

$$A_0 = \frac{P_f}{\sigma_f}$$

All we need do is look up the value of $\sigma_f$ for the material, and substitute it along with the value of 10,000 lb for $P_f$, and the problem is solved.

A number of materials' properties are listed in Appendix A, where we find the UTS of carbon steel to be 1200 MPa. We also note in Appendix A that many properties vary widely for given materials depending on their composition and processing, so the 1200 MPa value is only a preliminary design estimate. In light of that uncertainty and many other potential ones, it is common to include a "factor of safety" in the design. Selection of an appropriate factor is an often difficult choice, especially in cases where weight or cost restrictions place a great penalty on using excess material. In this case, steel is relatively inexpensive, and we don't have any special weight limitations, so we'll use a conservative 50 percent safety factor and assume that the ultimate tensile strength is $1200/2 = 600$ MPa.

We now have only to adjust the units before solving for area. Engineers must be very comfortable with unit conversions, especially given the mix of SI and older traditional units used today. Eventually, we'll likely be ordering steel rod using inches rather than meters, so we'll convert the MPa to psi rather than convert the pounds to newtons. Also using $A = \pi d^2/4$ to compute the diameter rather than the area, we have

$$d = \sqrt{\frac{4A}{\pi}} = \sqrt{\frac{4P_f}{\pi\sigma_f}} = \left[\frac{4 \times 10000(\text{lb})}{\pi \times 600 \times 10^6(\text{N/m}^2) \times 1.449 \times 10^{-4}\left(\frac{\text{lb/in}^2}{\text{N/m}^2}\right)}\right]^{\frac{1}{2}}$$
$$= 0.38 \text{ in}$$

We probably wouldn't order rod of exactly $0.38''$, which would be an oddball size and thus too expensive. However, $\frac{3}{8}''$ $(0.375'')$ would likely be a standard size and would be acceptable in light of our conservative safety factor.

If the specimen is loaded by an axial force $P$ less than the breaking load $P_f$, the *tensile stress* is defined by analogy with Eq. 1.1 as

$$\boxed{\sigma = \frac{P}{A_0}} \tag{1.2}$$

The tensile stress, the force per unit area acting on a plane transverse to the applied load, is a fundamental measure of the internal forces within the material. Much of this text will be concerned with elaborating this concept to include higher orders of dimensionality, working out methods of determining the stress for various geometries and loading conditions, and predicting what the material's response to the stress will be.

EXAMPLE 1.2

Many engineering applications, notably aerospace vehicles, require materials that are both strong and lightweight. One measure of this combination of properties is provided by computing how long a rod of the material can be that when suspended from its top will break under its own weight (see Fig. 1.7). Here the stress is not uniform along the rod: The material at the very top bears the weight of the entire rod, but that at the bottom carries no load at all.

To compute the stress as a function of position, let $y$ denote the distance from the bottom of the rod and let the weight density of the material, for instance in N/m³, be denoted by $\gamma$. (The weight density is related to the mass density $\rho$ [kg/m³] by $\gamma = \rho g$, where $g = 9.8$ m/s² is the acceleration due to gravity.) The weight supported by the cross section at $y$ is just the weight density $\gamma$ times the volume of material $V$ below $y$:

$$W(y) = \gamma V = \gamma A y$$

The tensile stress is then given as a function of $y$ by Eq. 1.2 as

$$\sigma(y) = \frac{W(y)}{A} = \gamma y$$

Note that the area cancels, leaving only the material density $\gamma$ as a design variable.

The length of rod that is just on the verge of breaking under its own weight can now be found by letting $y = L$ (the highest stress occurs at the top), setting $\sigma(L) = \sigma_f$, and solving for $L$:

$$\sigma_f = \gamma L \Rightarrow L = \frac{\sigma_f}{\gamma}$$

In the case of steel, we find the mass density $\rho$ in Appendix A to be $7.85 \times 10^3$ (kg/m³); then

$$L = \frac{\sigma_f}{\rho g} = \frac{1200 \times 10^6 (\text{N/m}^2)}{7.85 \times 10^3 (\text{kg/m}^3) \times 9.8 (\text{m/s}^2)} = 15.6 \text{ km}$$

This would be a long rod indeed; the purpose of such a calculation is not so much to design superlong rods as to provide a vivid way of comparing materials (see Problem 4 in this chapter).

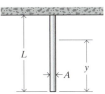

**FIGURE 1.7** Circular rod suspended from the top and bearing its own weight.

## 1.3 STIFFNESS

It is important to distinguish *stiffness*, which is a measure of the *load* needed to induce a given *deformation* in the material, from the *strength*, which usually refers to the material's resistance to failure by fracture or excessive deformation. The stiffness is usually measured by applying relatively small loads, well short of fracture, and measuring the resulting deformation. Since the deformations in most materials are very small for these loading conditions, the experimental problem is largely one of measuring small changes in length accurately.

Hooke[2] made a number of such measurements on long wires under various loads and observed that, to a good approximation, the load $P$ and its resulting deformation $\delta$ were related linearly as long as the loads were sufficiently small. This relation, generally known as *Hooke's law,* can be written algebraically as

$$P = k\delta \tag{1.3}$$

where $k$ is a constant of proportionality called the *stiffness* and having units of lb/in or N/m. The stiffness as defined by $k$ is not a function of the material alone, but is also influenced by the specimen shape. A wire gives much more deflection for a given load if coiled up like a watch spring, for instance.

A useful way to adjust the stiffness so as to be a purely materials property is to normalize the load by the cross-sectional area; that is, to use the tensile stress rather than the load. Further, the deformation $\delta$ can be normalized by noting that an applied load stretches all parts of the wire uniformly, so that a reasonable measure of "stretching" is the deformation per unit length:

$$\epsilon = \frac{\delta}{L_0} \tag{1.4}$$

Here $L_0$ is the original length, and $\epsilon$ is a dimensionless measure of stretching called the *strain.* Using these more general measures of load per unit area and displacement per unit length,[3] Hooke's law becomes

$$\frac{P}{A_0} = E\frac{\delta}{L_0} \tag{1.5}$$

or

$$\sigma = E\epsilon \tag{1.6}$$

The constant of proportionality $E$, called *Young's modulus*[4] or the *modulus of elasticity*, is one of the most important mechanical descriptors of a material. It has the same units as stress: Pa or psi. As shown in Fig. 1.9, Hooke's law can refer to either of Eqs. 1.3 or 1.6.

**FIGURE 1.8** Thomas Young. (E. Scott Barr Collection/AIP Niels Bohr Library)

---

[2] Robert Hooke (1635–1703) was a contemporary and rival of Isaac Newton. Hooke was a great pioneer in mechanics, but competing with Newton isn't easy.

[3] It was apparently the Swiss mathematician Jakob Bernoulli (1655–1705) who first realized the correctness of this form, published in the final paper of his life.

[4] After the English physicist Thomas Young (1773–1829), shown in Fig. 1.8, who also made notable contributions to the understanding of the interference of light as well as being a noted physician and Egyptologist.

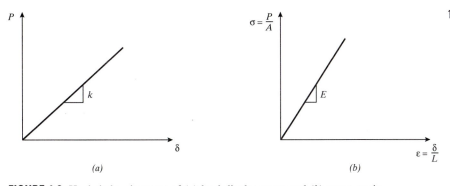

**FIGURE 1.9** Hooke's law in terms of (*a*) load-displacement and (*b*) stress-strain.

The Hookean stiffness $k$ is now recognizable as being related to the Young's modulus $E$ and the specimen geometry as

$$k = \frac{AE}{L} \tag{1.7}$$

where here the 0 subscript is dropped from the area $A$; it will be assumed from here on (unless stated otherwise) that the change in area during loading can be neglected. Another useful relation is obtained by solving Eq. 1.5 for the deflection in terms of the applied load as

$$\delta = \frac{PL}{AE} \tag{1.8}$$

Note that the stress $\sigma = P/A$ developed in a tensile specimen subjected to a fixed load is independent of the material properties, whereas the deflection depends on the material property $E$. Hence the stress $\sigma$ in a tensile specimen at a given load is the same whether the specimen is made of steel or polyethylene, but the strain $\epsilon$ will be different: The polyethylene will exhibit much larger strain and deformation, since its modulus is two orders of magnitude less than steel's.

### EXAMPLE 1.3

In Example 1.1 we found that a steel rod 0.38″ in diameter would safely bear a load of 10,000 lb. Now suppose that we have been given a second design goal, namely that the geometry requires that we use a rod 15 ft in length but that the loaded end cannot be allowed to deflect downward more than 0.3″ when the load is applied. Replacing $A$ in Eq. 1.8 by $\pi d^2/4$ and solving for $d$, the diameter for a given $\delta$ is

$$d = 2\sqrt{\frac{PL}{\pi \delta E}}$$

From Appendix A, the modulus of carbon steel is 210 GPa; using this along with the given load, length, and deflection, the required diameter is

$$d = 2 \sqrt{\frac{10^4(\text{lb}) \times 15(\text{ft}) \times 12(\text{in/ft})}{\pi \times 0.3(\text{in}) \times 210 \times 10^9(\text{N/m}^2) \times 1.449 \times 10^{-4}\left(\frac{\text{lb/in}^2}{\text{N/m}^2}\right)}}$$

$$= 0.5 \text{ in}$$

This diameter is larger than the 0.38″ computed earlier; therefore a larger rod must be used if the deflection as well as the strength goals are to be met. Clearly, using the larger rod makes the tensile stress in the material less and thus lowers the likelihood of fracture. This is an example of a *stiffness-critical* design, in which deflection rather than fracture is the governing constraint. As it happens, many structures throughout the modern era have been designed for stiffness rather than strength and thus wound up being "overdesigned" with respect to fracture. This has undoubtedly lessened the incidence of fracture-related catastrophes, which will be addressed in Chapter 7.

## EXAMPLE 1.4

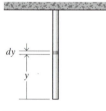

**FIGURE 1.10** Deformation of a column under its own weight.

When very long columns are suspended from the top, as in a cable hanging down the hole of an oil well, the deflection due to the weight of the material itself can be important. The solution for the total deflection is a minor extension of Eq. 1.8, in that now we must consider the increasing weight borne by each cross section as the distance from the bottom of the cable increases. As shown in Fig. 1.10, the total elongation of a column of length $L$, cross-sectional area $A$, and weight density $\gamma$ due to its own weight can be found by considering the incremental deformation $d\delta$ of a slice $dy$ a distance $y$ from the bottom. The weight borne by this slice is $\gamma A y$, so

$$d\delta = \frac{(\gamma A y)\, dy}{AE}$$

$$\delta = \int_0^L d\delta = \frac{\gamma}{E} \left.\frac{y^2}{2}\right|_0^L = \frac{\gamma L^2}{2E}$$

Note that $\delta$ is independent of the area $A$, so that finding a fatter cable won't help to reduce the deformation; the critical parameter is the *specific modulus, $E/\gamma$.* Since the total weight is $W = \gamma A L$, the result can also be written

$$\delta = \frac{WL}{2AE}$$

The deformation is the same as in a bar being pulled with a tensile force equal to half its weight; this is just the average force experienced by cross sections along the column.

In Example 1.2 we computed the length of a steel rod that would be just on the verge of breaking under its own weight if suspended from its top; we obtained $L = 15.6$ km. Were such a rod to be constructed, our analysis predicts the deformation at the bottom would be

$$\delta = \frac{\gamma L^2}{2E} = \frac{7.85 \times 10^3(\text{kg/m}^3) \times 9.8(\text{m/s}^2) \times [15.6 \times 10^3(\text{m})]^2}{2 \times 210 \times 10^9(\text{N/m}^2)} = 44.6 \text{ m}$$

However, this analysis assumes that Hooke's law holds over the entire range of stresses from zero to fracture. This is not true for many materials, including carbon steel, and later sections of this text will address materials response at high stresses.

The linear proportionality between stress and strain given by Hooke's law is not nearly as general as, say, Einstein's general theory of relativity, or even Newton's law of gravitation. It's really just an approximation that is observed to be reasonably valid for many materials as long the applied stresses are not too large. As the stresses are increased, eventually more complicated material response will be observed. Some of these effects will be outlined in Section 1.7, which introduces the stress-strain curve as an experimental means of plotting the strain response of materials over a range of stresses up to and including fracture.

If we were to push on the specimen rather than pulling on it, the loading would be described as *compressive* rather than tensile. In the range of relatively low loads, Hooke's law holds for this case as well. By convention, compressive stresses and strains are negative, so the expression $\sigma = E\epsilon$ holds for both tension and compression.

# 1.4 ATOMISTIC BASIS OF ELASTICITY

So far, we've introduced two very important material properties: the ultimate tensile strength $\sigma_f$ and Young's modulus $E$. To the effective mechanical designer, these aren't just numerical parameters that are looked up in tables and plugged into equations. The very nature of the material is reflected in these properties, and designers who try to function without a sense of how the material really works are very apt to run into trouble. Whenever practical in this text, we will make an effort to put the material's mechanical properties in context with its processing and microstructure. This section will describe how for most engineering materials the modulus is controlled by the atomic bond energy function.

For most materials the amount of stretching experienced by a tensile specimen under a small fixed load is controlled in a relatively simple way by the tightness of the chemical bonds at the atomic level, and this makes it possible to relate stiffness to the chemical architecture of the material. By contrast, more complicated mechanical properties such as fracture are controlled by a diverse combination of microscopic as well as molecular aspects of the material's internal structure and surface. Further, the stiffness of some materials—notably rubber—arises not from bond stiffness but from disordering or entropic factors. We will defer consideration of fracture and entropic elasticity effects to later sections, and concentrate initially on the bond energy basis of elasticity.

## 1.4.1 Energetic Effects

Chemical bonding between atoms can be viewed as arising from the electrostatic attraction between regions of positive and negative electronic charge. Materials can be classified based on the nature of these electrostatic forces, the three principal classes being

1. *Ionic materials,* such as NaCl, in which an electron is transferred from the less electronegative element (Na) to the more electronegative (Cl). The ions

therefore differ by one electronic charge and are thus attracted to one another. Further, the two ions are attracted not only to each other but also to other oppositely charged ions in their vicinity; they are also repelled from nearby ions of the same charge. Some ions may gain or lose more than one electron.

2. *Metallic materials,* such as iron and copper, in which one or more loosely bound outer electrons are released into a common pool, which then acts to bind the positively charged atomic cores.

3. *Covalent materials,* such as diamond and polyethylene, in which atomic orbitals overlap to form a region of increased electronic charge to which both nuclei are attracted. This bond is directional, with each of the nuclear partners in the bond being attracted to the negative region between them but not to any of the other atoms nearby.

In the case of ionic bonding, Coulomb's law of electrostatic attraction can be used to develop simple but effective relations for the bond stiffness. For ions of equal charge $e$ the attractive force $f_{attr}$ can be written:

$$f_{attr} = \frac{Ce^2}{r^2} \tag{1.9}$$

Here $C$ is a conversion factor; for $e$ in coulombs (C), $C = 8.988 \times 10^9$ N-m$^2$/C$^2$. For singly ionized atoms, $e = 1.602 \times 10^{-19}$ C is the charge on an electron. The *energy* associated with the Coulombic attraction is obtained by integrating the force, which shows that the bond energy varies inversely with the separation distance:

$$U_{attr} = \int f_{attr}\, dr = \frac{-Ce^2}{r} \tag{1.10}$$

where the energy of atoms at infinite separation is taken as zero.

If the material's atoms are arranged as a perfect crystal, it is possible to compute the electrostatic binding energy field in considerable detail. In the interpenetrating cubic lattice of the ionic NaCl structure shown in Fig. 1.11, for instance, each ion experiences attraction to oppositely charged neighbors and repulsion from equally charged ones. A particular sodium atom is surrounded by 6 Cl$^-$ ions at a distance $r$, 12 Na$^+$ ions at a distance $r\sqrt{2}$, 8 Cl$^-$ ions at a distance $r\sqrt{3}$, and so forth. The total electronic field sensed by the first sodium ion is then

$$U_{attr} = -\frac{Ce^2}{r}\left(\frac{6}{\sqrt{1}} - \frac{12}{\sqrt{2}} + \frac{8}{\sqrt{3}} - \frac{6}{\sqrt{4}} + \frac{24}{\sqrt{5}} - \cdots\right) \tag{1.11}$$

$$= \frac{-ACe^2}{r}$$

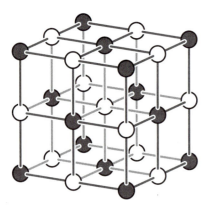

**FIGURE 1.11** The interpenetrating cubic NaCl lattice.

where $A = 1.747558\cdots$ is the result of the previous series, called the *Madelung constant.*[5] Note that it is not sufficient to consider only nearest-neighbor attractions in computing the bonding energy; in fact, the second term in the series is larger in magnitude than the first. The specific value for the Madelung constant is determined by the crystal structure, being 1.763 for CsCl and 1.638 for cubic ZnS.

---

[5]C. Kittel, *Introduction to Solid State Physics,* John Wiley & Sons, New York, 1966. The Madelung series does not converge smoothly, and this text includes some approaches to computing the sum.

At close separation distances the attractive electrostatic force is balanced by mutual repulsion forces that arise from interactions between overlapping electron shells of neighboring ions; this force varies much more strongly with the distance and can be written

$$U_{\text{rep}} = \frac{B}{r^n} \qquad (1.12)$$

Compressibility experiments have determined the exponent $n$ to be 7.8 for the NaCl lattice, so this is a much steeper function than $U_{\text{attr}}$.

As shown in Fig. 1.12, the total binding energy of one ion due to the presence of all others is then the sum of the attractive and repulsive components:

$$U = -\frac{ACe^2}{r} + \frac{B}{r^n} \qquad (1.13)$$

Note that the curve is *anharmonic* (not shaped like a sine curve), being more flattened out at larger separation distances. The system will adopt a configuration near the position of lowest energy, computed by locating the position of zero slope in the energy function:

$$\left(\frac{dU}{dr}\right)_{r=r_0} = \left(\frac{ACe^2}{r^2} - \frac{nB}{r^{n+1}}\right)_{r=r_0} = 0$$

$$r_0 = \left(\frac{nB}{ACe^2}\right)^{1/(n-1)} \qquad (1.14)$$

The range for $n$ is generally 5 to 12, increasing with the number of outer-shell electrons that cause the repulsive force.

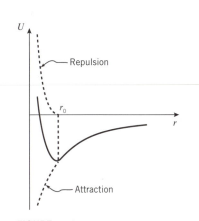

**FIGURE 1.12** The bond energy function.

## EXAMPLE 1.5

In practice the $n$ and $B$ parameters in Eq. 1.13 are determined from experimental measurements, for instance by using a combination of X-ray diffraction to measure $r_0$ and elastic modulus to infer the slope of the $U(r)$ curve. As an illustration of this process, picture a tensile stress $\sigma$ applied to a unit area of crystal ($A = 1$), as shown in Fig. 1.13, in a direction perpendicular to the crystal cell face (the [100] direction on the (100) face, using crystallographic notation; see Appendix E). The total force on this unit area is numerically equal to the stress: $F = \sigma A = \sigma$.

If the interionic separation is $r_0$, there will be $1/r_0^2$ ions on the unit area, each being pulled by a force $f$. Since the total force $F$ is just $f$ times the number of ions, the stress can then be written

$$\sigma = F = f\frac{1}{r_0^2}$$

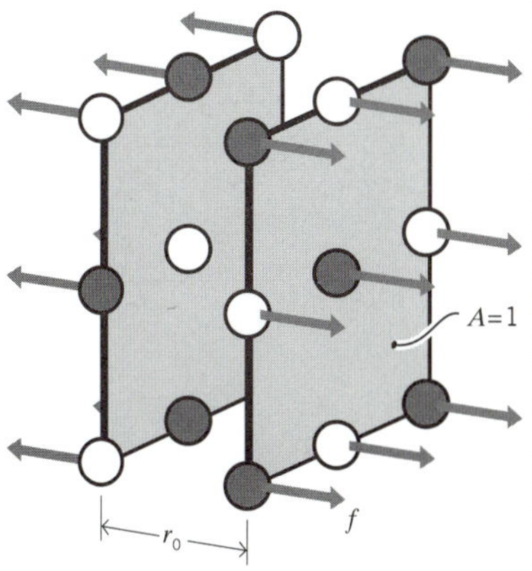

When the separation between two adjacent ions is increased by an amount $\delta$, the strain is $\epsilon = \delta/r_0$. The differential strain corresponding to a differential

**FIGURE 1.13** Simple tension applied to crystal face.

displacement is then

$$d\epsilon = \frac{dr}{r_0}$$

The elastic modulus $E$ is now the ratio of stress to strain, in the limit as the strain approaches zero:

$$E = \left.\frac{d\sigma}{d\epsilon}\right|_{\epsilon \to 0} = \left.\frac{1}{r_0}\frac{df}{dr}\right|_{r \to r_0} = \left.\frac{1}{r_0}\left(\frac{ACe^2}{r^2} - \frac{nB}{r^{n+1}}\right)\right|_{r \to r_0}$$

Using $B = ACe^2 r_0^{n-1}/n$ from Eq. 1.14 and simplifying,

$$E = \frac{(n-1)ACe^2}{r_0^4}$$

Note the very strong dependence of $E$ on $r_0$, which in turn reflects the tightness of the bond. If $E$ and $r_0$ are known experimentally, $n$ can be determined. For NaCl, $E = 3 \times 10^{10}$ N/m$^2$; using this along with the X-ray diffraction value of $r_0 = 2.82 \times 10^{-10}$ m, we find $n = 1.47$.

Using simple tension in this calculation is not really appropriate, because when a material is stretched in one direction, it will *contract* in the transverse directions. This is the *Poisson effect,* which will be treated in Chapter 2. Our tension-only example does not consider the transverse contraction, and the resulting value of $n$ is too low. A better but slightly more complicated approach is to use hydrostatic compression, which moves all the ions closer to one another. Problem 16 in this chapter outlines this procedure, which yields values of $n$ in the range of 5 to 12 as mentioned earlier.

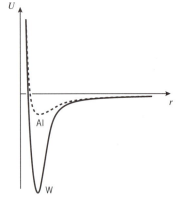

**FIGURE 1.14** Bond energy functions for aluminum and tungsten (approximate).

The stiffnesses of metallic and covalent systems will be calculated differently than those of ionic crystals, but the concept of electrostatic attraction applies to these nonionic systems as well. As a result, bond energy functions of a qualitatively similar nature result from all these materials. In general, the "tightness" of the bond, and hence the elastic modulus $E$, is related to the slope of the bond energy function. Steeper bond functions will also be deeper as a rule, so that within similar classes of materials the modulus tends to correlate with the energy needed to rupture the bonds, for instance by melting. Materials such as tungsten that fill many bonding and few antibonding orbitals have very deep bonding functions,[6] with correspondingly high stiffnesses and melting temperatures, as illustrated in Fig. 1.14. This correlation is obvious in Table 1.1, which lists the values of modulus for a number of metals, along with the values of melting temperature $T_m$ and melting energy $\Delta H$.

The system will generally have sufficient thermal energy to reside at a level somewhat above the minimum in the bond energy function and will oscillate between the two positions labeled $A$ and $B$ in Fig. 1.15, with an average position near $r_0$. This

---

[6]A detailed analysis of the cohesive energies of materials is an important topic in solid-state physics; see, for example, F. Seitz, *The Modern Theory of Solids,* McGraw-Hill, 1940.

| **TABLE 1.1** Modulus and bond strengths for transition metals | | | | |
|---|---|---|---|---|
| **Material** | **E** GPa (Mpsi) | **$T_m$** °C | **$\Delta H$** kJ/mol | **$\alpha_L$** $\times 10^{-6}, °C^{-1}$ |
| Pb | 14  (2) | 327 | 5.4 | 29 |
| Al | 69 (10) | 660 | 10.5 | 22 |
| Cu | 117 (17) | 1084 | 13.5 | 17 |
| Fe | 207 (30) | 1538 | 15.3 | 12 |
| W | 407 (59) | 3410 | 32 | 4.2 |

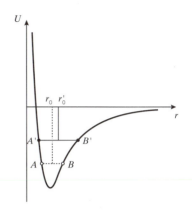

**FIGURE 1.15** Anharmonicity of the bond energy function.

simple idealization provides a rationale for why materials expand when the temperature is raised. As the internal energy is increased by the addition of heat, the system oscillates between the positions labeled $A'$ and $B'$ with an average separation distance $r_0'$. Since the curve is anharmonic, the average separation distance is now greater than before, so the material has expanded or stretched. To a reasonable approximation, the relative thermal expansion $\Delta L/L$ is often related linearly to the temperature rise $\Delta T$, and we can write

$$\boxed{\frac{\Delta L}{L} = \epsilon_T = \alpha_L \Delta T} \qquad (1.15)$$

where $\epsilon_T$ is a *thermal strain* and the constant of proportionality $\alpha_L$ is the *coefficient of linear thermal expansion*. The expansion coefficient $\alpha_L$ will tend to correlate with the depth of the energy curve, as is seen in Table 1.1.

## EXAMPLE 1.6

A steel bar of length $L$ and cross-sectional area $A$ is fitted snugly between rigid supports as shown in Fig. 1.16. We wish to find the compressive stress in the bar when the temperature is raised by an amount $\Delta T$.

   If the bar were free to expand, it would increase in length by an amount given by Eq. 1.15. Clearly, the rigid supports have to push on the bar—that is, put it into compression—to suppress this expansion. The magnitude of this thermally induced compressive stress could be found by imagining the material free to expand, then solving Eq. 1.6 for the stress needed to "push the material back" to its unstrained state. Equivalently, we could simply set the sum of a thermally induced strain and a mechanical strain $\epsilon_\sigma$ to zero:

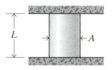

**FIGURE 1.16** Bar between rigid supports.

$$\epsilon = \epsilon_\sigma + \epsilon_T = \frac{\sigma}{E} + \alpha_L \Delta T = 0$$

$$\sigma = -\alpha_L E \Delta T$$

The minus sign in this result reminds us that a negative (compressive) stress is induced by a positive temperature change (temperature rises).

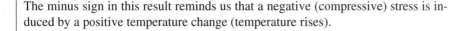

EXAMPLE 1.7

A glass container of stiffness $E$ and thermal expansion coefficient $\alpha_L$ is removed from a hot oven and plunged suddenly into cold water. We know from experience that this "thermal shock" could fracture the glass, and we would like to see what materials parameters control this phenomenon. The analysis is very similar to that of the previous example.

In the time period just after the cold-water immersion, before significant heat transfer by conduction can take place, the outer surfaces of the glass will be at the temperature of the cold water, but the interior will still be at the temperature of the oven. The outer surfaces will try to contract but are kept from doing so by the still-hot interior; this causes a tensile stress to develop on the surface. As before, the stress can be found by setting the total strain to zero:

$$\epsilon = \epsilon_\sigma + \epsilon_T = \frac{\sigma}{E} + \alpha_L \Delta T = 0$$

$$\sigma = -\alpha_L E \Delta T$$

Here the temperature change $\Delta T$ is negative if the glass is going from hot to cold, so the stress is positive (tensile). If the glass is not to fracture by thermal shock, this stress must be less than the ultimate tensile strength $\sigma_f$; hence, the maximum allowable temperature difference is

$$-\Delta T_{\max} = \frac{\sigma_f}{\alpha_L E}$$

To maximize the resistance to thermal shock, the glass should have as low a value of $\alpha_L E$ as possible. Pyrex® glass was developed specifically for improved thermal shock resistance by using boron rather than soda and lime as process modifiers; this yields a much reduced value of $\alpha_L$.

Material properties for a number of important structural materials are listed in Appendix A, and these data are also given in the computer file `props.csv` included with the text software diskette. This file is in comma-separated form, so that it can be read into most spreadsheets. When the column holding Young's modulus is plotted against the column containing the thermal expansion coefficients (using log-log coordinates), the graph shown in Fig. 1.17 is obtained. Here we see again the general inverse relationship between stiffness and thermal expansion, and the distinctive nature of polymers is apparent as well.

Not all types of materials can be described by these simple bond-energy concepts. *Intra*molecular polymer covalent bonds have energies comparable with ionic or metallic bonds, but most common polymers have substantially lower moduli than most metals or ceramics, because the *inter*molecular bonding in polymers is due to secondary bonds, which are much weaker than the strong intramolecular covalent bonds. Polymers can also have substantial entropic contributions to their stiffness, as will be described in the next section, and these effects do not necessarily correlate with bond energy functions.

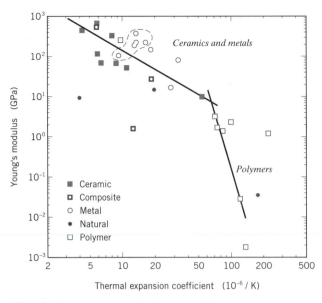

**FIGURE 1.17**  Correlation of stiffness and thermal expansion for materials of various types.

## 1.4.2  Entropic Effects

The internal energy as given by the function $U(r)$ is sufficient to determine the atomic positions in many engineering materials; the material "wants" to minimize its internal energy, and it does so by optimizing the balance of attractive and repulsive electrostatic bonding forces. When the absolute temperature is greater than approximately two-thirds of the melting temperature, however, there can be sufficient molecular mobility that entropic or disordering effects must be considered as well. This is often the case for polymers, even at room temperature, because of their weak intermolecular bonding.

When the temperature is high enough, polymer molecules can be viewed as an interpenetrating mass of (extremely long) wriggling worms, constantly changing their positions by rotation about carbon-carbon single bonds. This wriggling does not require straining the bond lengths or angles, and large changes in position are possible with no change in internal bonding energy.

The shape, or "conformation," of a polymer molecule can range from a fully extended chain to a randomly coiled sphere (see Fig. 1.18). Statistically, the coiled shape is much more likely than the extended one, simply because there are so many ways the chain can be coiled and only one way it can be fully extended. In thermodynamic terms, the entropy of the coiled conformation is very high (many possible "microstates"), and the entropy of the extended conformation is very low (only one possible microstate). If the chain is extended and then released, there will be more wriggling motions tending to the most probable state than to even more highly stretched states; the material would therefore shrink back to its unstretched and highest-entropy state. Equivalently, a person holding the material in the stretched state would feel a tensile force as the material tries to unstretch and is prevented from doing so. These effects are due to entropic factors, not to internal bond energy.

It is possible for materials to exhibit both internal-energy and entropic elasticity. Energy effects dominate in most materials, but rubber is much more dependent on entropic effects. An *ideal rubber* is one in which the response is completely entropic, with the internal energy changes being negligible.

**FIGURE 1.18**  Conformational change in polymers.

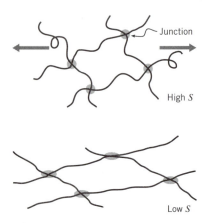

**FIGURE 1.19** Stretching of crosslinked or entangled polymers.

When we stretch a rubber band, the molecules in its interior become extended because they are *crosslinked* by chemical or physical junctions, as shown in Fig. 1.19. Without these links the molecules could simply slide past one another with little or no uncoiling. Bouncing putty is an example of an uncrosslinked polymer, and its lack of junction connections causes it to be a viscous fluid rather than a useful elastomer that can bear sustained loads without continuing flow. The crosslinks provide a means by which one molecule can pull on another and thus establish load transfer within the material. They also have the effect of limiting how far the rubber can be stretched before breaking, because the extent of the entropic uncoiling is limited by how far the material can extend before pulling up tight against the network of junction points. We will see below that the stiffness of a rubber can be controlled directly by adjusting the crosslink density, and this is an example of process-structure-property control in materials.

As the temperature is raised, the Brownian-type wriggling of the polymer is intensified, so that the material seeks more vigorously to assume its random high-entropy state. This means that the force needed to hold a rubber band at fixed elongation *increases* with increasing temperature. Similarly, if the band is stretched by hanging a fixed weight on it, the band will *shrink* as the temperature is raised. In some thermodynamic formalisms it is convenient to model this behavior by letting the coefficient of thermal expansion be a variable parameter, with the ability to become negative for sufficiently large tensile strains. This is a little tricky, however; for instance, the stretched rubber band will contract only along its long axis when the temperature is raised, and it will become thicker in the transverse directions. The coefficient of thermal expansion would have to be made not only stretch-dependent but also dependent on direction ("anisotropic").

## EXAMPLE 1.8

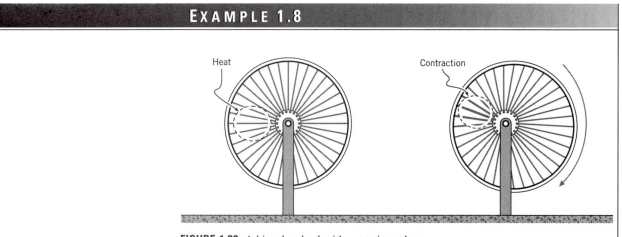

**FIGURE 1.20** A bicycle wheel with entropic spokes.

An interesting demonstration of the unusual thermal response of stretched rubber bands involves replacing the spokes of a bicycle wheel with stretched rubber bands, as seen in Fig. 1.20, then mounting the wheel so that a heat lamp shines on the bands to the right or left of the hub. As the bands warm up, they contract. This pulls the rim closer to the hub, causing the wheel to become unbalanced. It will then rotate under gravity, causing the warmed bands to move out from under the heat lamp and be replaced by other bands. The process continues, and the wheel rotates in a direction opposite to what would be expected were the spokes to expand rather than contract on heating.

The bicycle-wheel trick produces a rather weak response, and it is easy to stop the wheel with only a light touch of the finger. However, the same idea, using very highly stretched urethane bands and employing superheated geothermal steam as a heat source, becomes a viable route for converting the heat into mechanical energy.

It is worthwhile to study the response of rubbery materials in some depth, partly because doing so provides a broader view of the elasticity of materials. This isn't a purely academic exercise, however. Rubbery materials are being used in increasingly demanding mechanical applications (in addition to their use in tires, a very demanding application itself). Elastomeric bearings, vibration-control supports, and biomedical prostheses are but a few examples. We will outline what is known as the "kinetic theory of rubber elasticity," which treats the entropic effect using concepts of statistical thermodynamics. This theory stands as one of the very most successful atomistic theories of mechanical response. It leads to a result of good accuracy without the need for adjustable parameters or other fudge factors.

When pressure-volume changes are not significant, the competition between internal energy and entropy can be expressed by the Helmholtz free energy $A = U - TS$, where $T$ is the temperature and $S$ is the entropy. The system will move toward configurations of lowest free energy, which it can do either by reducing its internal energy or by increasing its entropy. Note that the influence of the entropic term increases explicitly with increasing temperature. With certain thermodynamic limitations in mind (see Problem 18 in this chapter), the mechanical work $dW = F\,dL$ done by a force $F$ acting through a differential displacement $dL$ will produce an increase in free energy given by

$$F\,dL = dW = dU - T\,dS \qquad (1.16)$$

or

$$F = \frac{dW}{dL} = \left(\frac{\partial U}{\partial L}\right)_{T,V} - T\left(\frac{\partial S}{\partial L}\right)_{T,V} \qquad (1.17)$$

For an ideal rubber, the energy change $dU$ is negligible, so the force is related directly to the temperature and the change in entropy $dS$ produced by the displacement. To determine the force-deformation relationship, we obviously need to consider how $S$ changes with deformation. We begin by writing an expression for the conformation, or shape, of the segment of polymer molecule between junction points as a statistical probability distribution. Here the length of the *segment* is the important molecular parameter, not the length of the entire molecule. In the simple form of this theory, each covalently bonded segment is idealized as a freely-jointed sequence of $n$ rigid links, each having length $a$.

A reasonable model for the end-to-end distance of a randomly wriggling segment is that of a "random walk" Gaussian distribution, treated in elementary statistics. One end of the chain is visualized at the origin of an $xyz$ coordinate system, as shown in Fig. 1.21, and each successive link in the chain is attached with a random orientation relative to the previous link. (An elaboration of the theory would constrain the orientation so as to maintain the 109.5° covalent bonding angle.) The probability $\Omega_1(r)$ that the other end of the chain is at a position $r = \left(x^2 + y^2 + z^2\right)^{1/2}$ can be shown to be

$$\Omega_1(r) = \frac{\beta^3}{\sqrt{\pi}} \exp(-\beta^2 r^2) = \frac{\beta^3}{\sqrt{\pi}} \exp\left[-\beta^2\left(x^2 + y^2 + z^2\right)\right]$$

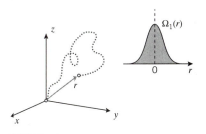

**FIGURE 1.21** Random-walk model of polymer conformation.

The parameter $\beta$ is a scale factor related to the number of units $n$ in the polymer segment and the bond length $a$; specifically it turns out that $\beta = \sqrt{3/2n}/a$. This is the "bell-shaped curve" well known to seasoned test-takers. The most probable end-to-end distance is seen to be zero, which is expected because the chain will end up a given distance to the left (or up, or back) of the origin exactly as often as it ends up the same distance to the right.

When the molecule is now stretched or otherwise deformed, the relative positions of the two ends are changed. Deformation in elastomers is usually described in terms of *extension ratios,* which are the ratios of stretched to original dimensions, $L/L_0$. Stretches in the $x$, $y$, and $z$ directions are denoted by $\lambda_x$, $\lambda_y$, and $\lambda_z$, respectively, The deformation is assumed to be *affine*; that is, the end-to-end distances of each molecular segment increase by these same ratios. Hence, if we continue to view one end of the chain as the origin, the other end will have moved to $x_2 = \lambda_x x$, $y_2 = \lambda_y y$, $z_2 = \lambda_z z$. The configurational probability of a segment being found in this stretched state is then

$$\Omega_2 = \frac{\beta^3}{\sqrt{\pi}} \exp\left[-\beta^2\left(\lambda_x^2 x^2 + \lambda_y^2 y^2 + \lambda_z^2 z^2\right)\right]$$

The relative change in probabilities between the perturbed and unperturbed states can now be written as

$$\ln \frac{\Omega_2}{\Omega_1} = -\beta^2\left[\left(\lambda_x^2 - 1\right)x^2 + \left(\lambda_y^2 - 1\right)y^2 + \left(\lambda_z^2 - 1\right)z^2\right]$$

Several stratagems have been used in the literature to simplify this expression. One simple approach is to let the initial position of the segment end $x, y, z$ be such that $x^2 = y^2 = z^2 = r_0^2/3$, where $r_0^2$ is the initial mean square end-to-end distance of the segment. (This is not zero, because when squares are taken the negative values no longer cancel the positive ones.) It can also be shown (see Problem 21 in this chapter) that the distance $r_0^2$ is related to the number of bonds $n$ in the segment and the bond length $a$ by $r_0^2 = na^2$. Making these substitutions and simplifying, we have

$$\ln \frac{\Omega_2}{\Omega_1} = -\frac{1}{2}\left(\lambda_x^2 + \lambda_y^2 + \lambda_z^2 - 3\right) \tag{1.18}$$

As is taught in subjects in statistical thermodynamics, changes in configurational probability are related to corresponding changes in thermodynamic entropy by the *Boltzmann relation:*

$$\Delta S = k \ln \frac{\Omega_2}{\Omega_1}$$

where $k = 1.38 \times 10^{-23}$ J/K is Boltzmann's constant. Substituting Eq. 1.18 in this relation:

$$\Delta S = -\frac{k}{2}\left(\lambda_x^2 + \lambda_y^2 + \lambda_z^2 - 3\right)$$

This is the entropy change for one segment. If there are $N$ chain segments per unit volume, the total entropy change per unit volume $\Delta S_V$ is just $N$ times this quantity:

$$\Delta S_V = -\frac{Nk}{2}\left(\lambda_x^2 + \lambda_y^2 + \lambda_z^2 - 3\right) \tag{1.19}$$

The associated work (per unit volume) required to change the entropy by this amount is

$$\Delta W_V = -T\Delta S_V = +\frac{NkT}{2}\left(\lambda_x^2 + \lambda_y^2 + \lambda_z^2 - 3\right) \qquad (1.20)$$

The quantity $\Delta W_V$ is therefore the strain energy per unit volume contained in an ideal rubber stretched by $\lambda_x, \lambda_y, \lambda_z$.

To illustrate the use of this expression for a simple but useful case, consider a rubber band, initially of length $L_0$, that is stretched to a new length $L$. Hence $\lambda = \lambda_x = L/L_0$. To a very good approximation, rubbery materials maintain a constant volume during deformation, and this lets us compute the transverse contractions $\lambda_y$ and $\lambda_z$ that accompany the stretch $\lambda_x$. An expression for the change $\Delta V$ in a cubical volume of initial dimensions $a_0, b_0, c_0$ that is stretched to new dimensions $a, b, c$ is

$$\Delta V = abc - a_0b_0c_0 = (a_0\lambda_x)(b_0\lambda_y)(c_0\lambda_z) - a_0b_0c_0 = a_0b_0c_0(\lambda_x\lambda_y\lambda_z - 1)$$

Setting this to zero gives

$$\lambda_x\lambda_y\lambda_z = 1 \qquad (1.21)$$

Hence, the contractions in the $y$ and $z$ directions are

$$\lambda_y^2 = \lambda_z^2 = \frac{1}{\lambda}$$

Using this in Eq. 1.20, the force $F$ needed to induce the deformation can be found by differentiating the total strain energy according to Eq. 1.17:

$$F = \frac{dW}{dL} = \frac{d(V\,\Delta W_V)}{L_0\,d\lambda} = A_0\,\frac{NkT}{2}\left(2\lambda - \frac{2}{\lambda^2}\right)$$

Here $A_0 = V/L_0$ is the original area. Dividing by $A_0$, we obtain the engineering stress:

$$\sigma = NkT\left(\lambda - \frac{1}{\lambda^2}\right) \qquad (1.22)$$

Clearly, the parameter $NkT$ is related to the stiffness of the rubber, as it gives the stress $\sigma$ needed to induce a given extension $\lambda$. It can be shown (see Problem 24 in this chapter) that the initial modulus—the slope of the stress-strain curve at the origin—is controlled by the temperature and the crosslink density according to $E = 3NkT$.

Crosslinking in rubber is usually produced by the "vulcanizing" process invented by Charles Goodyear (Fig. 1.22) in 1839. In this process, sulfur abstracts reactive hydrogens adjacent to the double bonds in the rubber molecule and forms permanent bridges between adjacent molecules. When crosslinking is done by using approximately 5 percent sulfur, a conventional rubber is obtained. When the sulfur is increased to $\approx$ 30 to 50 percent, a hard and brittle material named ebonite (or simply "hard rubber") is produced instead.

**FIGURE 1.22** Charles Goodyear. (Wiley Archive.)

The volume density of chain segments $N$ is also the density of junction points. This quantity is related to the specimen density $\rho$ and the molecular weight between crosslinks $M_c$ as $M_c = \rho N_A/N$, where $N$ is the number of crosslinks per unit volume and $N_A = 6.023 \times 10^{23}$ is Avogadro's number. When $N$ is expressed in terms of moles per unit volume, we have simply $M_c = \rho/N$ and the quantity $NkT$ in Eq. 1.22 is replaced by $NRT$, where $R = kN_A = 8.314$ J/mol-K is the gas constant.

## EXAMPLE 1.9

Young's modulus of a rubber is measured at $E = 3.5$ MPa for a temperature of $T = 300$ K. The molar crosslink density is then

$$N = \frac{E}{3RT} = \frac{3.5 \times 10^6 \text{ N/m}^2}{3 \times 8.314 \frac{\text{N·m}}{\text{mol·K}} \times 300 \text{ K}} = 468 \text{ mol/m}^3$$

The molecular weight per segment is

$$M_c = \frac{\rho}{N} = \frac{1100 \text{ kg/m}^3}{468 \text{ mol/m}^3} = 2350 \text{ g/mol}$$

## EXAMPLE 1.10

A person with more entrepreneurial zeal than caution wishes to start a bungee-jumping company and naturally wants to know how far the bungee cord will stretch; the clients sometimes complain if the cord fails to stop them before they reach the asphalt. It is probably easiest to obtain a first estimate from an energy point of view. Say the unstretched length of the cord is $L_0$, and let this be also the distance the jumper free-falls before the cord begins to stretch. Just as the cord begins to stretch, the jumper has lost an amount of potential energy $wL_0$, where $w$ is the jumper's weight. The jumper's velocity at this time could then be calculated from $(mv^2)/2 = wL_0$ if desired, where $m = w/g$ is the jumper's mass and $g$ is the acceleration of gravity. When the jumper's velocity has been brought to zero by the cord (assuming the cord doesn't break first, and the ground doesn't intervene), this energy will now reside as entropic strain energy within the cord. (See Fig. 1.23.) Using Eq. 1.20, we can equate the initial and final energies to obtain

$$wL_0 = \frac{A_0 L_0 \cdot NRT}{2}\left(\lambda^2 + \frac{2}{\lambda} - 3\right)$$

Here $A_0 L_0$ is the total volume of the cord; the entropic energy per unit volume $\Delta W_V$ must be multiplied by the volume to give total energy. Dividing out the initial length $L_0$ and using $E = 3NRT$, this result can be written in the dimensionless form

$$\frac{w}{A_0 E} = \frac{1}{6}\left(\lambda^2 + \frac{2}{\lambda} - 3\right)$$

**FIGURE 1.23** The elasticity of bungee cords arises from entropic rather than energetic molecular mechanisms. (Ed Bock/The Stock Market.)

The closed-form solution for $\lambda$ is messy, but the variable $w/A_0 E$ can easily be plotted versus $\lambda$ (see Fig. 1.24). Note that the length $L_0$ has canceled from the result, although it is still present implicitly in the extension ratio $\lambda = L/L_0$.

Taking a typical design case for illustration, say the desired extension ratio is taken at $\lambda = 3$ for a rubber cord of initial modulus $E = 100$ psi; this stops the jumper safely above the pavement and is verified to be well below the breaking extension of the cord. The value of the parameter $w/AE$ corresponding to $\lambda = 3$ is read from the graph to be 1.11. For a jumper weight of 150 lb, this corresponds to $A = 1.35$ in$^2$, or a cord diameter of $1.31''$.

If ever there was a strong case for field testing, this is it. An analysis such as this is nothing more than a crude starting point, and many tests such as drops with sandbags are obviously called for. Even then, the insurance costs would likely be very substantial.

Note that the stress-strain response for rubber elasticity is nonlinear and that the stiffness, as given by the stress needed to produce a given deformation, is predicted to increase with increasing temperature. This is in accord with the concept of more vigorous wriggling with a statistical bias toward the more disordered state. The rubber elasticity equation works well at lower extensions but tends to deviate from experimental values at high extensions, where the segment configurations become non-Gaussian.

Deviations from Eq. 1.22 can also occur due to crystallization at high elongations. (Rubbers are normally noncrystalline, and in fact polymers such as polyethylene that crystallize readily are not elastomeric, because of the rigidity imparted by the crystallites.) However, the decreased entropy that accompanies stretching in rubber increases the crystalline melting temperature according to the thermodynamic relation

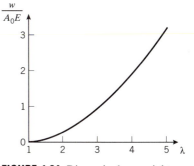

**FIGURE 1.24** Dimensionless weight versus cord extension.

$$T_m = \frac{\Delta U}{\Delta S} \tag{1.23}$$

where $\Delta U$ and $\Delta S$ are the change in internal energy and entropy on crystallization. The quantity $\Delta S$ is reduced if stretching has already lowered the entropy, so the crystallization temperature rises. If it rises above room temperature, the rubber develops crystallites that stiffen it considerably and cause further deviation from the rubber elasticity equation. (Since the crystallization is exothermic, the material will also increase in temperature; this can often be sensed by stretching a rubber band and then touching it to the lips.) Strain-induced crystallization also helps inhibit crack growth, and the excellent abrasion resistance of natural rubber is related to the ease with which it crystallizes upon stretching.

## 1.5 VISCOELASTIC RESPONSE OF POLYMERS

### 1.5.1 Molecular Mechanisms

When polymers are stretched, the stretching of bond lengths and angles that underlies the change in internal energy on mechanical loading occurs essentially instantaneously. However, the configurational changes involved in the entropic response take place by thermally activated rate processes that can occur very quickly or with glacial slowness. These rates are very temperature-dependent and can usually be described by Arrhenius-type expressions of the form

$$\text{Rate} \propto \exp \frac{-E^*}{RT} \tag{1.24}$$

where $E^*$ is an apparent activation energy of the process.

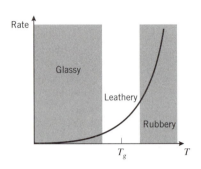

**FIGURE 1.25** Temperature dependence of entropic response rate.

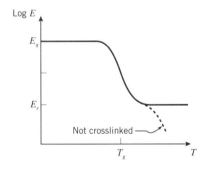

**FIGURE 1.26** A generic modulus-temperature map for polymers.

At temperatures much above the *glass transition temperature,* labeled $T_g$ in Fig. 1.25, the rates are so fast as to be essentially instantaneous, and the polymer acts in a rubbery manner, in which it exhibits large, instantaneous, and fully reversible strains in response to an applied stress. Conversely, at temperatures much less than $T_g$, the rates are so slow as to be negligible. Here the chain uncoiling process is essentially "frozen out," so the polymer is able to respond only by bond stretching. It now responds in a "glassy" manner, responding instantaneously and reversibly but incapable of being strained beyond a few percent before fracturing in a brittle manner.

In the range near $T_g$ the material is midway between the glassy and rubbery regimes. Its response is a combination of viscous fluidity and elastic solidity, and this region is termed "leathery" or, more technically, "viscoelastic." The value of $T_g$ is an important descriptor of polymer thermomechanical response and is a fundamental measure of the material's propensity for mobility. Factors that enhance mobility, such as absorbed diluents, expansive stress states, and lack of bulky molecular groups, all tend to produce lower values of $T_g$. The transparent polyvinyl butyral film used in automobile windshield laminates is an example of a material that is used in the viscoelastic regime; viscoelastic response can be a source of substantial energy dissipation during impact.

At temperatures well below $T_g$, when entropic motions are frozen and only elastic bond deformations are possible, polymers exhibit a relatively high modulus, called the "glassy modulus" $E_g$, which is on the order of 3 GPa (400 kpsi). As the temperature is increased through $T_g$, the stiffness drops dramatically, by perhaps two orders of magnitude, to a value called the "rubbery modulus" $E_r$. In elastomers that have been permanently crosslinked by sulfur vulcanization or other means, the value of $E_r$ is determined by the crosslink density according to Eq. 1.22. If the material is not crosslinked, the stiffness exhibits a short plateau, because molecular entanglements can act as network junctions; at still higher temperatures the entanglements slip, and the material becomes a viscous liquid. Neither the glassy nor the rubbery modulus depends strongly on time, but in the vicinity of the transition near $T_g$, time effects can be very important. Clearly, a plot of modulus versus temperature, such as is shown in Fig. 1.26, is a vital tool in polymer materials science and engineering. It provides a map of a vital engineering property and also is a fingerprint of the molecular motions available to the material.

In the viscoelastic range there is an observable time dependence in the response to an applied load or displacement. When a constant stress $\sigma(t) = \sigma_0$ is applied, an instantaneous strain will occur as the atoms move elastically to new positions, just as with a metal or ceramic material. As time continues, however, additional strain will develop due to conformational uncoiling, and this continuing strain is called *creep.* The converse of creep is *stress relaxation,* in which an abrupt strain $\epsilon(t) = \epsilon_0$ is imposed and the resulting time-dependent stress $\sigma(t)$ is measured. One important objective of viscoelastic theory is to develop reasonable mathematical models for these time-dependent functions.

### 1.5.2 Spring-Dashpot Models

The time dependence of viscoelastic response is analogous to the time dependence of reactive electrical circuits, and both can be described by identical ordinary differential equations in time. A convenient way of developing these relations while also helping to visualize molecular motions employs "spring-dashpot" models. These mechanical analogs use "Hookean" springs, depicted in Fig. 1.27 and described by

$$\sigma = k\epsilon$$

**FIGURE 1.27** Hookean spring (left) and Newtonian dashpot (right).

where $\sigma$ and $\epsilon$ are analogous to the spring force and displacement, and the spring constant $k$ is analogous to Young's modulus $E$. The spring models the instantaneous bond deformation of the material, and its magnitude will be related to the fraction of mechanical energy stored reversibly as strain energy $U$.

The entropic uncoiling process is fluidlike in nature and can be modeled by a "Newtonian[7] dashpot," also shown in Fig. 1.23, in which the stress produces not a strain but a strain *rate*:

$$\sigma = \eta \dot{\epsilon}$$

Here the overdot denotes time differentiation, and $\eta$ is a viscosity with units of Pa-s. In many of the relations to follow, it will be convenient to employ the ratio of viscosity to stiffness:

$$\tau = \frac{\eta}{k}$$

The unit of $\tau$ is time, and it will be seen that this ratio is a useful measure of the response time of the material's viscoelastic response.

The "Maxwell"[8] solid shown in Fig. 1.28 is a mechanical model in which a Hookean spring and a Newtonian dashpot are connected in series. The spring should be visualized as representing the elastic or energetic component of the response, while the dashpot represents the conformational or entropic component. In a series connection such as the Maxwell model, the stress on each element is the same and equal to the imposed stress, while the total strain is the sum of the strain in each element:

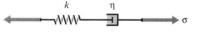

**FIGURE 1.28** The Maxwell model.

$$\sigma = \sigma_s = \sigma_d$$
$$\epsilon = \epsilon_s + \epsilon_d$$

Here the subscripts $s$ and $d$ represent the spring and dashpot, respectively. In seeking a single equation relating the stress to the strain, it is convenient to differentiate the strain equation and then write the spring and dashpot strain rates in terms of the stress:

$$\dot{\epsilon} = \dot{\epsilon}_s + \dot{\epsilon}_d = \frac{\dot{\sigma}}{k} + \frac{\sigma}{\eta}$$

Multiplying by $k$ and using $\tau = \eta/k$:

$$k\dot{\epsilon} = \dot{\sigma} + \frac{1}{\tau}\sigma \qquad (1.25)$$

This expression is the "constitutive" equation for the material, an equation that relates the stress to the strain. Note that it contains time derivatives, so that a simple

[7] After Isaac Newton (1642–1727), arguably the greatest scientist who ever lived.
[8] After the prolific Scottish physicist James Clerk Maxwell (1831–1879).

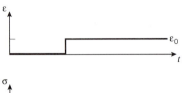

**FIGURE 1.29** Strain and stress histories in the stress relaxation test.

constant of proportionality between stress and strain does not exist. The concept of "modulus"—the ratio of stress to strain—must be broadened to account for this more complicated behavior.

One may define several different moduli appropriate for various circumstances. In a stress relaxation test, for instance, we view the constant strain as the "input" to the material and seek an expression for the resulting time-dependent stress; this is depicted in Fig. 1.29. The *relaxation modulus* is then defined as

$$E_{rel} = \frac{\sigma(t)}{\epsilon_0} \qquad (1.26)$$

Because in stress relaxation $\dot{\epsilon} = 0$, Eq. 1.25 becomes

$$\frac{d\sigma}{dt} = -\frac{1}{\tau}\sigma$$

Separating variables and integrating:

$$\int_{\sigma_0}^{\sigma} \frac{d\sigma}{\sigma} = -\frac{1}{\tau}\int_0^t dt$$

$$\ln\sigma - \ln\sigma_0 = -\frac{t}{\tau}$$

$$\sigma(t) = \sigma_0 \exp(-t/\tau)$$

Here the significance of $\tau \equiv \eta/k$ as a characteristic "relaxation time" is evident; it is physically the time needed for the stress to fall to $1/e$ of its initial value.

The relaxation modulus $E_{rel}$ may be obtained from this relation directly, noting that initially only the spring will deform and the initial stress and strain are related by $\sigma_0 = k\epsilon_0$. So

$$E_{rel}(t) = \frac{\sigma(t)}{\epsilon_0} = \frac{\sigma_0}{\epsilon_0}\exp(-t/\tau)$$

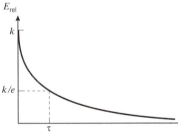

**FIGURE 1.30** Relaxation modulus for the Maxwell model.

$$\boxed{E_{rel}(t) = k\exp(-t/\tau)} \qquad (1.27)$$

This important function is plotted schematically in Fig. 1.30. The two adjustable parameters in the model, $k$ and $\tau$, can be used to force the model to match an experimental plot of the relaxation modulus at two points. The spring stiffness $k$ would be set to the initial or glass modulus $E_g$, and $\tau$ would be chosen to force the value $k/e$ to match the experimental data at $t = \tau$.

## EXAMPLE 1.11

The Laplace transformation is very convenient in viscoelasticity problems, and Appendix C lists some transform pairs encountered often in these problems. Rather than integrating the constitutive differential equation as in the previous example, let us start by transforming Eq. 1.25.

Since the stress and strain are zero as the origin is approached from the left, the transforms of the time derivatives are just the Laplace variable $s$ times the transforms of the functions; denoting the transformed functions with an overline, we have $\mathcal{L}(\dot{\epsilon}) = s\bar{\epsilon}$ and $\mathcal{L}(\dot{\sigma}) = s\bar{\sigma}$. Then writing the transform of an expression such as Eq. 1.25 is done simply by placing a line over the time-dependent functions and replacing the time-derivative overdot by an $s$ coefficient:

$$k\dot{\epsilon} = \dot{\sigma} + \frac{1}{\tau}\sigma \longrightarrow ks\bar{\epsilon} = s\bar{\sigma} + \frac{1}{\tau}\bar{\sigma}$$

Since now we don't have to worry about mixtures of ordinary functions and their time derivatives, we can simply solve this for $\bar{\sigma}$:

$$\bar{\sigma} = \frac{ks}{s + \frac{1}{\tau}}\bar{\epsilon} \qquad (1.28)$$

This is the Laplace-plane counterpart of the material stress-strain law.

In the case of stress relaxation, the strain function $\epsilon(t)$ is treated as a constant $\epsilon_0$ times the "Heaviside" or "unit step" function $u(t)$:

$$\epsilon(t) = \epsilon_0 u(t), \qquad u(t) = \begin{cases} 0, & x < 0 \\ 1, & x \geq 0 \end{cases}$$

This has the Laplace transform

$$\bar{\epsilon} = \frac{\epsilon_0}{s}$$

Using this in Eq. 1.28 and dividing through by $\epsilon_0$, we have

$$\frac{\bar{\sigma}}{\epsilon_0} = \frac{k}{s + \frac{1}{\tau}}$$

Since $\mathcal{L}^{-1} 1/(s + a) = e^{-at}$, this can be inverted directly to give

$$\frac{\sigma(t)}{\epsilon_0} \equiv E_{\text{rel}}(t) = k \exp(-t/\tau)$$

Later sections will make use of the Laplace transform approach to extend the treatment of viscoelasticity to more general cases. The reader should avoid the natural tendency to cringe when "advanced" mathematical techniques such as this are used. They're really not difficult, and in this case they're a *lot* easier to use than grinding problems out with the direct elementary methods.

The relaxation time $\tau$ is strongly dependent on temperature and other factors that affect the mobility of the material, and it is roughly inverse to the rate of molecular motion. Above $T_g$, $\tau$ is very short; below $T_g$, it is very long. More detailed consideration of the temperature dependence will be given in Chapter 6, in the context of creep effects in "thermorheologically simple" materials.

Most polymers do not exhibit the unrestricted flow permitted by the Maxwell model, although it would be a reasonable model for bouncing putty or warm tar.

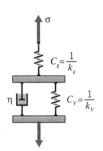

**FIGURE 1.31** The Standard Linear Solid model for creep.

Therefore, Eq. 1.27 is valid only for a very limited set of materials. For more typical polymers, whose conformational change is eventually limited by the network of entanglements or other types of junction points, more elaborate spring-dashpot models can be used effectively.

Figure 1.31 shows a spring-dashpot parallel arrangement, called the *Voigt model,* placed in series with a single spring; the overall model is called the *Standard Linear Solid* or also the "*Zener*[9] *solid.*" (Living during the development of the models seems to have been helpful in getting your name attached to something.) This model is convenient for describing creep, the time-dependent strain under a given constant stress.

Rather than working with spring stiffnesses, it will be convenient to develop an expression for creep using the spring *compliances.* A compliance is just the reciprocal of a stiffness or modulus; in other words, the compliance $C$ is the strain divided by the stress rather than the other way around, so $C = 1/E = \epsilon/\sigma$. The compliance of the spring in the Voigt part of the model is denoted $C_V = 1/k_V$, and that of the other spring is $C_g = 1/k_g$.

In the Voigt part of the model, the strain $\epsilon_V$ in both the spring and the dashpot is the same, while the total transmitted stress is the sum of the stress in the spring plus that in the dashpot:

$$\sigma = \sigma_s + \sigma_d = \frac{\epsilon_V}{C_V} + \eta \dot{\epsilon}_V$$

Taking transforms,

$$\bar{\sigma} = \frac{\bar{\epsilon}_V}{C_V} + \eta s \bar{\epsilon}_V$$

Multiplying by $C_V$ and using $\tau = \eta C_V$, this gives

$$\bar{\epsilon}_V = \frac{C_V \bar{\sigma}}{\tau \left( s + \dfrac{1}{\tau} \right)}$$

The strain in the series spring is just $\epsilon_s = \sigma C_g$, with transform $\bar{\epsilon}_s = \bar{\sigma} C_g$, so the transform of the total strain is

$$\bar{\epsilon} = \bar{\epsilon}_s + \bar{\epsilon}_V = \left[ C_g + \frac{C_V}{\tau \left( s + \dfrac{1}{\tau} \right)} \right] \bar{\sigma}$$

In a creep test $\sigma(t) = \sigma_0 = $ constant, so $\bar{\sigma} = \sigma_0/s$. Using this and inverting, the creep response of the Standard Linear Solid is

$$\frac{\epsilon(t)}{\sigma_0} \equiv C_{crp}(t) = C_g + C_V \left( 1 - e^{-t/\tau} \right)$$

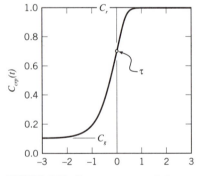

**FIGURE 1.32** Creep response of the Standard Linear Solid. Here $C_g = 0.1$, $C_r = 1$, $\tau = 1$.

This function is plotted in Fig. 1.32. The model parameter $C_g$ is the short-term, or glassy, compliance of the material. After loading, the strain rises and exponentially

---

[9] After the American engineer and physicist Clarence Melvin Zener (1905–1993).

parameter $C_V$ is the difference between $C_r$ and $C_g$, so the creep law can be written

$$C_{\text{crp}} = C_g + \left(C_r - C_g\right)\left(1 - e^{-t/\tau}\right) \qquad (1.29)$$

## 1.6  COMPOSITE MATERIALS

The term *composite* could mean almost anything if taken at face value, because all materials are composed of dissimilar subunits if examined at close enough detail. In modern materials engineering, however, the term *composite* usually refers to a "matrix" material that is reinforced with fibers. For instance, the term *FRP* (for Fiber-Reinforced Plastic) usually indicates a thermosetting polyester matrix containing glass fibers, and this particular composite has the lion's share of today's commercial market. Many composites used today are at the leading edge of materials technology, with performance and costs appropriate to ultrademanding applications such as spacecraft. However, heterogeneous materials combining the best aspects of dissimilar constituents have been used by nature for millions of years. Ancient society, imitating nature, used this approach as well. The Book of Exodus speaks of using straw to reinforce mud in brickmaking, without which the bricks would have almost no strength (Fig. 1.33).

As seen in Table 1.2,[10] the fibers used in modern composites have strengths and stiffnesses far above those of traditional bulk materials. The high strengths of

**FIGURE 1.33** Composites using straw and mud have been used since ancient times for dwelling construction; this method continues in use even today. (Nicholas DeVore/Tony Stone Images/New York, Inc.)

---

[10]F.P. Gerstle, "Composites," *Encyclopedia of Polymer Science and Engineering,* Wiley, New York, 1991. Copyright © 1991, John Wiley & Sons. Reprinted by permission of John Wiley & Sons, Inc.

**TABLE 1.2** Properties of composite reinforcing fibers

| Material | $E$, GPa | $\sigma_b$, GPa | $\rho$, kg/m³ | $E/\rho$, MJ/kg | $\sigma_b/\rho$, MJ/kg | Cost, $/kg |
|---|---|---|---|---|---|---|
| E-glass | 72.4 | 2.4 | 2,540 | 28.5 | 0.95 | 1.1 |
| S-glass | 85.5 | 4.5 | 2,490 | 34.3 | 1.8 | 22–33 |
| Aramid | 124 | 3.6 | 1,440 | 86 | 2.5 | 22–33 |
| Boron | 400 | 3.5 | 2,450 | 163 | 1.43 | 330–440 |
| HS graphite | 253 | 4.5 | 1,800 | 140 | 2.5 | 66–110 |
| HM graphite | 520 | 2.4 | 1,850 | 281 | 1.3 | 220–660 |

the glass fibers are due to processing that avoids the internal or surface flaws that normally weaken glass, and the strength and stiffness of the polymeric aramid fiber is a consequence of the nearly perfect alignment of the molecular chains with the fiber axis.

Of course, these materials are not generally usable as fibers alone. Typically they are impregnated with a matrix material that acts to transfer loads to the fibers and also to protect the fibers from abrasion and environmental attack. The matrix dilutes the properties to some degree, but even so very high specific (weight-adjusted) properties are available from these materials. Metal and glass are available as matrix materials, but these are currently very expensive and largely restricted to R&D laboratories. Polymers are much more commonly used, with unsaturated styrene-hardened polyesters having the majority of low- to medium-performance applications and epoxy or more sophisticated thermosets having the higher end of the market. Thermoplastic matrix composites are increasingly attractive materials, with processing difficulties being perhaps their principal limitation.

### 1.6.1 Stiffness

The fibers may be oriented randomly within the material, but it is also possible to arrange for them to be oriented preferentially in the direction expected to have the highest stresses. Such a material is said to be *anisotropic* (having different properties in different directions), and control of the anisotropy is an important means of optimizing the material for specific applications. At a microscopic level the properties of these composites are determined by the orientation and distribution of the fibers as well as by the properties of the fiber and matrix materials. The topic known as *composite micromechanics* is concerned with developing estimates of the overall material properties from these parameters.

Consider a typical region of material of unit dimensions, containing a volume fraction $V_f$ of fibers all oriented in a single direction. The matrix volume fraction is then $V_m = 1 - V_f$. This region can be idealized as shown in Fig. 1.34 by gathering all the fibers together, leaving the matrix to occupy the remaining volume. If a stress $\sigma_1$ is applied along the fiber direction, the fiber and matrix phases act in *parallel* to support the load. In these parallel connections, just as with the spring-dashpot models discussed earlier, the strains in each phase must be the same. Hence the strain $\epsilon_1$ in the fiber direction can be written as

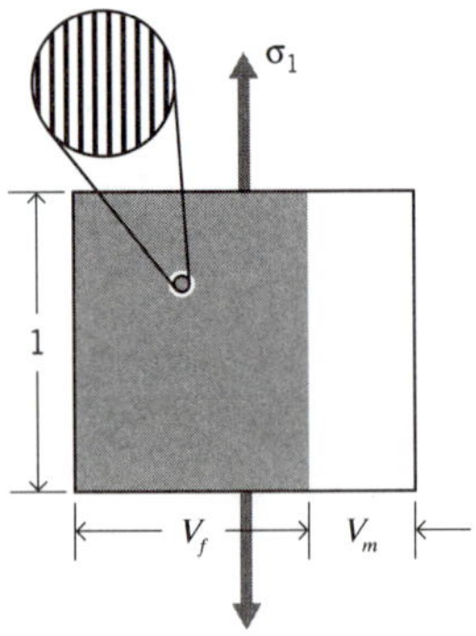

$$\epsilon_f = \epsilon_m = \epsilon_1$$

**FIGURE 1.34** Loading parallel to the fibers.

The forces in each phase must add to balance the total load on the material. Since the forces in each phase are the phase stresses times the area (here numerically equal to

$$\sigma_1 = \sigma_f V_f + \sigma_m V_m = E_f \epsilon_1 V_f + E_m \epsilon_1 V_m$$

The stiffness in the fiber direction is found by dividing by the strain:

$$E_1 = \frac{\sigma_1}{\epsilon_1} = V_f E_f + V_m E_m \qquad (1.30)$$

This relation is known as a *rule of mixtures* prediction of the overall modulus in terms of the moduli of the constituent phases and their volume fractions.

    If the stress is applied in the direction transverse to the fibers, as depicted in Fig. 1.35, the fiber and matrix materials can be modeled as acting *in series*. In this case the stress in the fiber and matrix are equal (an idealization), but the deflections add to give the overall transverse deflection. In this case it can be shown (see Problem 44 in this chapter) that

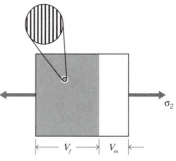

**FIGURE 1.35** Loading perpendicular to the fibers.

$$\frac{1}{E_2} = \frac{V_f}{E_f} + \frac{V_m}{E_m} \qquad (1.31)$$

    In more complicated composites, for instance those with fibers in more than one direction or those having particulate or other nonfibrous reinforcements, Eq. 1.30 provides an *upper bound* to the composite modulus, and Eq. 1.31 is a *lower bound* (see Fig. 1.36). Most practical cases will be somewhere between these two values, and the search for reasonable models for these intermediate cases has occupied considerable attention in the composites research community. Perhaps the most popular model is an empirical one known as the *Halpin-Tsai* equation,[11] which can be written in the form

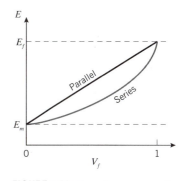

**FIGURE 1.36** Bounds on composite modulus.

$$E = \frac{E_m[E_f + \xi(V_f E_f + V_m E_m)]}{V_f E_m + V_m E_f + \xi E_m} \qquad (1.32)$$

Here $\xi$ is an adjustable parameter that results in series coupling for $\xi = 0$ and parallel averaging for very large $\xi$.

## 1.6.2  Strength

Rule of mixtures estimates for strength proceed along lines similar to those for stiffness. For instance, consider a unidirectionally reinforced composite that is strained up to the value at which the fibers begin to break. Denoting this value $\epsilon_{fb}$, the stress transmitted by the composite is given by multiplying the stiffness (Eq. 1.30):

$$\sigma_b = \epsilon_{fb} E_1 = V_f \sigma_{fb} + (1 - V_f)\sigma^*$$

The stress $\sigma^*$ is the stress in the matrix, which is given by $\epsilon_{fb} E_m$. This relation is linear in $V_f$, rising from $\sigma^*$ to the fiber breaking strength $\sigma_{fb} = E_f \epsilon_{fb}$. However, this relation is not realistic at low fiber concentration, since the breaking strain of the matrix $\epsilon_{mb}$ is usually substantially greater than $\epsilon_{fb}$. If the matrix had *no* fibers in it, it would fail at a stress $\sigma_{mb} = E_m \epsilon_{mb}$. If the fibers were considered to carry no load at

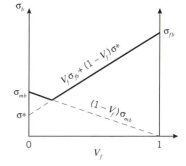

**FIGURE 1.37** Strength of unidirectional composite in fiber direction.

[11]Cf. J.C. Halpin and J.L. Kardos, *Polymer Engineering and Science,* Vol. 16, May 1976, pp. 344–352.

all, having broken at $\epsilon = \epsilon_{fb}$ and leaving the matrix to carry the remaining load, the strength of the composite would fall off with fiber fraction according to

$$\sigma_b = (1 - V_f)\sigma_{mb}$$

Since the breaking strength actually observed in the composite is the greater of these two expressions, there will be a range of fiber fraction in which the composite is *weakened* by the addition of fibers. These relations are depicted in Fig. 1.37.

## 1.7 STRESS-STRAIN CURVES

Most of the experiments outlined in the preceding section involve placing a relatively small fixed load at the end of a clamped rod and measuring the resulting displacement. A more generally useful experiment involves measuring the load needed to increase the displacement continuously, so that a full *stress-strain curve* is obtained as shown in Fig. 1.38.[12] In modern servo-controlled testing machines, such as that seen in Fig. 1.39, either the load or the displacement can be selected as the controlled variable, which can be viewed as the "input" to the material. The resulting displacement or load is then the "output" variable. It is impossible to specify both the load and the displacement simultaneously; the material has some say in the matter, and can be viewed as the "transfer function" between the input and the output.

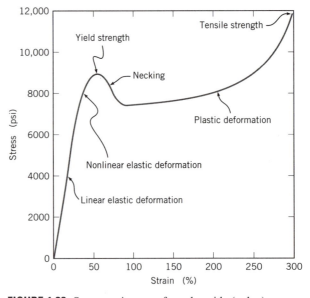

**FIGURE 1.38** Stress-strain curve for polyamide (nylon) thermoplastic.

---

[12]Stress-strain testing as well as almost all of the other experimental procedures mentioned in this text are detailed by standards-setting organizations, notably the American Society for Testing and Materials (ASTM). Tensile testing of metals in prescribed by ASTM Test E8, plastics by ASTM D638, and composite materials by ASTM D3039.

The specific shape of the stress-strain curve is a property of the material, and many variations are possible. However, they do tend to exhibit certain characteristic features, some of which can be described in the paragraphs to follow.

### 1.7.1 Young's Modulus

This important parameter, defined earlier as the ratio of stress to strain in the region of linear elasticity, is the slope of the stress-strain curve at the origin (see Fig. 1.40). A related term, the *tangent modulus,* is the slope of the curve at any arbitrary value of strain. Still another term, the *secant modulus,* is the slope of a straight line drawn from the origin to intersect the curve at an arbitrary value of strain. Only when the material response is linear do these three moduli coincide.

### 1.7.2 Proportional Limit

Stress-strain curves typically deviate from linearity at some value of stress, called the *proportional limit.* The material nonlinearity may arise from the anharmonic nature of the bond energy function itself or, more probably, because the material is beginning to change due to stress-induced rearrangements in its internal structure.

Linearity implies that an increase in strain, say by a factor of $a$, will increase the stress by this same factor. (Double the strain, double the stress.) Mathematically, this can be written $\sigma(a\epsilon) = a\sigma(\epsilon)$. When the material is time-dependent, as in viscoelastic response, this simple multiplicative scaling is not a sufficient criterion. To be complete, we require that the stress response to the sum of two different strain histories $\epsilon_1(t)$ and $\epsilon_2(t)$ be just the sum of what would have been the stress responses to each strain applied by itself. Combining this with the multiplicative requirement, the necessary and sufficient condition for linearity is

$$\sigma(a\epsilon_1 + b\epsilon_2) = a\sigma(\epsilon_1) + b\sigma(\epsilon_2) \tag{1.33}$$

### 1.7.3 Elastic Limit

Elasticity is the property of complete and immediate recovery from an imposed displacement on release of the load, and the *elastic limit* is the value of stress at which the material experiences a *permanent* deformation that is not lost on unloading. This is the beginning of the *yielding* process, in which molecular slip causes atoms to move to new equilibrium positions. This elastic limit is not necessarily the same as the proportional limit, although the two points may coincide.

### 1.7.4 Yield Stress

For many materials, there is a stress—denoted by $\sigma_Y$ in Fig. 1.41—at which the stress-strain curve becomes horizontal ($d\sigma/d\epsilon = 0$). This zero slope indicates that an increment of strain may take place without a corresponding incremental increase in stress. This is fluidlike behavior, taking place by molecular slip of the sort initiated at the elastic limit. It is also termed *plastic flow.* In materials that slip without exhibiting this point of zero slope, a yield stress is often defined by drawing a straight line parallel to the initial portion of the stress-strain curve but offset to the right by an arbitrary amount, usually 0.2 percent. The stress at which this line intersects the curve as it bends over is called the *offset yield stress.*

**FIGURE 1.39** Tensile testing machine. (Courtesy Instron Corp.)

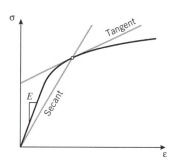

**FIGURE 1.40** Definition of common engineering moduli.

Not all materials yield; the ones that yield significantly are termed *ductile*. Materials such as porcelain are resistant to slip, and in these materials slip would require such a high stress that the material fractures first; these materials are *brittle*. Brittleness is usually a very dangerous property in load-bearing materials, and the relative dominance of yield versus fracture processes is a central aspect of materials engineering. Yield is also amenable to control by various chemical and processing variations, so that the degree of ductility can often be optimized for a given application. Modifications in yield behavior are sometimes termed *strengthening,* and this topic will be treated more fully in Chapter 6.

During slip, the material's atoms move plastically to new equilibrium positions, so that the original shape of the specimen is not recovered when the loads are removed. The effect is seen in the material's *unloading curve,* which is usually a straight line of slope equal to the original elastic modulus. The *residual strain* is then the strain at which the unloading line reaches the $\sigma = 0$ axis, as shown in Fig. 1.42.

### 1.7.5 Drawing

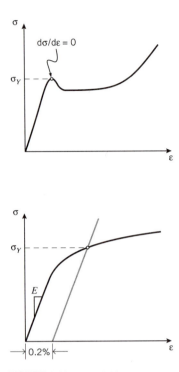

**FIGURE 1.41** The yield stress $\sigma_Y$.

Many materials, particularly the ductile metals such as aluminum, yield but experience fracture if strained further. In these cases yielding takes place initially at some location along the specimen length, and the flow at this position causes the cross-sectional area to diminish there. There is not as much area to hold the load, so the stress-strain curve drops to lower values beyond the yield point. The true stress at this location (the load divided by the current, reduced area) rises, so flow accelerates and the area reduces further. The process continues until the material fails.

This process can be observed without the need for a testing machine by stretching a polyethylene six-pack holder. Instead of failing after yielding, the material forms a "neck" that appears to propagate as the stretching is continued. The stretched-out material between the neck shoulders is said to be *drawn,* and if the stretching is done carefully, fracture will not take place until the entire length of plastic being gripped is drawn. The strain in the drawn region is very large compared with the yield strain, being on the order of 500 percent. The stress-strain curve is rather ambiguous once the yielding and drawing process begins, because the strain is no longer uniform along the specimen length. During drawing, only the material at the neck shoulders is deforming; the strain in the remainder of the material remains constant.

The key to the stable drawing process observed in polyethylene fibers and sheets, but not in aluminum, is *strain hardening.* Why doesn't the plastic break when its area is reduced and the true stress there rises? The material has been strengthened by orientation—more of the polymer's strong covalent bonds have been lined up with the load direction—and this strengthening process more than makes up for the reduction in area. This is depicted schematically in Fig. 1.43.

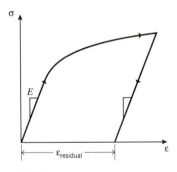

**FIGURE 1.42** Unloading and residual strain.

### 1.7.6 Strain Hardening

The term *hardness* is often taken to be the resistance of a material to being scratched or indented by a sharp probe. However, it is also used to indicate a material's resistance to deformation in general. As the specimen strains, it may become increasingly more resistant to deformation; for instance, the resistance may increase by dislocation interactions in crystalline materials or by molecular orientation in polymers. The stress-strain curve bends upward (the tangent modulus increases), and this is termed *strain hardening.* Each successive increment of increasing strain requires a larger increment of increasing stress.

**FIGURE 1.43** Necking and drawing in polymers.

## 1.7.7 Strain Energy

The area under the stress-strain curve up to a given value of strain is the total mechanical energy per unit volume consumed by the material in straining it to that value. This is easily shown as follows:

$$U^* = \frac{1}{V} \int P \, dL = \int_0^L \frac{P}{A_0} \frac{dL}{L_0} = \int_0^\epsilon \sigma \, d\epsilon \qquad (1.34)$$

In the absence of molecular slip and other mechanisms for energy dissipation, this mechanical energy is stored reversibly within the material as *strain energy*. When the stresses are low enough that the material remains in the elastic range, the strain energy is just the triangular area in Fig. 1.44:

$$U^* = \frac{1}{2}\sigma\epsilon = \frac{\sigma^2}{2E} = \frac{E\epsilon^2}{2} \qquad (1.35)$$

Note that the strain energy increases *quadratically* with the stress or strain; that is, as the strain increases, the total energy stored grows as the square of the strain. This has important consequences. For example, an archery bow, to work well, cannot be simply a curved piece of wood. A real bow is initially straight, then bent when it is strung; this stores substantial strain energy in it. When it is bent further on drawing the arrow back, the energy available to throw the arrow is very much greater than if the bow were simply carved in a curved shape without actually *bending* it (Fig. 1.45). Figure 1.46 shows schematically the amount of strain energy available for two equal increments of strain $\Delta\epsilon$, applied at different levels of existing strain.

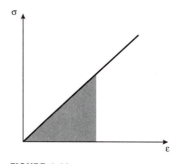

FIGURE 1.44 Strain energy = area under stress-strain curve.

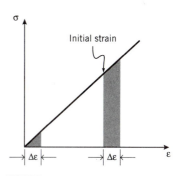

FIGURE 1.46 Energy associated with increments of strain.

FIGURE 1.45 A bow is bent as it is being strung, and this stores substantial strain energy in both the bow and the string. This increases greatly the energy made available when the string is drawn back prior to shooting. (Lori Adamski Peek/Tony Stone Images/New York, Inc.)

The area up to the yield point is termed the *modulus of resilience,* and the total area up to fracture is termed the *modulus of toughness;* these are shown in Fig. 1.47.

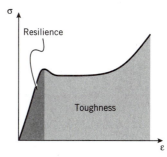

FIGURE 1.47 Moduli of resilience and toughness.

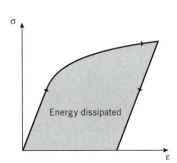

**FIGURE 1.48** Energy loss = area under stress-strain loop.

The term *modulus* is used because the units of strain energy per unit volume are N-m/m$^3$ or N/m$^2$, which are the same as those of stress or modulus of elasticity. The term *resilience* refers to the material being unaffected by the applied stress up to the point of yielding, so that upon unloading it returns to its original shape. When the strain exceeds the yield point, the material is deformed irreversibly so that some residual strain will persist even after unloading. The modulus of resilience is then the quantity of energy the material can absorb without suffering damage. Similarly, the modulus of toughness is the energy needed to fracture the material completely. Materials showing good impact resistance are generally those with high moduli of toughness. Table 1.3[13] lists energy absorption values for a number of common materials. Note that natural and polymeric materials can provide extremely high energy absorption per unit weight.

During loading, the area under the stress-strain curve is the strain energy per unit volume absorbed by the material. Conversely, the area under the *unloading* curve is the energy released by the material. In the elastic range, these areas are equal and no net energy is absorbed. If the material is loaded into the plastic range, however, as shown in Fig. 1.48, the energy absorbed exceeds the energy released, and the difference is dissipated as heat.

### 1.7.8 Fracture

The end point of the stress-strain curve occurs when the specimen breaks into two pieces. The value of stress at which fracture occurs is obviously an important aspect of engineering design and also of materials science. The strength of the material's chemical bonds is important in governing fracture processes, but many additional factors, such as the number and size of internal or surface flaws, may be even more important in practice. The important field of *fracture mechanics*, outlined in Chapter 7, deals with this problem in more detail.

## 1.8 COMPRESSION

### 1.8.1 Special Considerations

This chapter is concerned primarily with simple tension—that is, uniaxial loading that increases the interatomic spacing. However, as long as the loads are sufficiently small (stresses less than the proportional limit), in many materials the relations outlined in

| TABLE 1.3 Energy absorption of various materials | | | | | |
|---|---|---|---|---|---|
| **Material** | **Maximum strain, %** | **Maximum stress, MPa** | **Modulus of toughness, MJ/m$^3$** | **Density, kg/m$^3$** | **Max. energy, J/kg** |
| Ancient iron | 0.03 | 70 | 0.01 | 7,800 | 1.3 |
| Modern spring steel | 0.3 | 700 | 1.0 | 7,800 | 130 |
| Yew wood | 0.3 | 120 | 0.5 | 600 | 900 |
| Tendon | 8.0 | 70 | 2.8 | 1,100 | 2,500 |
| Rubber | 300 | 7 | 10.0 | 1,200 | 8,000 |

[13] J.E. Gordon, *Structures, or Why Things Don't Fall Down,* Plenum Press, New York, 1978.

the foregoing sections apply equally well if loads are placed so as to put the specimen in compression rather than tension. The expression $\delta = PL/AE$ applies just as before, with negative values for $\delta$ and $P$ indicating compression. Further, the modulus $E$ is the same in tension and compression to a good approximation. As seen in Fig. 1.49, the stress-strain curve simply extends as a straight line into the third quadrant.

There are some practical difficulties in performing stress-strain tests in compression. If excessively large loads are mistakenly applied in a tensile test, perhaps by wrong settings on the testing machine, the specimen simply breaks and the test must be repeated with a new specimen. In compression, however, a mistake can easily damage the load cell or other sensitive components, because even after specimen failure the loads are not necessarily relieved.

Specimens loaded cyclically so as to alternate between tension and compression can exhibit *hysteresis loops* if the loads are high enough to induce plastic flow (stresses above the yield stress). The enclosed area in the loop seen in Fig. 1.50 is the strain energy per unit volume, released as heat in each loading cycle; for example, the well-known tendency of a wire that is being bent back and forth to become quite hot at the region of plastic bending. The temperature of the specimen will rise according to the magnitude of this internal heat generation and the rate at which the heat can be removed by conduction within the material and convection from the specimen surface.

Specimen failure by cracking is inhibited in compression, because cracks will be closed up rather than opened by the stress state. A number of important materials are much stronger in compression than in tension for this reason. Concrete, for example, has good compressive strength and so finds extensive use in construction in which the dominant stresses are compressive, but it has essentially no strength in tension, as cracks in sidewalks and building foundations attest: Tensile stresses appear as these structures settle, and cracks begin at very low tensile strain in unreinforced concrete.

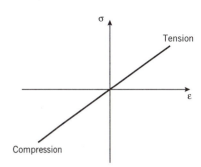

**FIGURE 1.49** Stress-strain curve in tension and compression.

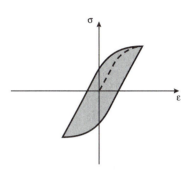

**FIGURE 1.50** Hysteresis loop.

## 1.8.2 Buckling and Structural Collapse

So far, we have discussed failure as being caused by an overstressing that induces fracture if in tension and crushing if in compression. However, it is important to realize that loaded structures can fail by means other than simple tensile or compressive overstressing. As an example, long and slender specimens loaded in compression can fail by a mode called *buckling* rather than by compressive crushing. This result is depicted in Fig. 1.51. The effect is somewhat like "trying to push on a rope," because the long specimen does not have sufficient bending resistance to keep it straight. This situation will be treated in some detail in Chapter 4, but we will touch on it here in order to show that the designer must be aware of a number of failure modes in addition to the obvious ones.

For now, we will state without proof that the critical load $P_{cr}$ at which a column of length $L$ will buckle is given by the "Euler[14] buckling formula" as

$$P_{cr} = \frac{\pi^2}{k} \frac{EI}{L^2} \tag{1.36}$$

where $k$ is a factor depending on the nature of the column end conditions and $I$ is the rectangular moment of inertia of the column's cross section with respect to its centroidal axis. ($I = bh^3/12$ for a column having a rectangular cross section of dimensions $b \times h$.) The product $EI$ arises frequently in bending problems and is known as the *section modulus* of the beam or column. Note that the load that the column can

---

[14] After the prolific Swiss mathematician and physicist Leonhard Euler (1707–1783).

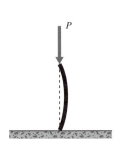

$P$

**FIGURE 1.51** Column buckling.

withstand without buckling increases linearly with the section modulus but decreases quadratically with increasing length. Whether the column fails by buckling or by crushing depends on the value of $P_{cr}$ relative to the compressive failure load.

Gordon[15] has noted that buckling considerations provide an interesting insight to the relative efficiencies of materials. Since a square column of width $h$ has $I \propto h^4$, a column of fixed length will have a width that scales with the required buckling load according to

$$h^2 \propto \sqrt{\frac{P}{E}}$$

The weight of such a column is $W = \rho h^2 L$, where $\rho$ is the density. The scaling law for the column weight can therefore be written in terms of the buckling load rather than the width as

$$W \propto \rho h^2 \propto \rho \sqrt{\frac{P}{E}}$$

The column *buckling efficiency,* the load per unit weight, therefore scales as

$$\frac{P}{W} \propto \sqrt{P} \cdot \frac{\sqrt{E}}{\rho} \tag{1.37}$$

Note that the buckling efficiency depends more strongly on the density than on the modulus. Nature makes extensive use of hollow cells in buckling-critical structures, such as trees and bones, largely for this reason.

When rectangular panels rather than columns are considered, the relevant moment of inertia scales with $h^3$ rather than $h^4$, where now $h$ is the panel thickness. A development similar to that just stated now gives the buckling efficiency as scaling with $\sqrt[3]{E/\rho}$, which increases further the importance of density in comparison with modulus. Table 1.4[16] lists these parameters for a number of common materials.

It is probably the case that for any given material some ranking of properties can be found that makes that material "look better" than its competitors. That's a big part of how advertising works. But it is also true that materials design optimization depends on the application at hand. If panel buckling is the name of the game, the designer might be able to save a lot of money specifying plywood rather than graphite composite.

**TABLE 1.4** Efficiencies of several materials

| Material | $E$, GPa | $\rho$, kg/m³ | $E/\rho$ | $\sqrt{E}/\rho$ | $\sqrt[3]{E}/\rho$ |
|---|---|---|---|---|---|
| Steel | 210 | 7,800 | 25,000 | 190 | 7.5 |
| Titanium | 120 | 4,500 | 25,000 | 240 | 11.0 |
| Aluminum | 73 | 2,800 | 25,000 | 310 | 15.0 |
| Magnesium | 42 | 1,700 | 24,000 | 380 | 20.5 |
| Carbon-fiber composite | 200 | 2,000 | 100,000 | 700 | 29.0 |
| Spruce | 14 | 500 | 25,000 | 750 | 48.0 |

[15] J.E. Gordon, op. cit.

[16] Ibid.

**FIGURE 1.52** "Flying" buttresses opposing the outward forces due to peaked roofs in Reims Cathedral, France. (Marburg—Art Reference Bureau.)

Buckling is an example of *structural collapse,* and it is important to remember that mechanical failures can take place by such events without the material failing by fracture or excessive deformation due to high stress. Putting a peaked roof on a house so the rain will run off, for instance, causes problems because the roof pushes outward on the walls, and it is necessary to counter this effect by providing a roof joist that acts in tension to hold the walls up. The beautiful domed roof on a large cathedral has the same outward-pushing effect: The walls have only a very low compressive stress, but there is a dangerous tendency for them to tip over. Here using a joist would be aesthetically objectionable, and medieval architects learned to use buttresses to push on the walls from the outside (see Fig. 1.52). Often these buttresses provide a special beauty of their own to the building.

The principal goal of this text will be to develop means of estimating stresses in existing or planned structures and of predicting the material's response to them. This is perhaps the most important aspect of mechanical design, but the designer must be alert to the possibility of disaster due to causes beyond simple (or not so simple) stress rupture. Structural collapse is one of the more important of these, and the history of such failures makes a fascinating if chilling study. The reader may be aware of the works of Robert Byrne, whose fictional accounts of technological disaster make both compelling and instructive reading. His work *Skyscraper*[17] tells of a tall building in Manhattan falling over because its piling system cannot withstand the tipping forces caused by wind.

It may seem an absurd fiction that a skyscraper might actually be able to tip over, but the famous Hancock Tower in Boston was found to be vulnerable to exactly this form of collapse (see Fig. 1.53). In its April 9, 1988 issue, *The Boston Globe* published an account that had up to then been suppressed by agreement among litigants in a number of suits and countersuits related to this building's many problems. Perhaps the most astounding of these issues was an analysis by Bruno Thurlimann of Zurich, whose studies (which included extensive wind tunnel testing) showed a substantial

---

[17]Atheneum Press, New York, 1984.

**FIGURE 1.53** Boston's Hancock Tower had many materials and structural problems in its early years, including a tendency of the original windows to break. For a time, many of the windows were replaced by plywood sheets. (The Boston Globe.)

danger of tipover. An extensive network of diagonal braces, costing several million dollars, was added to the building, doubling its stiffness along its long axis. Stories such as this, while no doubt embarrassing (and expensive!) to the engineers and architects involved, should be sought out by the novice designer and studied for their lessons.[18]

## 1.9 PROBLEMS

**1.** Determine the stress and total deformation of an aluminum wire, 30 m long and 5 mm in diameter, subjected to an axial load of 250 N.

**2.** Two rods, one of nylon and one of steel, are rigidly connected as shown in Fig. P.1.2. Determine the stresses and axial deformations when an axial load of $F = 1$ kN is applied.

**3.** A steel cable 10 mm in diameter and 1 km long bears a load in addition to its own weight of $W = 150$ N as shown in Fig. P.1.3. Find the total elongation of the cable.

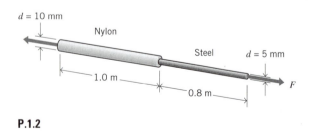

P.1.2

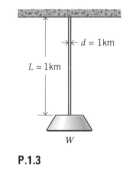

P.1.3

---

[18] See, for example H. Petroski, *To Engineer Is Human: the Role of Failure in Successful Design,* Vintage Press, New York, 1992.

**4.** Using the spreadsheet of material properties `props.csv` included on the text software diskette, rank the given materials in terms of the length of the rod that will just barely support its own weight.

**5.** Plot the maximum self-supporting rod lengths of the materials in Problem 4 versus the cost (per unit cross-sectional area) of the rod.

**6.** Augment the plot of Problem 5 by including the composite reinforcing fibers of Table 1.2.

**7.** Show that the effective stiffness of two springs connected in (*a*) series and (*b*) parallel (see Fig. P.1.7) is

(*a*) Series: $\dfrac{1}{k_{\text{eff}}} = \dfrac{1}{k_1} + \dfrac{1}{k_2}$

(*b*) Parallel: $k_{\text{eff}} = k_1 + k_2$

(Note that these are the *reverse* of the relations for the effective electrical resistance of two resistors connected in series and parallel—which use the same graphic symbols.)

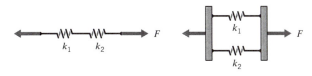

**P.1.7**

**8.** A tapered column of modulus $E$ and mass density $\rho$ varies linearly from a radius of $r_1$ to $r_2$ in a length $L$, as shown in Fig. P.1.8. Find the total deformation caused by an axial load $P$.

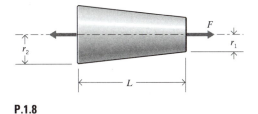

**P.1.8**

**9.** A tapered column of modulus $E$ and mass density $\rho$ varies linearly from a radius of $r_1$ to $r_2$ in a length $L$ and is hanging from its broad end, as shown in Fig. P.1.9. Find the total deformation due to the weight of the bar.

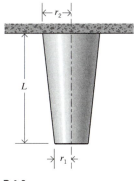

**P.1.9**

**10.** A rod of circular cross section hangs under the influence of its own weight and also has an axial load $P$ suspended from its free end, as shown in Fig. P.1.10. Determine the shape of the bar—that is, the function $r(y)$ such that the axial stress is constant along the bar's length.

**P.1.10**

**11.** A bolt with 20 threads per inch passes through a sleeve, and a nut is threaded over the bolt as shown in Fig. P.1.11. The nut is then tightened one half-turn beyond finger tightness; find the stresses in the bolt and the sleeve. All materials are steel, the cross-sectional area of the bolt is 0.5 in², and the area of the sleeve is 0.4 in².

**P.1.11**

**12.** Two bars, one of steel and one of aluminum, are fitted snugly between rigid supports. They have identical length $L$ and cross-sectional area $A$ (see Fig. P.1.12). Find the compressive stress in each bar and the movement of the interface between them when the temperature is raised by an amount $\Delta T$.

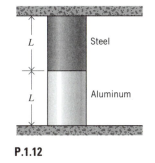

**P.1.12**

**13.** A tapered circular bar has a diameter that varies linearly from $d_1$ to $d_2$ over a distance $L$, as shown in Fig. P.1.13. The bar is fitted snugly between rigid supports, and the temperature is then changed by an amount $\Delta T$. Show that the force exerted by the bar on the supports is

$$F = -\frac{\pi \alpha E d_1 d_2 \Delta T}{4}$$

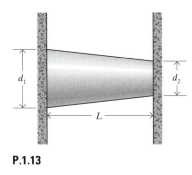

**P.1.13**

Note that $L$ does not appear in the result. Why? Evaluate this force for a copper bar with $d_1 = 25$ mm, $d_2 = 50$ mm, and $\Delta T = 5°C$.

**14.** Justify the first two terms of the Madelung series given in Eq. 1.11.

**15.** Using Eq. 1.14 to write the parameter $B$ in terms of the equilibrium interionic distance $r_0$, show that the binding energy of an ionic crystal, per bond pair, can be written as

$$U = -\frac{(n-1)ACe^2}{nr_0}$$

where $A$ is the Madelung constant, $C$ is the appropriate units conversion factor, and $e$ is the ionic charge.

**16.** Measurements of bulk compressibility are valuable for probing the bond energy function because, unlike simple tension, hydrostatic pressure causes the interionic distance to decrease uniformly. The *modulus of compressibility, $K$,* of a solid is the ratio of the pressure $p$ needed to induce a relative change in volume $dV/V$:

$$K = -\frac{dp}{(dV)/V}$$

The minus sign is needed because positive pressures induce reduced volumes (volume change negative).

(a) Use the relation $dU = p\,dV$ for the energy associated with pressure acting through a small volume change to show

$$\frac{K}{V_0} = \left(\frac{d^2U}{dV^2}\right)_{V=V_0}$$

where $V_0$ is the crystal volume at the equilibrium interionic spacing $r = a_0$.

(b) The volume of an ionic crystal containing $N$ negative and $N$ positive ions can be written as $V = cNr^3$, where $c$ is a constant dependent on the type of lattice (2 for NaCl). Use this to obtain the relation

$$\frac{K}{V_0} = \left(\frac{d^2U}{dV^2}\right)_{V=V_0} = \frac{1}{9c^2N^2r^2} \cdot \frac{d}{dr}\left(\frac{1}{r^2}\frac{dU}{dr}\right)$$

(c) Carry out the indicated differentiation of the expression for binding energy to obtain the expression

$$\frac{K}{V_0} = \frac{K}{cNr_0^3} = \frac{N}{9c^2Nr_0^2}\left[\frac{-4ACe}{r_0^5} + \frac{n(n+3)B}{r_0^{n+4}}\right]$$

Then, using the expression $B = ACe^2r_0^{n-1}/n$, obtain the formula for $n$ in terms of compressibility:

$$n = 1 + \frac{9cr_0^4K}{ACe^2}$$

**17.** Complete the following spreadsheet, filling in the values for repulsion exponent $n$ and lattice energy $U$.

| Type | $r_0$ (pm) | $K$ (GPa) | $A$ | $n$ | $U$(kJ/mol) | $U_{expt}$ |
|------|-----------|-----------|------|-----|-------------|-----------|
| LiF | 201.4 | 6.710e+01 | 1.750 | | | $-1014$ |
| NaCl | 282.0 | 2.400e+01 | 1.750 | | | $-764$ |
| KBr | 329.8 | 1.480e+01 | 1.750 | | | $-663$ |

The column labeled $U_{expt}$ lists experimentally obtained values of the lattice energy.

**18.** Given the definition of Helmholtz free energy,

$$A = U - TS$$

along with the first and second laws of thermodynamics,

$$dU = dQ + dW$$

$$dQ = T\,dS$$

where $U$ is the internal energy, $T$ is the temperature, $S$ is the entropy, $Q$ is the heat, and $W$ is the mechanical work, show that the force $F$ required to hold the ends of a tensile specimen a length $L$ apart is related to the Helmholtz energy as

$$F = \left(\frac{\partial A}{\partial L}\right)_{T,V}$$

**19.** Show that the temperature dependence of the force needed to hold a tensile specimen at fixed length as the temperature is changed (neglecting thermal expansion effects) is related to the dependence of the entropy on extension as

$$\left(\frac{\partial F}{\partial T}\right)_L = -\left(\frac{\partial S}{\partial L}\right)_T$$

**20.** Show that if an ideal rubber ($dU = 0$) of mass $M$ and specific head $c$ is extended adiabatically (that is, without addition or removal of heat), its temperature will change according to the relation

$$\frac{\partial T}{\partial L} = \frac{-T}{Mc}\left(\frac{\partial S}{\partial L}\right)$$

That is, if the entropy is reduced upon extension, the temperature will rise. This is known as the *thermoelastic effect.*

**21.** Use the expression found in Problem 20 to obtain the temperature change $dT$ in terms of a increase $d\lambda$ in the extension ratio as

$$dT = \frac{\sigma}{\rho c} d\lambda$$

where $\sigma$ is the engineering stress (load divided by original area) and $\rho$ is the mass density.

**22.** Show that the end-to-end distance $r_0$ of a chain composed of $n$ freely jointed links of length $a$ is given by $r_0 = na^2$.

**23.** Evaluate the temperature rise in a rubber specimen of $\rho = 1100$ kg/m³, $c = 2$ kJ/kg-K, $NkT = 500$ kPa, subjected to an axial extension $\lambda = 4$.

**24.** Show that the initial engineering modulus of a rubber whose stress-strain curve is given by Eq. 1.22 is $E = 3NRT$.

**25.** Calculate Young's modulus of a rubber of density 1100 g/mol and whose inter-crosslink segments have a molecular weight of 2500 g/mol. The temperature is 25°C.

**26.** Show that in the case of biaxial extension ($\lambda_x$ and $\lambda_y$ prescribed), the $x$-direction stress based on the original cross-sectional dimensions is

$$\sigma_x = NkT\left(\lambda_x - \frac{1}{\lambda_x^3 \lambda_y^2}\right)$$

and based on the deformed dimensions,

$$_t\sigma_x = NkT\left(\lambda_x^2 - \frac{1}{\lambda_x^2 \lambda_y^2}\right)$$

where the $t$ subscript indicates a "true" or current stress.

**27.** Estimate the initial elastic modulus $E$, at a temperature of 20°C, of an elastomer having a molecular weight of 7,500 g/mol between crosslinks and a density of 1.0 g/cm³. What is the percentage change in the modulus if the temperature is raised to 40°C?

**28.** Consider a line on a rubber sheet, originally oriented at an angle $\phi_0$ from the vertical. When the sheet is stretched in the vertical direction by an amount $\lambda_y = \lambda$, the line rotates to a new inclination angle $\phi'$. Show that

$$\tan \phi' = \frac{1}{\lambda^{3/2}} \tan \phi_0$$

**29.** Before stretching, the molecular segments in a rubber sheet are assumed to be distributed uniformly over all directions, so the fraction of segments $f(\phi)$ oriented in a particular range of angles $d\phi$ is

$$f(\phi) = \frac{dA}{A} = \frac{2\pi r^2 \sin \phi d\phi}{2\pi r}$$

The *Herrman orientation parameter* is defined in terms of the mean orientation as

$$f = \frac{1}{2}\left(3\langle\cos^2 \phi'\rangle - 1\right), \qquad \langle\cos^2 \phi'\rangle = \int_0^{\pi/2} \cos^2 \phi' f(\phi) d\phi$$

Using the result of the previous problem, plot the orientation function $f$ as a function of the extension ratio $\lambda$.

**30.** The *Voigt* model for linear viscoelastic response consists of a Hookean spring of constant $k$ and Newtonian dashpot of viscosity $\eta$ connected in parallel, as shown in Fig. P. 1.30. Why is this model not applicable to stress relaxation experiments?

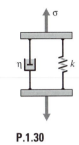

**P.1.30**

**31.** Give an intuitive description of the response of the Voigt model to a constant creep stress.

**32.** Determine the differential stress-strain law for the Voigt model.

**33.** Solve the stress-strain law for the Voigt model in the case of constant stress to give the model creep compliance $C_{crp} = \epsilon(t)/\sigma_0$.

**34.** Show that the differential stress-strain relation for the Maxwell form of the Standard Linear Solid, shown in Fig. P.1.34, can be written

$$(k_e + k_1)\dot{\epsilon} + \frac{k_e \epsilon}{\tau} = \dot{\sigma} + \frac{\sigma}{\tau}$$

where $\tau = \eta/k_1$.

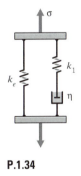

**P.1.34**

**35.** Show that the relaxation modulus of the Maxwell Standard Linear Solid is

$$E_{rel} = k_e + k_1 e^{-t/\tau}$$

Sketch the form of this relation versus time and also logarithmic time, and state how the glassy and rubbery moduli of the material are related to the constants $k_e$ and $k_1$.

**36.** Show the creep compliance of the Maxwell Standard Linear Solid to be

$$C_{crp} = \frac{\epsilon(t)}{\sigma_0} = \frac{1}{k_e} + \left[\frac{1}{k_e + k_1} - \frac{1}{k_e}\right]e^{-t/\tau_c}$$

where

$$\tau_c = \frac{k_e + k_1}{k_e}\tau$$

is the "retardation time." Note that creep response is slower (larger characteristic time) than relaxation, by the ratio of the glassy to rubbery moduli. See Fig. P.1.36.

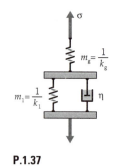

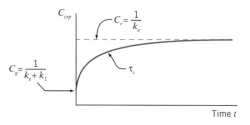

**P.1.36**

**37.** An formulation of the Standard Linear Solid uses a Voigt model in series with a spring (see Fig. P.1.37). Sketch the response of this Voigt stress relaxation using intuition alone, identifying the critical points on the curve (the glass modulus $E_g$, the rubbery modulus $E_r$, the relaxation time $\tau_{rel}$) in terms of the model parameters $k_g$, $k_v$, and $\tau = \eta/k_v$. Will the relaxation time $\tau_{rel}$ be the same as, greater than, or less than the model $\tau$? Why?

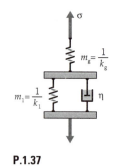

**P.1.37**

**38.** Derive an analytical expression for the relaxation modulus $E_{rel}$ of the Voigt Standard Linear Solid in the preceding problem. Show that the expression provides correct values for the glassy (instantaneous) and rubbery (long-time) response.

**39.** Plot the relaxation modulus $E_{rel} = \sigma(t)/\epsilon_0$ and the creep modulus $E_{crp} = 1/C_{crp} = \sigma_0/\epsilon(t)$ for the Maxwell form of the Standard Linear Solid, with $k_e = 10^3$ psi, $k_1 = 10^5 - 10^3$ psi, and $\tau = 1$ s. Use log-log coordinates, and let $\log t$ range from $-2$ to $3$. Also plot the difference $E_{crp} - E_{rel}$; nonzero values are an indication that viscoelastic effects are present.

**40.** Compute the longitudinal and transverse stiffness $(E_1, E_2)$ of an S-glass epoxy lamina for a fiber volume fraction $V_f = 0.7$, using the constituent properties in Appendix A and Table 1.2.

**41.** Plot the longitudinal stiffness $E_1$ of an E-glass/nylon unidirectionally reinforced composite as a function of the volume fraction $V_f$.

**42.** Plot the longitudinal tensile strength of an E-glass/epoxy unidirectionally reinforced composite as a function of the volume fraction $V_f$.

**43.** What is the maximum fiber volume fraction $V_f$ that could be obtained in a unidirectionally reinforced composite with optimal fiber packing?

**44.** Show the transverse modulus of a unidirectionally reinforced composite to be

$$\frac{1}{E_2} = \frac{v_f}{E_f} + \frac{v_m}{E_m}$$

or, in terms of compliances,

$$C_2 = C_f v_f + C_m v_m$$

**45.** *Computer spreadsheet exercise:* The file curve.csv in the computer diskette accompanying this text contains a comma-separated text file in which the first and second entries on each line are the displacement and load, respectively, recorded during a tensile test of an oriented polyethylene monofilament. The diameter of the filament was 0.254 mm, and the gage length was 762 mm.

Import the data into your own spreadsheet, and create columns for engineering strain and stress.

**46.** Plot the stress-strain curve for the specimen of Problem 45.

**47.** Perform a linear regression on the initial linear portion of the curve in Problem 46, and report the value of Young's modulus based on the slope.

**48.** Integrate the curve of Problem 44 numerically to obtain the modulus of rupture. (The trapezoidal rule is convenient in spreadsheet work:

$$U = \int \sigma\, d\epsilon \approx \sum_j \left(\frac{\sigma_j + \sigma_{j-1}}{2}\right)(\epsilon_j - \epsilon_{j-1})$$

where the $j$ subscript denotes the spreadsheet row.)

# SIMPLE TENSILE AND SHEAR STRUCTURES

## CHAPTER 2

In this and subsequent chapters, we seek to apply the concepts of materials response outlined in Chapter 1 to actual structures of the sort encountered in engineering design. We will restrict ourselves initially to situations having relatively simple geometries—specifically, trusses, pressure vessels, and shafts loaded in torsion. In part this is done to introduce new concepts without excessive detail, but these structural types are far from being of academic interest only. As it is very often true that the best design is the simplest design, these simple structures make up a large part of modern construction.

## 2.1 TRUSSES

A *truss* is an assemblage of long, slender structural elements that are connected at their ends. Trusses find substantial use in modern construction, for instance as derrick booms (see Fig. 2.1), bridges, or scaffolding. In addition to their practical importance

**FIGURE 2.1** Crane derrick truss. (Mitchell Funk/The Image Bank.)

as useful structures, truss elements have a dimensional simplicity that will help us extend further the concepts of mechanics introduced in the last chapter. This chapter will use trusses to introduce important concepts in statics and numerical analysis that will be extended in later chapters to more general problems.

## EXAMPLE 2.1

Trusses are often used to stiffen structures, and most people have seen the often very elaborate systems of cross-bracing used in bridges. The truss bracing used to stiffen the towers of suspension bridges against buckling is hard to miss, but not everyone notices the vertical truss panels on most such bridges, which serve to stiffen the deck against flexural and torsional deformation (see Fig. 2.2).

Many readers will have seen the very famous movie taken on November 7, 1940, by Barney Elliott of The Camera Shop in Tacoma, Washington. The wind, gusting up to 42 mph that day, induced a sequence of spectacular undulations and eventual collapse of the Tacoma Narrows bridge;[1] Fig. 2.3 shows a frame from the movie just before collapse. This bridge was built using relatively short I-beams for deck stiffening rather than truss panels, reportedly for aesthetic reasons; bridge designs of the period favored very slender and graceful-appearing structures. Even during construction the bridge became well known for its alarming tendency to sway in the wind, earning it the local nickname "Galloping Gertie."

Truss stiffeners were used when the bridge was rebuilt in 1950, and the new bridge was free of the oscillations that led to the collapse of its predecessor. This

**FIGURE 2.2** Bridge tower and deck trusses. (Larry Ulrich/Tony Stone Images/New York, Inc.)

**FIGURE 2.3** The Tacoma Narrows bridge just before collapse. (Wide World Photos, Inc.)

---

[1]An interactive instructional videodisc of the Tacoma Narrows Bridge collapse is available from Wiley Educational Software (ISBN 0-471-87320-9).

is a good example of one important use of trusses, but it is probably an even better example of the value of caution and humility in engineering. The glib answers often given for the original collapse—resonant wind gusts, von Kármán vortices, and so forth—are not really satisfactory beyond the obvious statement that the deck was not stiff enough. Even today, knowledgeable engineers argue about the very complicated structural dynamics involved. Ultimately, many uncertainties exist even in designs completed using very modern and elaborate techniques. A wise designer will never fully trust a theoretical result, computer-generated or not, and will take as much advantage of experience and intuition as possible.

## 2.1.1 Statics Analysis of Forces

Newton observed that a mass accelerates according to the vector sum of forces applied to it: $\sum \mathbf{F} = m\mathbf{a}$. (Vector quantities are indicated by boldface type.) In structures that are anchored so as to prevent motion, there is obviously no acceleration and the forces must sum to zero. This vector equation has as many scalar components as the dimensionality of the problem; for two-dimensional cases we have

$$\sum F_x = 0 \tag{2.1}$$

$$\sum F_y = 0 \tag{2.2}$$

where $F_x$ and $F_y$ are the components of $\mathbf{F}$ in the $x$ and $y$ Cartesian coordinate directions. These two equations, which we can interpret as constraining the structure against translational motion in the $x$ and $y$ directions, allow us to solve for at most two unknown forces in structural problems. If the structure is constrained against rotation as well as translation, we can add a *moment equation*, which states that the sum of moments or torques in the $x$-$y$ plane must also add to zero:

$$\sum M_{xy} = 0 \tag{2.3}$$

In two dimensions, then, we have three equations of static equilibrium that can be used to solve for unknown forces. In three dimensions, a third force equation and two more moment equations are added, for a total of six:

$$\begin{array}{ll} \sum F_x = 0 & \sum M_{xy} = 0 \\ \sum F_y = 0 & \sum M_{xz} = 0 \\ \sum F_z = 0 & \sum M_{yz} = 0 \end{array} \tag{2.4}$$

These equations can be applied to the structure as a whole, or we can (conceptually) remove a piece of the structure and consider the forces acting on the removed piece. A sketch of the piece, showing all forces acting on it, is called a *free-body diagram*. If the number of unknown forces in the diagram is equal to or less than the number of available static equilibrium equations, the unknowns can be solved in a straightforward manner; such problems are termed *statically determinate*. Note that these equilibrium equations do not assume anything about the material from which the structure is made, so the resulting forces are also independent of the material.

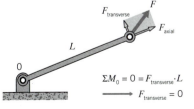

**FIGURE 2.4** Pinned elements cannot support transverse loads.

In the analyses to be considered here, the truss elements are assumed to be joined together by pins or other such connections that allow free rotation around the joint. As seen in the free-body diagram of Fig. 2.4, this inability to resist rotation implies that the force acting on a truss element's pin joint must be in the element's axial direction: Any transverse component would tend to cause rotation, and if the element is to be in static equilibrium, the moment equation forces the transverse component to vanish. If the element ends were to be welded or bolted rather than simply pinned, the end connection could transmit transverse forces and bending moments into the element. Such a structure would then be called a *frame* rather than a truss, and its analysis would have to include bending effects. Such structures will be treated in Chapter 4.

Knowing that the force in each truss element must be in the element's axial direction is the key to solving for the element forces in trusses that contain many elements. Each element meeting at a pin joint will pull or push on the pin depending on whether the element is in tension or compression, and since the pin must be in static equilibrium, the sum of all element forces acting on the pin must equal the force that is externally applied to the pin:

$$\sum_e \mathbf{F}_i^e = \mathbf{F}_i$$

Here the *e* superscript indicates the vector force supplied by the element on the *i*th pin in the truss, and $\mathbf{F}_i$ is the force externally applied to that pin. The summation is over all the elements connected to the pin.

## EXAMPLE 2.2

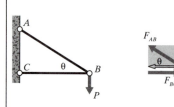

FIGURE 2.5 A two-element truss.

The very simple two-element truss often found in high school physics books and shown in Fig. 2.5 can be analyzed this way. Intuition tells us that the upper element, connecting joints *A* and *B*, is in tension, while element *BC* is in compression. In more complicated problems it is not always possible to determine the sign of the element force by inspection, but it doesn't matter. In sketching the free-body diagrams for the pins, the load can be drawn in either direction; if the guess turns out to be wrong, the solution will give a negative value for the force magnitude.

The unknown forces on the connecting pin *B* are in the direction of the elements attached to it, and because there are only two such forces, they may be determined from the two static equilibrium force equations:

$$\sum F_y = 0 = +F_{AB} \sin \theta - P \Rightarrow F_{AB} = \frac{P}{\sin \theta}$$

$$\sum F_x = 0 = -F_{AB} \cos \theta + F_{BC} \Rightarrow F_{BC} = F_{AB} \cos \theta = \frac{P}{\tan \theta}$$

In more complicated trusses the general approach is to start at a pin joint that contains no more than two elements having unknown forces and then to work from joint to joint, using the element forces from the previous step to reduce the number of unknowns. Consider the six-element truss shown in Fig. 2.6, in which the joints and elements are numbered as indicated, with the element numbers appearing in circles. Joint 3 is a natural starting point, since only forces $F_2$ and $F_5$ appear as unknowns.

Once $F_5$ is found, an analysis of joint 5 has only forces $F_4$ and $F_6$ as unknowns. Finally, the free-body diagram of node 2 can be completed, since only $F_1$ and $F_3$ are now unknown. The force analysis is then complete.

There are often many ways to complete problems such as this, perhaps with some being easier than others. Another approach might be to start at one of the joints at the wall: joint 1 or joint 4. The problem as originally stated gives these joints as having fixed (zero) displacements rather than specified forces. This is an example of a *mixed boundary value problem,* with some parts of the boundary having specified forces and the remaining parts having specified displacements. Such problems are generally more difficult, and require more mathematical information for their solution, than problems having only one or the other type of boundary condition. However, in the *statically determinate* problems, the structure can be converted to a load-only type by invoking static equilibrium on the structure as a whole. The fixed-displacement boundary conditions are then replaced by *reaction forces* that are set up at the points of constraint.

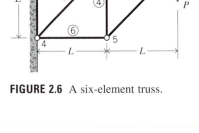

**FIGURE 2.6** A six-element truss.

Moment equilibrium equations were not useful in the joint-by-joint analysis described earlier, because individual elements cannot support moments. However, as seen in Fig. 2.7, we can consider the six-element truss as a whole and take moments around joint 4. With clockwise-tending moments taken as positive, this gives

$$\sum M_4 = 0 = -F_1 \times L + P \times 2L \Rightarrow F_1 = 2P$$

The force $F_1$ is the force applied by the wall to joint 1, and this is obviously equal to the tensile force in element 1. There can be no vertical component of this reaction force, because the element forces must be axial and only element 1 is connected to joint 1. At joint 4, reaction forces $R_x$ and $R_y$ can act in both the $x$ and $y$ directions, because element 3 is not perpendicular to the wall. These reaction forces can be found by invoking horizontal and vertical equilibrium:

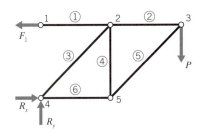

**FIGURE 2.7** Free-body diagram of six-element truss.

$$\sum F_x = 0 = -F_1 + R_x \Rightarrow R_x = F_1 = 2P$$

$$\sum F_y = 0 = +R_y - P \Rightarrow R_y = P$$

A joint-by-joint analysis can now be started from joint 4, since only two unknown forces act there (see Fig. 2.8). For vertical equilibrium, $F_3 \cos 45° = P$, so $F_3 = \sqrt{2}P$. Then, for horizontal equilibrium, $F_6 + F_3 \cos 45° = 2P$, so $F_6 = P$. Now moving to joint 5, horizontal equilibrium gives $F_5 \cos 45° = P$, so $F_5 = F_3 = \sqrt{2}P$, and vertical equilibrium gives $F_4 = F_5 \cos 45°$, so $F_4 = P$. Finally, at joint 3, horizontal equilibrium gives $F_2 = F_5 \cos 45°$, so $F_2 = P$.

In actual truss design, once each element's force is known, its cross-sectional area can then be calculated so as to keep the element stress, according to $\sigma = P/A$, safely less than the material's yield point. Elements in compression, however, must be analyzed for buckling as well, because their ratios of $EI$ to $L^2$ are generally low. The buckling load can be increased substantially by bracing the element against sideward deflection, and this bracing is evident in most bridges and cranes. Also, the truss elements are usually held together by welded or bolted joints rather than pins. These joints can carry some bending moments, which helps stiffen the truss against buckling.

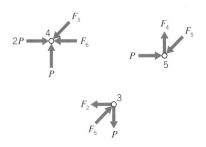

**FIGURE 2.8** Individual joint diagrams.

## 2.1.2 Deflections

It may be important in some applications that the truss be stiff enough to keep the deformations inside specified limits. Astronomical telescopes such as that in Fig. 2.9

**FIGURE 2.9** A truss-based astronomical telescope.

are an example, since deflection of the structure supporting the optical assemblies can degrade the focusing ability of the instrument. A typical derrick or bridge, however, is probably more likely to be strength- rather than stiffness-critical, so it might appear that deflections would be relatively unimportant. However, it will be seen that deflections must be considered to solve the great number of structures that are *not* statically determinate. The following sections treat truss deflections for both these reasons.

### Geometrical Approach

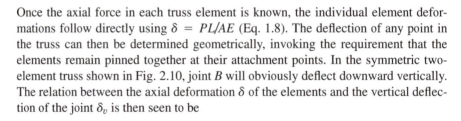

**FIGURE 2.10** Two-element truss.

Once the axial force in each truss element is known, the individual element deformations follow directly using $\delta = PL/AE$ (Eq. 1.8). The deflection of any point in the truss can then be determined geometrically, invoking the requirement that the elements remain pinned together at their attachment points. In the symmetric two-element truss shown in Fig. 2.10, joint B will obviously deflect downward vertically. The relation between the axial deformation $\delta$ of the elements and the vertical deflection of the joint $\delta_v$ is then seen to be

$$\delta_v = \frac{\delta}{\cos \theta}$$

It is assumed here that the deformation is small enough that the gross aspects of the geometry are essentially unchanged; in this case, that the angle $\theta$ is the same before and after the load is applied.

In geometrical analyses of more complicated trusses it is sometimes convenient to visualize unpinning the elements at a selected joint, letting the elements elongate or shrink according to the axial force they are transmitting, and then swinging them around the still-pinned joint until the pin locations match up again. The motions of the unpinned ends would trace out circular paths, but if the deflections are small, the path can be approximated as a straight line perpendicular to the element axis. The joint position can then be computed from Pythagorean relationships.

In the two-element truss shown earlier in Fig. 2.5, we had $P_{AB} = P/\sin\theta$ and $P_{BC} = P/\tan\theta$. If the pin at joint $B$ were removed, the element deflections would be

$$\delta_{AB} = \frac{P}{\sin\theta}\left(\frac{L}{AE}\right)_{AB} \quad \text{(tension)}$$

$$\delta_{BC} = \frac{P}{\tan\theta}\left(\frac{L}{AE}\right)_{BC} \quad \text{(compression)}$$

The total downward deflection of joint $B$ is then

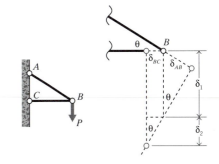

**FIGURE 2.11** Displacements in the two-element truss.

$$\delta_v = \delta_1 + \delta_2 = \frac{\delta_{AB}}{\sin\theta} + \frac{\delta_{BC}}{\tan\theta}$$

$$= \frac{P}{\sin^2\theta}\left(\frac{L}{AE}\right)_{AB} + \frac{P}{\tan^2\theta}\left(\frac{L}{AE}\right)_{BC}$$

These deflections are shown in Fig. 2.11.

The horizontal deflection $\delta_h$ of the pin is easier to compute, since it is just the contraction of element $BC$:

$$\delta_h = \delta_{BC} = \frac{P}{\tan\theta}\left(\frac{L}{AE}\right)_{BC}$$

### Energy Approach

The geometrical approach to truss deformation analysis can be rather tedious, especially as problems become larger. Many problems can be solved more easily using a strain energy rather than a force-at-a-point approach. The total strain energy $U$ in a single, elastically loaded truss element is

$$U = \int P\,d\delta$$

The increment of deformation $d\delta$ is related to a corresponding increment of load $dP$ as follows:

$$\delta = \frac{PL}{AE} \Rightarrow d\delta = \frac{L}{AE}dP$$

The strain energy is then

$$U = \int P\frac{L}{AE}dP = \frac{P^2 L}{2AE}$$

The *incremental* increase in strain energy corresponding to an increase in deformation $d\delta$ is just $dU = Pd\delta$. If the force-elongation curve is linear, this is identical to the increase in the quantity called the *complementary* strain energy: $dU^c = \delta\,dP$. These quantities are depicted in Fig. 2.12. Now consider a system subjected to a number of loads acting at different nodes. If we were to increase the $i$th load slightly

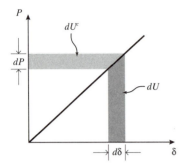

**FIGURE 2.12** Increments of strain and complementary strain energy.

while holding all the other loads constant, the increase in the complementary energy would be

$$dU^c = \delta_i \, dP_i$$

where $\delta_i$ is the displacement that would occur at the location of $P_i$, moving in the same direction as the force vector for $P_i$. Rearranging,

$$\delta_i = \frac{\partial U^c}{\partial P_i}$$

and since $U^c = U$:

$$\boxed{\delta_i = \frac{\partial U}{\partial P_i}} \tag{2.5}$$

Hence the displacement at a given point is the derivative of the total strain energy with respect to the load acting at that point. This provides the basis of an extremely useful method of displacement analysis known as *Castigliano's theorem*,[2] which can be stated for truss problems as the following recipe:

1. Let the load applied at the joint whose deformation is sought, in the direction of the desired deformation, be written as an algebraic variable, say $Q$. If the load is known numerically, replace the number with a letter. If there is no load at the desired location and direction, put an imaginary one there that will be set to zero at the end of the problem.

2. Solve for the forces $F_i(Q)$ in each truss element, each of which may be dependent on the load $Q$ assigned in the previous step.

3. Use these forces to compute the strain energy for each element, and sum the energies in each element to obtain the total strain energy for the truss:

$$U_{\text{tot}} = \sum_i U_i = \sum_i \frac{F_i^2 L_i}{2 A_i E_i} \tag{2.6}$$

Each term in this summation may contain the variable $Q$.

4. The deformation *congruent* to $Q$ (that is, the deformation at the point where $Q$ is applied and in the same direction as $Q$) is then

$$\delta_Q = \frac{\partial U_{\text{tot}}}{\partial Q} = \sum_i \frac{F_i L_i}{A_i E_i} \frac{\partial F_i(Q)}{\partial Q} \tag{2.7}$$

5. The load $Q$ is replaced by its numerical value, if known, or by zero, if it was an imaginary load in the first place.

---

[2]From the 1873 thesis of the Italian engineer Alberto Castigliano (1847–1884) at the Turin Polytechnical Institute.

Applying this method to the vertical deflection of the two-element truss of Fig.
2.5, the problem already has a force in the required direction: the applied downward
load $P$. The forces have already been shown to be $P_{AB} = P/\sin\theta$ and $P_{BC} = P/\tan\theta$,
so the vertical deflection can be written immediately as

$$\delta_v = P_{AB}\left(\frac{AE}{L}\right)_{AB}\frac{\partial P_{AB}}{\partial P} + P_{BC}\left(\frac{AE}{L}\right)_{BC}\frac{\partial P_{BC}}{\partial P}$$

$$= \frac{P}{\sin\theta}\left(\frac{AE}{L}\right)_{AB}\frac{1}{\sin\theta} + \frac{P}{\tan\theta}\left(\frac{AE}{L}\right)_{BC}\frac{1}{\tan\theta}$$

This is identical to the expression obtained from geometric considerations. The energy
method didn't save too many algebraic steps in this case, but it avoided having to
visualize and idealize the displacements geometrically.

If the horizontal displacement at joint $B$ is desired, the method requires that a
horizontal force exist at that point. One isn't given, so we place an imaginary one
there, say $Q$. The truss is then reanalyzed statically to find how the element forces are
influenced by this new force $Q$. The upper element force is $P_{AB} = P/\sin\theta$ as before,
and the lower element force becomes $P_{BC} = P/\tan\theta - Q$. Repeating the Castigliano
process, but now differentiating with respect to $Q$,

$$\delta_h = P_{AB}\left(\frac{AE}{L}\right)_{AB}\frac{\partial P_{AB}}{\partial Q} + P_{BC}\left(\frac{AE}{L}\right)_{BC}\frac{\partial P_{BC}}{\partial Q}$$

$$= \frac{P}{\sin\theta}\left(\frac{AE}{L}\right)_{AB}\cdot 0 + \left(\frac{P}{\tan\theta} - Q\right)\left(\frac{AE}{L}\right)_{BC}(-1)$$

The first term vanishes upon differentiation, because $Q$ did not appear in the expression for $P_{AB}$. This is the method's way of noticing that the horizontal deflection is
determined completely by the contraction of element $BC$. Upon setting $Q = 0$, the
final result is

$$\delta_h = -\frac{P}{\tan\theta}\left(\frac{AE}{L}\right)_{BC}$$

as before.

## EXAMPLE 2.3

Consider the six-element truss of Fig. 2.6 whose individual element forces were
found earlier by free-body diagrams. We are seeking the vertical deflection of node
6, which is congruent to the force $P$. Using Castigliano's method, this deflection
is the derivative of the total strain energy with respect to $P$. Equivalently, we can
differentiate the strain energy of each element with respect to $P$ individually and
then add the contributions of each element to obtain the final result:

$$\delta_P = \frac{\partial}{\partial P}\sum_i\frac{F_i^2 L_i}{2A_i E_i} = \sum_i\left(\frac{F_i L_i}{A_i E_i}\frac{\partial F_i}{\partial P}\right)$$

To systemize this approach, we can form a table of needed parameters as follows:

| $i$ | $F_i$ | $\dfrac{L_i}{A_i E_i}$ | $\dfrac{\partial F_i}{\partial P}$ | $\dfrac{F_i L_i}{A_i E_i}\dfrac{\partial F_i}{\partial P}$ |
|---|---|---|---|---|
| 1 | $2P$ | $L/AE$ | 2 | $4PL/AE$ |
| 2 | $P$ | $L/AE$ | 1 | $PL/AE$ |
| 3 | $\sqrt{2}P$ | $\sqrt{2}L/AE$ | $\sqrt{2}$ | $2.83PL/AE$ |
| 4 | $P$ | $L/AE$ | 1 | $PL/AE$ |
| 5 | $\sqrt{2}P$ | $\sqrt{2}L/AE$ | $\sqrt{2}$ | $2.83PL/AE$ |
| 6 | $P$ | $L/AE$ | 1 | $PL/AE$ |
| | | | | $\delta_P = \sum = 12.7PL/AE$ |

If for instance we have as numerical parameters $P = 1000$ lbs, $L = 100$ in, $E = 30$ Mpsi, and $A = 0.5$ in$^2$, then $\delta_p = 0.0844$ in.

### Statically Indeterminate Trusses

It has already been noted that the element forces in the truss problems treated up to now do not depend on the properties of the materials used in their construction, just as the stress in a simple tension test is independent of the material. This result, which certainly makes the problem easier to solve, is a consequence of the earlier problems being statically determinate—that is, solvable using only the equations of static equilibrium. Statical determinacy, then, is an important aspect of the difficulty we can expect in solving the problem. Not all problems are statically determinate, and one consequence of this indeterminacy is that the forces in the structure *may* depend on the material properties.

After performing a static analysis of the truss as a whole to find reaction forces at the supports, we typically try to find the element forces using the joint-at-a-time method described previously. However, there can be at most two unknown forces at a pin joint in a two-dimensional truss problem if the joint is to be solved using statics alone, because the moment equation does not provide usable information in this case. If more unknowns are present no matter in which order the truss joints are analyzed, then a number of additional equations equal to the remaining unknowns must be found. These extra equations enforce compatibility of the various joint displacements, each of which must be such as to keep the truss joints pinned together.

## EXAMPLE 2.4

A simple example—just two truss elements acting in parallel, as shown in Fig. 2.13—will show the approach needed. Here the compatibility condition is just

$$\delta_A = \delta_B$$

The individual element displacements are related to the element forces by $\delta = PL/AE$, which is material-dependent and can be termed a *constitutive* equation because it reflects the material's mechanical constitution. Combining this with the compatibility condition gives

$$\frac{P_A L}{A_A E_A} = \frac{P_B L}{A_B E_B} \Rightarrow P_B = P_A \frac{A_B E_B}{A_A E_A}$$

Finally, the individual element forces must add up to the total applied load $P$ in order to satisfy equilibrium:

$$P = P_A + P_B = P_A + P_A \frac{A_B E_B}{A_A E_A} \Rightarrow P_A = P\left(\frac{1}{1 + \left(\frac{A_B E_B}{A_A E_A}\right)}\right)$$

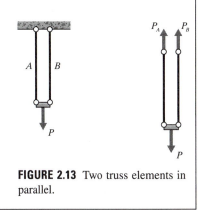

**FIGURE 2.13** Two truss elements in parallel.

Note that the final answer in the preceding example depends on the element dimensions and material stiffnesses, as promised. Here the geometrical compatibility condition was very simple and obvious, namely that the displacements of the two element end joints were identical. In more complex trusses these relations can be subtle, but they tend to become more evident with practice.

Three different types of relations were used in the foregoing problem: a *compatibility* equation, stating how the structure must deform kinematically in order to remain connected; a *constitutive* equation, embodying the stress-strain response of the material; and an *equilibrium* equation, stating that the forces must sum to zero if acceleration is to be avoided. These three concepts, made somewhat more general mathematically to handle geometrically more elaborate problems, underlie all of solid mechanics.

In Chapter 1 we noted that the stress in a tensile specimen is determined only by considerations of static equilibrium, being given by $\sigma = P/A$ independent of the material properties. We see now that the statical determinacy depends, among other things, on the material being *homogeneous* (identical throughout). If the tensile specimen is composed of two subunits each having different properties, the stresses will be allocated differently among the two units, and the stresses will not be uniform. Whenever a stress or deformation formula is copied out of a handbook, the user must be careful to note the limitations of the underlying theory. The handbook formulae are usually applicable only to homogeneous materials in their linear elastic range, and higher-order theories must be used when these conditions are not met.

## EXAMPLE 2.5

Figure 2.14*a* shows another statically indeterminate truss, with three elements having the same area and modulus, but different lengths, meeting at a common node. At a glance we can see that node 4 has three elements meeting there whose forces are unknown, and this is one more than the useful equations of static equilibrium will be able to handle. This is also evident in the free-body diagram of Fig. 2.14 *b*: Horizontal and vertical equilibrium gives

$$\sum F_x = 0 = -F_1 \sin\theta + F_3 \sin\theta \rightarrow F_1 = F_3$$

$$\sum F_y = 0 = -P + F_2 + F_1 \cos\theta + F_3 \cos\theta \rightarrow F_2 + 2F_3 \cos\theta = P \quad (2.8)$$

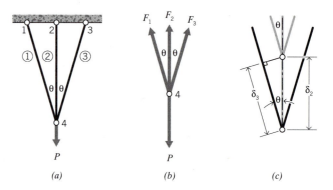

**FIGURE 2.14** (*a*) Three-element statically indeterminate truss. (*b*) Free-body diagram of node 4. (*c*) Deflections at node 4.

These two equations are clearly not sufficient to determine the unknowns $F_1, F_2, F_3$. We need another equation, which is provided by requiring that the deformation be such as to keep the truss pinned together at node 4. Since the symmetry of the problems tells us that the deflection there is straight downward, the diagram in Fig. 2.14*c* can be used. Also, since the deflection is small relative to the lengths of the elements, the angle of element 3 remains essentially unchanged after deformation. This lets us write

$$\delta_3 = \delta_2 \cos\theta$$

or

$$\frac{F_3 L_3}{A_3 E_3} = \frac{F_2 L_2}{A_2 E_2}\cos\theta$$

Using $A_2 = A_3$, $E_2 = E_3$, $L_3 = L$, and $L_2 = L\cos\theta$, this becomes

$$F_3 = F_2 \cos^2\theta$$

Solving this simultaneously with Eq. 2.8, we obtain

$$F_2 = \frac{P}{1 + 2\cos^3\theta}, \qquad F_3 = \frac{P\cos^2\theta}{1 + 2\cos^3\theta}$$

Note that the modulus $E$ does not appear in this result, even though the problem is statically indeterminate. If the elements had different stiffnesses, however, the cancellation of $E$ would not have occurred.

## 2.1.3 Matrix Analysis of Trusses

The joint-by-joint free-body analysis of trusses is tedious for large and complicated structures, especially if statical indeterminacy requires that displacement compatibility be considered along with static equilibrium. However, even statically indeterminate trusses can be solved quickly and reliably for both forces and displacements by a straightforward numerical procedure known as *matrix structural analysis*. This method is a forerunner of the more general computer method named *finite element analysis* (FEA), which has come to dominate much of engineering analysis in the past two decades. The foundations of matrix analysis will be outlined here, primarily as an introduction to the more general use of FEA in stress analysis.

Matrix analysis of trusses operates by considering the stiffness of each truss element one at a time and then using these stiffnesses to determine the forces that are set up in the truss elements by the displacements of the joints, usually called "nodes" in finite element analysis. Then, noting that the sum of the forces contributed by each element to a node must equal the force that is externally applied to that node, we can assemble a sequence of linear algebraic equations in which the nodal displacements are the unknowns and the applied nodal forces are known quantities. These equations are conveniently written in matrix form,[3] which gives the method its name:

$$\begin{bmatrix} K_{11} & K_{12} & \cdots & K_{1n} \\ K_{21} & K_{22} & \cdots & K_{2n} \\ \vdots & \vdots & \ddots & \vdots \\ K_{n1} & K_{n2} & \cdots & K_{nn} \end{bmatrix} \begin{Bmatrix} u_1 \\ u_2 \\ \vdots \\ u_n \end{Bmatrix} = \begin{Bmatrix} f_1 \\ f_2 \\ \vdots \\ f_n \end{Bmatrix}$$

Here $u_i$ and $f_j$ indicate the deflection at the $i$th node and the force at the $j$th node (these would actually be vector quantities, with subcomponents along each coordinate axis). The $K_{ij}$ coefficient array is called the *global stiffness matrix*, with the $ij$ component being physically the influence of the $j$th displacement on the $i$th force. The matrix equations can be abbreviated as

$$K_{ij}u_j = f_i \quad \text{or} \quad \mathbf{Ku} = \mathbf{f} \tag{2.9}$$

using either subscripts or boldface to indicate vector and matrix quantities.

Either the force externally applied or the displacement is known at the outset for each node, and it is impossible to specify simultaneously both an arbitrary displacement *and* a force on a given node. These prescribed nodal forces and displacements are the boundary conditions of the problem. The task of analysis is to determine

- The forces that accompany the imposed displacements
- The displacements at the nodes where known external forces are applied

### Stiffness Matrix for a Single Truss Element

As a first step in developing a set of matrix equations that describe truss systems, we need a relationship between the forces and displacements at each end of a single truss element. Consider such an element in the *x-y* plane, as shown in Fig. 2.15, attached to nodes numbered $i$ and $j$ and inclined at an angle $\theta$ from the horizontal.

Considering the elongation vector $\boldsymbol{\delta}$ to be resolved in directions along and transverse to the element, the elongation in the truss element can be written in terms of the differences in the displacements of its end points:

$$\delta = (u_j \cos\theta + v_j \sin\theta) - (u_i \cos\theta + v_i \sin\theta)$$

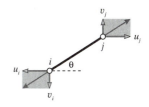

**FIGURE 2.15** Individual truss element.

---

[3] See Appendix D for a review of matrix notation.

where $u$ and $v$ are the horizontal and vertical components of the deflections, respectively. This relation can be written in matrix form as

$$\delta = [-c \ -s \ c \ s] \begin{Bmatrix} u_i \\ v_i \\ u_j \\ v_j \end{Bmatrix}$$

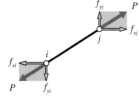

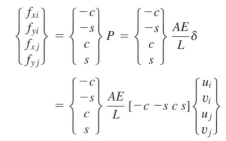

**FIGURE 2.16** Components of nodal force.

Here $c = \cos\theta$ and $s = \sin\theta$. The axial force $P$ that accompanies this elongation is given by Hooke's law for linear elastic bodies as $P = (AE/L)\delta$. The horizontal and vertical nodal forces are shown in Fig. 2.16; these can be written in terms of the total axial force as

$$\begin{Bmatrix} f_{xi} \\ f_{yi} \\ f_{xj} \\ f_{yj} \end{Bmatrix} = \begin{Bmatrix} -c \\ -s \\ c \\ s \end{Bmatrix} P = \begin{Bmatrix} -c \\ -s \\ c \\ s \end{Bmatrix} \frac{AE}{L} \delta$$

$$= \begin{Bmatrix} -c \\ -s \\ c \\ s \end{Bmatrix} \frac{AE}{L} [-c \ -s \ c \ s] \begin{Bmatrix} u_i \\ v_i \\ u_j \\ v_j \end{Bmatrix}$$

Carrying out the matrix multiplication:

$$\begin{Bmatrix} f_{xi} \\ f_{yi} \\ f_{xj} \\ f_{yj} \end{Bmatrix} = \frac{AE}{L} \begin{bmatrix} c^2 & cs & -c^2 & -cs \\ cs & s^2 & -cs & -s^2 \\ -c^2 & cs & c^2 & cs \\ -cs & -s^2 & cs & s^2 \end{bmatrix} \begin{Bmatrix} u_i \\ v_i \\ u_j \\ v_j \end{Bmatrix} \tag{2.10}$$

The quantity in brackets, multiplied by $AE/L$, is known as the element stiffness matrix $\mathbf{k}_{ij}$. Each of its terms has a physical significance, representing the contribution of one of the displacements to one of the forces. The global system of equations is formed by combining the element stiffness matrices from all of the truss elements, so their computation is central to the method of matrix structural analysis. The principal difference between the matrix truss method and the general finite element method is in how the element stiffness matrices are formed; most of the other computer operations are the same.

### Assembly of Multiple Element Contributions

The next step is to consider an assemblage of many truss elements connected by pin joints. Each element meeting at a joint, or node, will contribute a force there as dictated by the displacements of both that element's nodes (see Fig. 2.17). To maintain static equilibrium, all element force contributions $f_i^{\text{elem}}$, at the node and in the direction identified together by the index $i$, must sum to the force component $f_i^{\text{ext}}$ that is externally applied at that node in that direction:

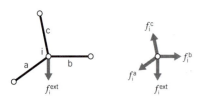

**FIGURE 2.17** Element contributions to total nodal force.

$$f_i^{\text{ext}} = \sum_{\text{elem}} f_i^{\text{elem}} = \left( \sum_{\text{elem}} k_{ij}^{\text{elem}} u_j \right) = \left( \sum_{\text{elem}} k_{ij}^{\text{elem}} \right) u_j = K_{ij} u_j$$

Each element stiffness matrix $k_{ij}^{elem}$ is added to the appropriate location $K_{ij}$ of the over-all, or *global,* stiffness matrix, which relates all of the truss displacements and forces. This process is called "assembly." The index numbers in the foregoing relation must be the "global" numbers assigned to the truss structure as a whole. However, it is generally convenient to compute the individual element stiffness matrices using a local scheme and then to have the computer convert to global numbers when assembling the individual matrices.

## EXAMPLE 2.6

The assembly process is at the heart of the finite element method, and it is worthwhile to do a simple case by hand to see how it really works. Consider the two-element truss problem of Fig. 2.10, with the nodes being assigned arbitrary "global" numbers from 1 to 3. Each node can in general move in two directions, so there are $3 \times 2 = 6$ total degrees of freedom in the problem. The global stiffness matrix will then be a $6 \times 6$ array relating the six displacements to the six externally applied forces. Only one of the displacements is unknown in this case, since all but the vertical displacement of node 2 (degree of freedom number 4) is constrained to be zero. Figure 2.18 shows a workable listing of the global numbers as well as "local" numbers for each individual element.

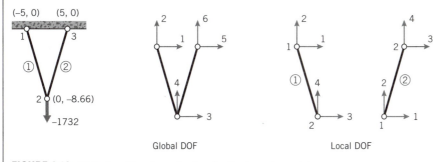

Global DOF                      Local DOF

**FIGURE 2.18** Global and local numbering for the two-element truss.

Using the local numbers, the $4 \times 4$ element stiffness matrix of each of the two elements can be evaluated according to Eq. 2.10. The inclination angle is calculated from the nodal coordinates as

$$\theta = \tan^{-1} \frac{y_2 - y_1}{x_2 - x_1}$$

The resulting matrix for element 1 is

$$\mathbf{k}^{(1)} = \begin{bmatrix} 25.00 & -43.30 & -25.00 & 43.30 \\ -43.30 & 75.00 & 43.30 & -75.00 \\ -25.00 & 43.30 & 25.00 & -43.30 \\ 43.30 & -75.00 & -43.30 & 75.00 \end{bmatrix} \times 10^3$$

and for element 2,

$$\mathbf{k}^{(2)} = \begin{bmatrix} 25.00 & 43.30 & -25.00 & -43.30 \\ 43.30 & 75.00 & -43.30 & -75.00 \\ -25.00 & -43.30 & 25.00 & 43.30 \\ -43.30 & -75.00 & 43.30 & 75.00 \end{bmatrix} \times 10^3$$

(It is important the units be consistent; here lengths are in inches, forces in pounds, and moduli in psi. The modulus of both elements is $E = 10$ Mpsi, and both have area $A = 0.1$ in$^2$.) These matrices have rows and columns numbered from 1 to 4, corresponding to the local degrees of freedom of the element. However, each of the local degrees of freedom can be matched to one of the global degrees of the overall problem. By inspection, we can form the following table that maps local to global numbers:

| Local | Global, element 1 | Global, element 2 |
|:-----:|:-----------------:|:-----------------:|
| 1 | 1 | 3 |
| 2 | 2 | 4 |
| 3 | 3 | 5 |
| 4 | 4 | 6 |

Using this table, we see for instance that the second degree of freedom for element 2 is the fourth degree of freedom in the global numbering system, and the third local degree of freedom corresponds to the fifth global degree of freedom. Hence the value in the second row and third column of the element stiffness matrix of element 2, denoted $k_{23}^{(2)}$, should be added into the position in the fourth row and fifth column of the $6 \times 6$ global stiffness matrix. We write this as

$$k_{23}^{(2)} \longrightarrow K_{4,5}$$

Each of the sixteen positions in the stiffness matrix of each of the two elements must be added into the global matrix according to the mapping given by the table. This gives the result

$$\mathbf{K} = \begin{bmatrix} k_{11}^{(1)} & k_{12}^{(1)} & \cdot & k_{13}^{(1)} & k_{14}^{(1)} & 0 & 0 \\ k_{21}^{(1)} & k_{22}^{(1)} & k_{23}^{(1)} & k_{24}^{(1)} & 0 & 0 \\ k_{31}^{(1)} & k_{32}^{(1)} & k_{33}^{(1)}+k_{11}^{(2)} & k_{34}^{(1)}+k_{12}^{(2)} & k_{13}^{(2)} & k_{14}^{(2)} \\ k_{41}^{(1)} & k_{42}^{(1)} & k_{43}^{(1)}+k_{21}^{(2)} & k_{44}^{(1)}+k_{22}^{(2)} & k_{23}^{(2)} & k_{24}^{(2)} \\ 0 & 0 & k_{31}^{(2)} & k_{32}^{(2)} & k_{33}^{(2)} & k_{34}^{(1)} \\ 0 & 0 & k_{41}^{(2)} & k_{42}^{(2)} & k_{43}^{(2)} & k_{44}^{(2)} \end{bmatrix}$$

This matrix premultiplies the vector of nodal displacements according to Eq. 2.9 to yield the vector of externally applied nodal forces. The full system of equations, taking into account the known forces and displacements, is then

$$10^3 \begin{bmatrix} 25.0 & -43.3 & -25.0 & 43.3 & 0.0 & 0.00 \\ -43.3 & 75.0 & 43.3 & -75.0 & 0.0 & 0.00 \\ -25.0 & 43.3 & 50.0 & 0.0 & -25.0 & -43.30 \\ 43.3 & -75.0 & 0.0 & 150.0 & -43.3 & -75.00 \\ 0.0 & 0.0 & -25.0 & -43.3 & 25.0 & 43.30 \\ 0.0 & 0.0 & -43.3 & -75.0 & 43.3 & 75.00 \end{bmatrix} \begin{Bmatrix} 0 \\ 0 \\ 0 \\ u_4 \\ 0 \\ 0 \end{Bmatrix} = \begin{Bmatrix} f_1 \\ f_2 \\ f_3 \\ -1732 \\ f_5 \\ f_5 \end{Bmatrix}$$

Note that either the force or the displacement for each degree of freedom is known, with the accompanying displacement or force being unknown. Here only one of the displacements ($u_4$) is unknown, but in most problems the unknown displacements far outnumber the unknown forces. Note also that only those elements that are physically connected to a given node can contribute a force to that node. In most cases this results in the global stiffness matrix containing many zeroes, corresponding to nodal pairs that are not spanned by an element. Effective computer implementations will take advantage of the matrix sparseness to conserve memory and reduce execution time.

In larger problems the matrix equations are solved for the unknown displacements and forces by Gaussian reduction or other techniques. In this two-element problem the solution for the single unknown displacement can be written down almost from inspection. Multiplying out the fourth row of the system, we have

$$0 + 0 + 0 + 150 \times 10^3 u_4 + 0 + 0 = -1732$$

$$u_4 = -1732/150 \times 10^3 = -0.01155 \text{ in}$$

Now any of the unknown forces can be obtained directly. Multiplying out the first row, for instance, gives

$$0 + 0 + 0 + (43.4)(-0.0115) + 0 + 0 = f_1$$

$$f_1 = -500 \text{ lb}$$

The negative sign here indicates the horizontal force on global node 1 is to the left, opposite the direction assumed in Fig. 2.18.

---

The process of cycling through each element to form the element stiffness matrix, assembling the element matrix into the correct positions in the global matrix, solving the equations for displacements, back-multiplying to compute the forces, and printing the results can be automated to make a very versatile computer code.

The software diskette accompanying this text contains a DOS-executable version of a recently developed code named `felt`, authored by Jason Gobat and Darren Atkinson. `felt` is intended for educational use and incorporates a number of novel features to promote user-friendliness. Complete information describing this code, as well as the C-language source and a number of trial runs and auxiliary code modules, is available via anonymous ftp from `cs.ucsd.edu` in the `/pub/felt` locker. If you have access to X Window workstations, a graphical shell named `velvet` is available as well.

## EXAMPLE 2.7

To illustrate how this code operates for a somewhat larger problem, consider the six-element truss of Fig. 2.6, analyzed earlier both by the joint-at-a-time free-body analysis approach and by Castigliano's method. The truss is redrawn in Fig. 2.19 by the `velvet` graphical interface.

The input dataset, which can be written manually or developed graphically in `velvet`, employs parsing techniques to simplify what can be a very tedious and error-prone step in finite element analysis. The dataset for this 6-element truss is

```
problem description
nodes=5 elements=6
```

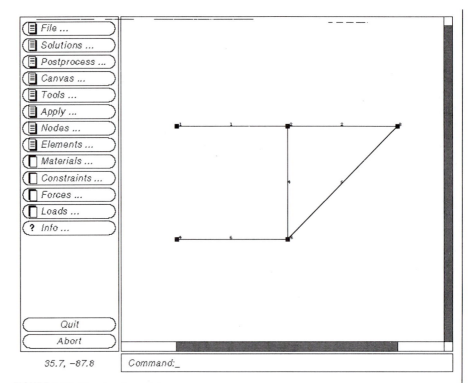

**FIGURE 2.19** The six-element truss.

```
nodes
1   x=0 y=100 z=0 constraint=pin
2   x=100 y=100 z=0 constraint=planar
3   x=200 y=100 z=0 force=P
4   x=0 y=0 z=0 constraint=pin
5   x=100 y=0 z=0 constraint=planar

truss elements
1   nodes=[1,2] material=steel
2   nodes=[2,3]
3   nodes=[4,2]
4   nodes=[2,5]
5   nodes=[5,3]
6   nodes=[4,5]

material properties
steel E=3e+07 A=0.5
distributed loads
constraints
free Tx=u Ty=u Tz=u Rx=u Ry=u Rz=u
pin Tx=c Ty=c Tz=c Rx=u Ry=u Rz=u
planar Tx=u Ty=u Tz=c Rx=u Ry=u Rz=u

forces
P Fy=-1000

end
```

The meaning of these lines should be fairly evident on inspection, although the `felt` documentation should be consulted for more detail. The output produced by `felt` for these data is

```
**    **

Nodal Displacements
-----------------------------------------------------------------
Node #       DOF 1       DOF 2   DOF 3   DOF 4   DOF 5   DOF 6
-----------------------------------------------------------------
  1              0           0       0       0       0       0
  2       0.013333    -0.03219      0       0       0       0
  3           0.02   -0.084379      0       0       0       0
  4              0           0       0       0       0       0
  5     -0.0066667   -0.038856      0       0       0       0


Element Stresses
-----------------------------------------------------------------
  1:          4000
  2:          2000
  3:        -2828.4
  4:          2000
  5:        -2828.4
  6:         -2000


Reaction Forces
------------------------------------------
Node #     DOF      Reaction Force
------------------------------------------
  1        Tx           -2000
  1        Ty               0
  1        Tz               0
  2        Tz               0
  3        Tz               0
  4        Tx            2000
  4        Ty            1000
  4        Tz               0
  5        Tz               0

Material Usage Summary
---------------------------
Material: steel
Number:   6
Length:   682.8427
Mass:       0.0000

Total mass:     0.0000
```

Note that the vertical displacement of node 3 (the DOF 2 value) is $-0.0844$, the same value obtained earlier in Example 2.3. Figure 2.20 shows the `velvet` graphical output for the truss deflections (greatly magnified).

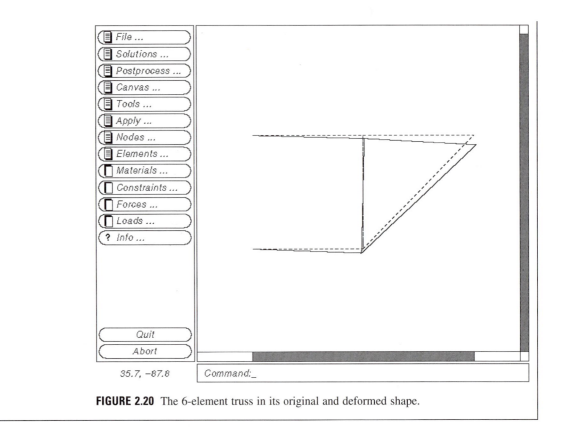

| File ... |
| Solutions ... |
| Postprocess ... |
| Canvas ... |
| Tools ... |
| Apply ... |
| Nodes ... |
| Elements ... |
| Materials ... |
| Constraints ... |
| Forces ... |
| Loads ... |
| ? Info ... |

| Quit |
| Abort |

35.7, −87.8        Command:_

**FIGURE 2.20** The 6-element truss in its original and deformed shape.

## 2.2 PRESSURE VESSELS

A good deal of the mechanics of materials can be introduced entirely within the confines of uniaxially stressed structural elements; that was the goal of the previous sections. Of course, the real world is three-dimensional, and we need to extend these concepts accordingly. Now we take the next step and consider those structures in which the loading is still simple, but where the stresses and strains require a second dimension for their description. Both for their value in demonstrating two-dimensional effects and for their practical use in mechanical design we turn to two slightly more complicated structural types: the thin-walled pressure vessel and the circular shaft loaded in torsion.

Structures such as pipes or bottles capable of holding internal pressure have been very important in the history of science and technology. Although the ancient Romans had developed municipal engineering to a high order in many ways, the very need for their impressive system of large aqueducts for carrying water was due to their not yet having pipes that could maintain internal pressure. Water *can* flow uphill when driven by the hydraulic pressure of a reservoir at a higher elevation, but without a pressure-containing pipe an aqueduct must be constructed so that the water can run downhill all the way from the reservoir to the destination.

Airplane cabins are another familiar example of pressure-containing structures. They illustrate very dramatically the importance of proper design, because the atmosphere in the cabin has enough energy associated with its relative pressurization compared to the thin air outside that catastrophic crack growth is a real possibility.

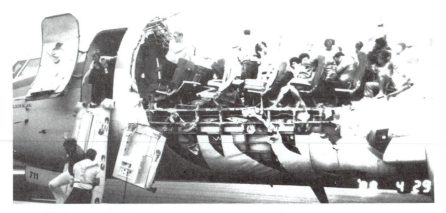

**FIGURE 2.21**  Failure of the Aloha Airlines cabin. (Robert Nichols/Black Star.)

A number of fatal commercial tragedies have resulted from this, particularly famous ones being the Comet aircraft that disintegrated in flight in the 1950s[4] and the loss of a 5-meter section of the roof in the first-class section of an Aloha Airlines B737 in April, 1988[5] (see Fig. 2.21).

In the sections to follow we will outline the means of determining stresses and deformations in structures such as these, because this determination is a vital first step in designing against failure.

## 2.2.1  Stresses

In two dimensions the state of stress at a point is conveniently illustrated by drawing four perpendicular lines, which we can view as representing four adjacent planes of atoms taken from an arbitrary position within the material. The planes on this "stress square" shown in Fig. 2.22 can be identified by the orientations of their normals; the upper horizontal plane is a $+y$ plane, since its normal points in the $+y$ direction. The vertical plane on the right is a $+x$ plane. Similarly, the left vertical and lower horizontal planes are $-y$ and $-x$, respectively.

The sign convention in common use regards tensile stresses as positive and compressive stresses as negative. A positive tensile stress acting in the $x$ direction is drawn on the $+x$ face as an arrow pointed in the $+x$ direction. However, for the stress square to be in equilibrium, this arrow must be balanced by another acting on the $-x$ face and pointed in the $-x$ direction. Of course, these are not two separate stresses but simply indicate that the stress state is one of uniaxial tension. A positive stress is therefore indicated by a + arrow on a + face or a − arrow on a − face. Compressive stresses are the reverse: a − arrow on a + face or a + arrow on a − face. A stress state with both

**FIGURE 2.22**  State of stress in two dimensions: the stress square.

[4]T. Bishop, "Fatigue and the Comet Disasters," *Metal Progress,* Vol. 67, pp. 79–85, May 1955.

[5]E. E. Murphy, "Aging Aircraft: Too Old to Fly?" *IEEE Spectrum,* pp. 28–31, June 1989.

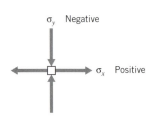

$\sigma_y$ Negative

$\sigma_x$ Positive

**FIGURE 2.23** The sign convention for normal stresses.

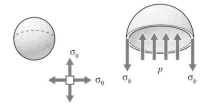

**FIGURE 2.24** Wall stresses in a spherical pressure vessel.

**FIGURE 2.25** Free-body diagram for axial stress in a closed-end vessel.

**FIGURE 2.26**
Hoop stresses in a cylindrical pressure vessel.

positive and negative components is shown in Fig. 2.23. These are termed "normal" stresses because they act normally (perpendicular) to the faces.

Consider now a simple spherical vessel of radius $r$ and wall thickness $b$, such as a round balloon. An internal pressure $p$ induces equal biaxial tangential tensile stresses in the walls, which can be denoted using spherical $r\theta\phi$ coordinates as $\sigma_\theta$ and $\sigma_\phi$.

The magnitude of these stresses can be determined by considering a free-body diagram of half the pressure vessel, including its pressurized internal fluid (see Fig. 2.24). The fluid itself is assumed to have negligible weight. The internal pressure generates a force of $pA = p(\pi r^2)$ acting on the fluid, which is balanced by the force obtained by multiplying the wall stress times its area, $\sigma_\phi(2\pi rb)$. Equating these,

$$p(\pi r^2) = \sigma_\phi(2\pi rb)$$

$$\boxed{\sigma_\phi = \frac{pr}{2b}} \tag{2.11}$$

Note that this is a statically determined result, with no dependence on the material properties. Further, note that the stresses in any two orthogonal circumferential directions are the same; in other words, $\sigma_\phi = \sigma_\theta$.

The accuracy of this result depends on the vessel being "thin-walled"; that is, $r \gg b$. At the surfaces of the vessel wall a radial stress $\sigma_r$ must be present to balance the pressure there. However, the inner-surface radial stress is equal to $p$, whereas the circumferential stresses are $p$ times the ratio $(r/2b)$. When this ratio is large, the radial stresses can be neglected in comparison with the circumferential stresses.

The stresses $\sigma_z$ in the axial direction of a cylindrical pressure vessel with closed ends are found using this same approach as seen in Fig. 2.25, and yielding the same answer:

$$p(\pi r^2) = \sigma_z(2\pi r)b$$

$$\boxed{\sigma_z = \frac{pr}{2b}} \tag{2.12}$$

However, a different view is needed to obtain the circumferential or "hoop" stresses $\sigma_\theta$. Considering an axial section of unit length, the force balance for Fig. 2.26 gives

$$2\sigma_\theta(b \cdot 1) = p(2r \cdot 1)$$

$$\boxed{\sigma_\theta = \frac{pr}{b}} \tag{2.13}$$

Note that the hoop stress is twice the axial stress. This result— different stresses in different directions—occurs more often than not in engineering structures and shows one of the compelling advantages of engineered materials that can be made stronger in one direction than another (the property of *anisotropy*). If a pressure vessel constructed of conventional isotropic material is made thick enough to keep the hoop stresses below yield, it will be twice as strong as it needs to be in the axial direction. In applications placing a premium on weight this may well be something to avoid.

EXAMPLE 2.8

Consider a cylindrical pressure vessel to be constructed by *filament winding*, in which fibers are laid down at a prescribed helical angle $\alpha$ (see Fig. 2.27). Taking a free body of unit axial dimension along which $n$ fibers transmitting tension $T$ are present, the circumferential distance cut by these same $n$ fibers is then $\tan \alpha$. To balance the hoop and axial stresses, the fiber tensions must satisfy the relations

$$\text{Hoop}: \quad nT \sin \alpha = \frac{pr}{b}(1)(b)$$

$$\text{Axial}: \quad nT \cos \alpha = \frac{pr}{2b}(\tan \alpha)(b)$$

Dividing the first of these expressions by the second and rearranging, we have

$$\tan^2 \alpha = 2, \qquad \alpha = 54.7°$$

This is the "magic angle" for filament-wound vessels, at which the fibers are inclined just enough toward the circumferential direction to make the vessel twice as strong circumferentially as it is axially. Firefighting hoses are also braided at this same angle, because otherwise the nozzle would jump forward or backward when the valve is opened and the fibers try to align themselves along the correct direction.

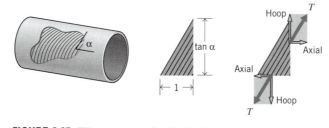

**FIGURE 2.27** Filament-wound cylindrical pressure vessel.

## 2.2.2 Deformation: The Poisson Effect

When a pressure vessel has open ends, like a pipe connecting one chamber with another, there will be no axial stress, because there are no end caps for the fluid to push against. Then only the hoop stress $\sigma_\theta = pr/b$ exists, and the corresponding hoop strain is given by Hooke's law as

$$\epsilon_\theta = \frac{\sigma_\theta}{E} = \frac{pr}{bE}$$

This strain is the change in circumference $\delta_C$ divided by the original circumference $C = 2\pi r$, so we can write

$$\delta_C = C \epsilon_\theta = 2\pi r \frac{pr}{bE}$$

The change in circumference and the corresponding change in radius $\delta_r$ are related by $\delta_r = \delta_C/2\pi$, so the radial expansion is

$$\delta_r = \frac{pr^2}{bE} \qquad (2.14)$$

This is analogous to the expression $\delta = PL/AE$ for the elongation of a uniaxial tensile specimen.

## EXAMPLE 2.9

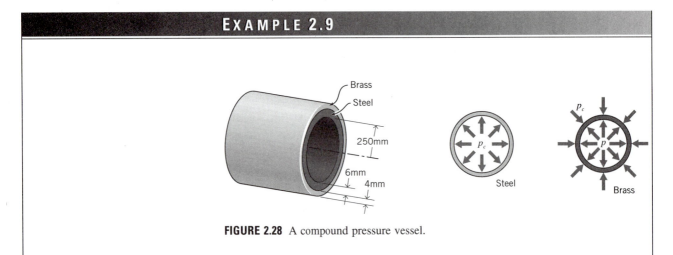

**FIGURE 2.28** A compound pressure vessel.

Consider a *compound cylinder*, in which a cylinder of brass fits snugly inside another of steel, as shown in Fig. 2.28, subjected to an internal pressure $p = 2$ MPa.

When the pressure is put inside the inner cylinder, it will naturally try to expand. But the outer cylinder pushes back so as to limit this expansion, and a "contact pressure" $p_c$ develops at the interface between the two cylinders. The inner cylinder now expands according to the difference $p - p_c$, while the outer cylinder expands as demanded by $p_c$ alone. However, because the two cylinders are obviously going to remain in contact, it should be clear that the radial expansions of the inner and outer cylinders must be the same, and we can write

$$\delta_b = \delta_s \longrightarrow \frac{(p - p_c)r_b^2}{E_b b_b} = \frac{p_c r_s^2}{E_s b_s}$$

where the $b$ and $s$ subscripts refer to the brass and steel cylinders, respectively.

Substituting numerical values and solving for the unknown contact pressure $p_c$,

$$p_c = 976 \text{ kPa}$$

Now, knowing $p_c$, we can calculate the radial expansions and the stresses if desired. For instance, the hoop stress in the inner brass cylinder is

$$\sigma_{\theta,b} = \frac{(p - p_c)r_b}{b_b} = 62.5 \text{ MPa} = 906 \text{ psi}$$

Note that the stress is no longer independent of the material properties ($E_b$ and $E_s$), depending as it does on the contact pressure $p_c$, which in turn depends on the material stiffnesses. This loss of static determinacy occurs here because the problem has a mixture of some load boundary values (the internal pressure) and some displacement boundary values (the constraint that both cylinders have the same radial displacement).

If a cylindrical vessel has closed ends, both axial and hoop stresses appear together, as given by Eqs. 2.12 and 2.13. Now the deformations are somewhat subtle, because a positive (tensile) strain in one direction will also contribute a negative (compressive) strain in the other direction, just as stretching a rubber band to make it longer in one direction makes it thinner in the other directions (see Fig. 2.29). This lateral contraction accompanying a longitudinal extension is called the *Poisson effect*,[6] and *Poisson's ratio* is a material property defined as

$$\nu = \frac{-\epsilon_{\text{lateral}}}{\epsilon_{\text{longitudinal}}} \qquad (2.15)$$

where the minus sign accounts for the sign change between the lateral and longitudinal strains. The stress-strain, or "constitutive," law of the material must be extended to include these effects, because the strain in any given direction is influenced not only by the stress in that direction but also by the Poisson strains contributed by the stresses in the other two directions.

A material subjected only to a stress $\sigma_x$ in the $x$ direction will experience a strain in that direction given by $\epsilon_x = \sigma_x/E$. A stress $\sigma_y$ acting alone in the $y$ direction will induce an $x$-direction strain given from the definition of Poisson's ratio of $\epsilon_x = -\nu\epsilon_y = -\nu(\sigma_y/E)$. If the material is subjected to both stresses $\sigma_x$ and $\sigma_y$ at once, the effects can be superimposed (since the governing equations are *linear*) to give

$$\epsilon_x = \frac{\sigma_x}{E} - \frac{\nu\sigma_y}{E} = \frac{1}{E}(\sigma_x - \nu\sigma_y) \qquad (2.16)$$

Similarly, for a strain in the $y$ direction,

$$\epsilon_y = \frac{\sigma_y}{E} - \frac{\nu\sigma_x}{E} = \frac{1}{E}(\sigma_y - \nu\sigma_x) \qquad (2.17)$$

The material is in a state of *plane stress* if no stress components act in the third dimension (the $z$ direction, here). This occurs commonly in thin sheets loaded in their planes. The $z$ components of stress vanish at the surfaces because there are no forces acting externally in that direction to balance them, and these components do not have sufficient specimen distance in the thin through-thickness dimension to build up to appreciable levels. However, a state of plane stress is *not* a state of plane strain. The sheet will experience a strain in the $z$ direction equal to the Poisson strain contributed by the $x$ and $y$ stresses:

$$\epsilon_z = -\frac{\nu}{E}(\sigma_x + \sigma_y) \qquad (2.18)$$

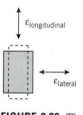

**FIGURE 2.29** The Poisson effect.

---

[6]After the French mathematician Simeon Denis Poisson (1781–1840).

In the case of a closed-end cylindrical pressure vessels, Eq. 2.16 or 2.17 can be used directly to give the hoop strain as

$$\epsilon_\theta = \frac{1}{E}(\sigma_\theta - \nu\sigma_z) = \frac{1}{E}\left(\frac{pr}{b} - \nu\frac{pr}{2b}\right)$$

$$= \frac{pr}{bE}\left(1 - \frac{\nu}{2}\right)$$

The radial expansion is then

$$\delta_r = r\epsilon_\theta = \frac{pr^2}{bE}\left(1 - \frac{\nu}{2}\right) \tag{2.19}$$

Note that the radial expansion is reduced by the Poisson term; the axial deformation contributes a shortening in the radial direction.

---

## EXAMPLE 2.10

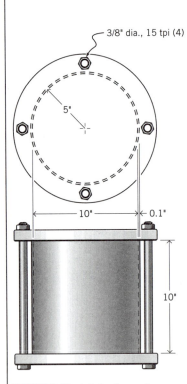

**FIGURE 2.30** A bolt-clamped pressure vessel.

It is common to build pressure vessels by using bolts to hold end plates on an open-ended cylinder, as shown in Fig. 2.30. Consider for example a cylinder made of copper alloy, with radius $R = 5''$, length $L = 10''$, and wall thickness $b_c = 0.1''$. Rigid plates are clamped to the ends by nuts threaded on four $\frac{3}{8}''$ diameter steel bolts, each having 15 threads per inch. Each of the nuts is given an additional $\frac{1}{2}$ turn beyond the just-snug point, and we wish to estimate the internal pressure that will just cause incipient leakage from the vessel.

As pressure $p$ inside the cylinder increases, a force $F = p(\pi R^2)$ is exerted on the end plates, and this is reacted equally by the four restraining bolts; each is thus subjected to a force $F_b$ given by

$$F_b = \frac{p(\pi R^2)}{4}$$

The bolts then stretch by an amount $\delta_b$ given by Eq. 1.8 as

$$\delta_b = \frac{F_b L}{A_b E_b}$$

It is tempting to say that the vessel will start to leak when the bolts have stretched by an amount equal to the original tightening, that is, $\frac{1}{2}$ turn/15 turns per inch. But as $p$ increases, the cylinder itself deforms as well; it experiences a radial expansion according to Eq. 2.14. The radial expansion by itself doesn't cause leakage, but it is accompanied by a Poisson contraction $\delta_c$ in the axial direction. This means the bolts don't have to stretch as far before the restraining plates are lifted clear. (As leakage begins, the plates cease to push on the cylinder, so the axial loading of the plates on the cylinder becomes zero and is not needed in the analysis.)

The relations governing leakage, in addition to the foregoing expressions for $\delta_b$ and $F_b$, are therefore

$$\delta_b + \delta_c = \frac{1}{2} \times \frac{1}{15}$$

where here the subscripts $b$ and $c$ refer to the bolts and the cylinder respectively. The axial deformation $\delta_c$ of the cylinder is just $L$ times the axial strain $\epsilon_z$, which in turn is given by an expression analogous to Eq. 2.17:

$$\delta_c = \epsilon_z L = \frac{L}{E_c}[\sigma_z - \nu\sigma_\theta]$$

Since $\sigma_z$ becomes zero just as the plate lifts off and $\sigma_\theta = pR/b_c$, this becomes

$$\delta_c = \frac{L}{E_c}\frac{\nu pR}{b_c}$$

Combining the foregoing relations and solving for $p$, we have

$$p = \frac{2A_b E_b E_c b_c}{15RL(\pi RE_c b_c + 4\nu A_b E_b)}$$

On substituting the geometrical and materials numerical values, this gives

$$p = 496 \text{ psi}$$

Poisson's ratio is a dimensionless parameter that provides a good deal of insight into the nature of the material. The major classes of engineered structural materials fall neatly into order when ranked by Poisson's ratio, shown to the right. (The values here are approximate.) It will be noted that the most brittle materials have the lowest Poisson's ratio, and that the materials appear to become generally more flexible as Poisson's ratio increases. The ability of a material to contract laterally as it is extended longitudinally is related directly to its molecular mobility, with rubber being liquidlike and ceramics being very tightly bonded.

| Material class | Poisson's ratio $\nu$ |
|---|---|
| Ceramics | 0.2 |
| Metals | 0.3 |
| Plastics | 0.4 |
| Rubber | 0.5 |

Poisson's ratio is also related to the compressibility of the material. The *bulk modulus K*, also called the modulus of compressibility, is the ratio of the hydrostatic pressure $p$ needed for a unit relative decrease in volume $\Delta V/V$:

$$K = \frac{-p}{\Delta V/V} \qquad (2.20)$$

where the minus sign indicates that a compressive pressure (traditionally considered positive) produces a negative volume change. It can be shown (see Chapter 3, Problem 6) that for isotropic materials the bulk modulus is related to the elastic modulus and Poisson's ratio as

$$K = \frac{E}{3(1 - 2\nu)} \qquad (2.21)$$

This expression becomes unbounded as $\nu$ approaches 0.5, so that rubber is essentially incompressible. Further, $\nu$ cannot be larger than 0.5, since that would mean volume would *increase* on the application of positive pressure. A ceramic at the lower end of Poisson's ratios, by contrast, is so tightly bonded that it is unable to rearrange itself to "fill the holes" that are created when a specimen is pulled in tension; it has no choice but to suffer a volume increase. Paradoxically, the tightly bonded ceramics have lower bulk moduli than the very mobile elastomers.

### 2.2.3 Viscoelastic Response

Polymers such as polybutylene and polyvinyl chloride are finding increasing use in plumbing and other liquid delivery systems, and these materials exhibit measurable viscoelastic time dependency in their mechanical response. It is common to ignore these rate effects in design of simple systems by using generous safety factors. However, in more critical situations the designer may wish to extend the elastic theory outlined in the foregoing sections to include material viscoelasticity. In the section to follow we will describe how this can be done, both for its practical use in pressure vessel analysis and also to demonstrate a general method by which viscoelastic effects can be incorporated into stress analyses.

One important point to repeat at the outset is that in the case of statically determined stresses, the material properties do not enter the problem and the stresses are not influenced by viscoelasticity. The results of Eqs. 2.12 and 2.13 apply without change. However, the displacements (for instance, $\delta_r$) will most certainly be affected, increasing with time as the strain in the material increases via molecular conformational change. For an open-ended cylindrical vessel the radial expansion was shown earlier to be

$$\delta_r = \frac{pr^2}{bE}$$

The elastic modulus in the denominator indicates that the radial expansion will increase as material loses stiffness through viscoelastic response; we denoted this time-dependent increase in strain in Section 1.5.2 by the term *creep*. In quantifying this behavior, it is convenient to replace the modulus $E$ by the *compliance*, $C = 1/E$. The expression for radial expansion now has the material constant in the numerator:

$$\delta_r = \frac{pr^2}{b}C$$

If the pressure $p$ is constant, viscoelasticity enters the problem only through the material compliance $C$, which must be made a suitable time-dependent function. (Here we assume that values of $r$ and $b$ can be treated as constant, which will usually be valid to a good approximation.) The value of $\delta_r$ at time $t$ is then simply the factor $(pr^2/b)$ times the value of $C(t)$ at that time.

The function $C(t)$ needed here is the material's *creep compliance,* the time-dependent strain exhibited by the material in response to an imposed unit tensile stress: $C_{crp} = \epsilon(t)/\sigma_0$. The standard linear solid provides a simple model for the creep compliance in which $C_{crp}$ rises exponentially from a glassy value $C_g$ to a higher, rubbery value $C_r$ with a characteristic single relaxation time $\tau$, as given by Eq. 1.29:

$$C_{crp} = C_g + (C_r - C_g)(1 - e^{-t/\tau}) \tag{2.22}$$

where here it is assumed that the stress is applied at time $t = 0$. The radial expansion of a pressure vessel, subjected to a constant internal pressure $p_0$ and constructed of a material for which the Standard Linear Solid is a reasonable model, is then

$$\delta_r(t) = \frac{p_0 r^2}{b}\left[C_g + (C_r - C_g)(1 - e^{-t/\tau})\right] \tag{2.23}$$

This function is shown schematically in Fig. 2.31.

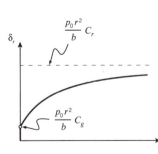

**FIGURE 2.31** Creep of open-ended pressure vessel.

The situation is a bit more complicated if both the internal pressure and the material compliance are time-dependent. It is incorrect simply to use the Eq. (2.23) with the value of $p_0$ replaced by the value of $p(t)$ at an arbitrary time, because the radial expansion at time $t$ is influenced by the pressure at previous times as well as the pressure at the current time.

The correct procedure is to "fold" the pressure and compliance functions together in the mathematical operation called *convolution*. To illustrate the physical significance of this operation, consider the radial expansion at the *current time*, indicated by $t$, due to the application of a small constant pressure $\Delta p_1$ at some *previous* time, indicated by $\xi_1$. This is illustrated in Fig. 2.32. The pressure will have been applied for a total time duration $t - \xi_1$, so the radial expansion is given by the creep compliance using this time duration as its argument:

$$(\Delta\delta_r)_1 = \frac{\Delta p_1 r^2}{b} C_{\mathrm{crp}}(t - \xi_1)$$

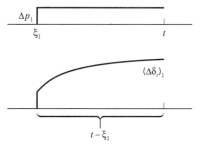

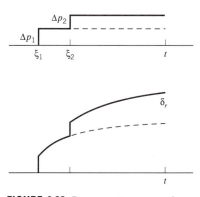

**FIGURE 2.32** Response to an initial pressure increment.

Now consider the effect of a *second* pressure increment $\Delta p_2$ applied at a different time $\xi_2$ as shown in Fig. 2.33. If the material is linear, the effect of both of these pressure inputs acting together is the sum of the expansions that would have been generated by each acting alone:

$$\delta_r(t) = (\Delta\delta_r)_1 + (\Delta\delta_r)_2$$

$$= \frac{\Delta p_1 r^2}{b} C_{\mathrm{crp}}(t - \xi_1) + \frac{\Delta p_2 r^2}{b} C_{\mathrm{crp}}(t - \xi_2)$$

Generalizing this to a large number of stress increments and then passing to the limit of an arbitrary continuous pressure function, we have

$$\delta_r(t) = \sum_j \frac{\Delta p_j r^2}{b} C_{\mathrm{crp}}(t - \xi_j) \longrightarrow \frac{r^2}{b} \int_{-\infty}^{t} C_{\mathrm{crp}}(t - \xi) \, dp$$

Since $dp = (dp/d\xi) \, d\xi$, this can be written

$$\delta_r(t) = \frac{r^2}{b} \int_{-\infty}^{t} C_{\mathrm{crp}}(t - \xi) \dot{p}(\xi) \, d\xi \tag{2.24}$$

**FIGURE 2.33** Response to a second pressure increment.

The parameter $\xi$ is a dummy variable, representing the history of previous times leading up to $t$. The lower limit of integration must be set to include all previous pressure inputs, where the lower limit of $-\infty$ here is a pessimistic guess.

## EXAMPLE 2.11

Let the internal pressure be a constantly increasing "ramp" function, so that $p = Rt$, with $R$ being the rate of increase; then we have $\dot{p}(\xi) = R$. Using the Standard Linear Solid of Eq. 2.22 for the creep compliance, the stress is calculated from the convolution integral as

$$\delta_r(t) = \frac{r^2}{b} \int_0^t \left[ C_g + (C_r - C_g)(1 - e^{-(t-\xi)/\tau}) \right] R \, d\xi$$

$$= \frac{r^2}{b} \left[ RtC_r - R\tau \left( C_r - C_g \right) \left( 1 - e^{-t/\tau} \right) \right]$$

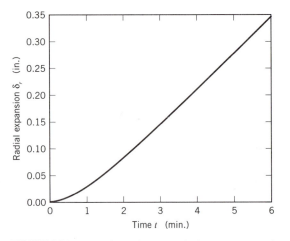

**FIGURE 2.34** Creep $\delta_r(t)$ of hypothetical pressure vessel for constantly increasing internal pressure.

This function is plotted in Fig. 2.34 for a hypothetical material with parameters $C_g = \frac{1}{3} \times 10^5$ psi$^{-1}$, $C_r = \frac{1}{3} \times 10^4$ psi$^{-1}$, $b = 0.2$ in, $r = 2$ in, $\tau = 1$ min, and $R = 100$ psi/min. Note that the creep rate increases from an initial value $(r^2/b)RC_g$ to a final value $(r^2/b)RC_r$ as the glassy elastic components relax away.

When the pressure vessel has closed ends and must therefore resist axial as well as hoop stresses, the radial expansion is $\delta_r = (pr^2/bE)\,[1 - (\nu/2)]$ as given by Eq. 2.19. The extension of this relation to viscoelastic material response *and* a time-dependent pressure is another step up in complexity. Now *two* material descriptors, $E$ and $\nu$, must be modeled by suitable time-dependent functions and then folded into the pressure function. The superposition approach just described could be used here as well, but with more algebraic complexity. The viscoelastic correspondence principle, to be presented in Chapter 5, is often more tractable in practice, but the superposition concept is very important in understanding time-dependent materials response.

## 2.3 SHEARING STRESSES AND STRAINS

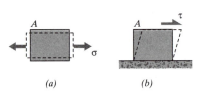

**FIGURE 2.35** (*a*) Normal and (*b*) shearing deformations.

Not all deformation is elongational or compressive, and we need to extend our concept of strain to include "shearing," or "distortional," effects. To illustrate the nature of shearing distortions, first consider a square grid inscribed on a tensile specimen as depicted in Fig. 2.35*a*. Upon uniaxial loading, the grid would be deformed so as to increase the length of the lines in the tensile loading direction and contract the lines perpendicular to the loading direction. However, the lines remain perpendicular to one another. These are termed *normal* strains because planes normal to the loading direction are moving apart.

Now consider the case illustrated in Fig. 2.35*b*, in which the load $P$ is applied *transversely* to the specimen. Here the horizontal lines tend to *slide* relative to one another, with line lengths of the originally square grid remaining unchanged. The vertical lines tilt to accommodate this motion, so the originally right angles between the lines are distorted. Such a loading is termed *direct shear.* Analogously to our definition of normal stress as force per unit area, or $\sigma = P/A$, we write the *shear stress* $\tau$ as

$$\tau = \frac{P}{A}$$

This expression is identical to Eq. 1.2 for normal stress, but the different symbol $\tau$ reminds us that the loading is transverse rather than extensional.

## EXAMPLE 2.12

Two timbers, of cross-sectional dimension $b \times h$, are to be glued together using a tongue-and-groove joint as shown in Fig. 2.36, and we wish to estimate the depth $d$ of the glue joint so as to make the joint approximately as strong as the timber itself.

The axial load $P$ on the timber acts to shear the glue joint, and the shear stress in the joint is just the load divided by the total glue area:

$$\tau = \frac{P}{2bd}$$

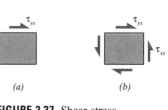

**FIGURE 2.36**  Tongue-and-groove adhesive joint.

If the bond fails when $\tau$ reaches a maximum value $\tau_f$, the load at failure will be $P_f = (2bd)\tau_f$. The load needed to fracture the timber in tension is $P_f = bh\sigma_f$, where $\sigma_f$ is the ultimate tensile strength of the timber. Hence, if the glue joint and the timber are to be equally strong, we have

$$(2bd)\tau_f = bh\sigma_f \rightarrow d = \frac{h\sigma_f}{2\tau_f}$$

Normal stresses act to pull parallel planes within the material apart or push them closer together, whereas shear stresses act to slide planes along one another. Normal stresses promote crack formation and growth, and shear stresses underlie yield and plastic slip. The shear stress can be depicted on the stress square used earlier in Fig. 2.22, as shown in Fig. 2.37a; it is traditional to use a half-arrowhead to distinguish shear stress from normal stress. The $yx$ subscript indicates that the stress is on the $y$ plane in the $x$ direction.

The $\tau_{yx}$ arrow on the $+y$ plane must be accompanied by one in the opposite direction on the $-y$ plane, in order to maintain horizontal equilibrium. However, these two arrows by themselves would tend to cause a clockwise rotation. To maintain moment equilibrium we must also add two vertical arrows as shown in Fig. 2.37b; these are labeled $\tau_{xy}$ because they are on $x$ planes in the $y$ direction. For rotational equilibrium, the magnitudes of the horizontal and vertical stresses must be equal:

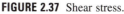

(a)                    (b)

**FIGURE 2.37**  Shear stress.

$$\tau_{yx} = \tau_{xy} \tag{2.25}$$

Hence, any shearing that tends to cause tangential sliding of horizontal planes is accompanied by an equal tendency to slide vertical planes as well. Note that all of these are positive by our earlier convention of + arrows on + faces being positive. A positive state of shear stress, then, has arrows meeting at the upper right and lower left of the stress square. Conversely, arrows in a negative state of shear meet at the lower right and upper left.

The strain accompanying the shear stress $\tau_{xy}$ is a *shear strain*, denoted $\gamma_{xy}$. This quantity is a deformation per unit length, just as was the normal strain $\epsilon$, but now the

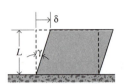

**FIGURE 2.38** Shear strain.

displacement is transverse to the length over which it is distributed (see Fig. 2.38). This is also the distortion or change in the right angle:

$$\frac{\delta}{L} = \tan \gamma \approx \gamma \qquad (2.26)$$

This angular distortion is found experimentally to be linearly proportional to the shear stress at sufficiently small loads, and the shearing counterpart of Hooke's law can be written as

$$\tau_{xy} = G\gamma_{xy} \qquad (2.27)$$

where $G$ is a material property called the *shear modulus*. There is no Poisson-type effect to consider in shear, so the shear strain is not influenced by the presence of normal stresses. Similarly, application of a shearing stress has no influence on the normal strains. For plane stress situations (no normal or shearing stress components in the $z$ direction), the constitutive equations as developed so far can be written:

$$\epsilon_x = \frac{1}{E}\left(\sigma_x - \nu\sigma_y\right)$$
$$\epsilon_y = \frac{1}{E}\left(\sigma_y - \nu\sigma_x\right) \qquad (2.28)$$
$$\gamma_{xy} = \frac{1}{G}\tau_{xy}$$

It will be shown later that for isotropic materials only two of the material constants here are independent, and that

$$G = \frac{E}{2(1 + \nu)} \qquad (2.29)$$

Hence, if any two of the three properties $E$, $G$, and $\nu$ are known, the other is determined.

## 2.4 TORSION OF CIRCULAR SHAFTS

Torsionally loaded shafts are among the most commonly used structures in engineering. For instance, the drive shaft of a standard rear-wheel-drive automobile, depicted in Fig. 2.39, serves primarily to transmit torsion. These shafts, which are almost always hollow and circular in cross section, transmit power from the transmission to the differential joint, at which the rotation is diverted to the drive wheels. As in the case of pressure vessels, it is important to be aware of design methods for such structures purely for their inherent usefulness. However, we study them here because they illustrate the role of shearing stresses and strains.

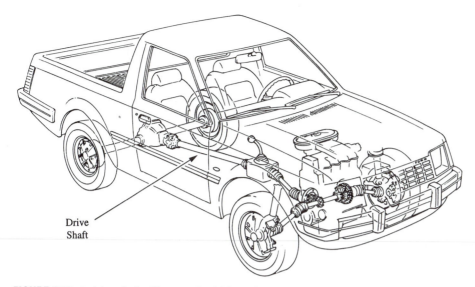

Drive
Shaft

**FIGURE 2.39** A drive shaft. (Courtesy Ford Motor Company.)

## 2.4.1 Statics Considerations

Twisting moments, or *torques*, are forces acting through distances (*lever arms*) so as to promote rotation. A simple example is that of using a wrench to tighten a nut on a bolt as shown in Fig. 2.40. If the bolt, wrench, and force are all perpendicular to one another, the moment is just the force $F$ times the length $l$ of the wrench: $T = F \cdot l$.

**FIGURE 2.40** Simple torque. (International Museum of Photography, George Eastman House.)

This relation will suffice when the geometry of torsional loading is simple, as in this case, when the torque is applied "straight."

Often, however, the geometry of the applied moment is a bit more complicated. Consider a not uncommon instance where a spark plug must be loosened and there just isn't room to put a wrench on it properly. Here a swiveled socket wrench might be needed, which can result in the lever arm not being perpendicular to the spark plug axis and the applied force from the mechanic's hand not being perpendicular to the lever arm. Vector algebra can make the geometrical calculations easier in such cases. Here the *moment vector* around a point $O$ is obtained by crossing the vector representation of the lever arm **r** from $O$ with the force vector **F**:

$$\mathbf{T} = \mathbf{r} \times \mathbf{F} \tag{2.30}$$

This vector is in a direction given by the right-hand rule and is normal to the plane containing the point $O$ and the force vector. The torque tending to loosen the spark plug is then the component of this moment vector along the plug axis:

$$T = \mathbf{i} \cdot (\mathbf{r} \times \mathbf{F}) \tag{2.31}$$

where **i** is a unit vector along the axis. The result, a torque or twisting moment *around an axis,* is a *scalar* quantity.

## EXAMPLE 2.13

We wish to find the effective twisting moment on a spark plug, where the force applied to a swivel wrench that is skewed away from the plug axis as shown in Fig. 2.41. An $x'y'z'$ Cartesian coordinate system is established in which $z'$ is the spark plug axis; the free end of the wrench is $2''$ above the $x'y'$ plane perpendicular to the plug axis, and $12''$ away from the plug along the $x'$ axis. A 15 lb force is applied to the free end at a skewed angle of $25°$ vertical and $20°$ horizontal.

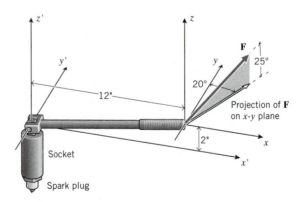

**FIGURE 2.41** "Working on your good old car"—trying to get the spark plug out.

The force vector applied to the free end of the wrench is

$$\mathbf{F} = 15(\cos 25 \sin 20\mathbf{i} + \cos 25 \cos 20\mathbf{j} + \sin 25\mathbf{k})$$

The vector from the axis of rotation to the applied force is

$$\mathbf{r} = 12\mathbf{i} + 0\mathbf{j} + 2\mathbf{k}$$

where $\mathbf{i}$, $\mathbf{j}$, $\mathbf{k}$, are the unit vectors along the $x$, $y$, $z$ axes. The moment vector around the point $O$ is then

$$\mathbf{T}_O = \mathbf{r} \times \mathbf{F} = (-25.55\mathbf{i} - 66.77\mathbf{j} + 153.3\mathbf{k})$$

and the scalar moment along the $z'$ axis is

$$T_{z'} = \mathbf{k} \cdot (\mathbf{r} \times \mathbf{F}) = 153.3 \text{ in-lb}$$

This is the torque that will loosen the spark plug, for someone luckier than the author is with cars.

Shafts in torsion are used in almost all rotating machinery, as in our earlier example of a drive shaft transmitting the torque of an automobile engine to the wheels. When the car is operating at constant speed (not accelerating), the torque on a shaft is related to its rotational speed $\omega$ and the power $W$ being transmitted:

$$W = T\omega \tag{2.32}$$

Geared transmissions are usually necessary to keep the engine speed in reasonable bounds as the car speeds up, and the gearing must be considered in determining the torques applied to the shafts.

## EXAMPLE 2.14

Consider a simple two-shaft gearing as shown in Fig. 2.42, with one end of shaft $A$ clamped and the free end of shaft $B$ loaded with a moment $T$. Drawing free-body diagrams for the two shafts separately, we see that the force $F$ transmitted at the gear periphery is just the force that keeps shaft $B$ in rotational equilibrium:

$$F \cdot r_B = T$$

This same force acts on the periphery of gear $A$, so the torque $T_A$ experienced by shaft $A$ is

$$T_A = F \cdot r_A = T \cdot \frac{r_A}{r_B}$$

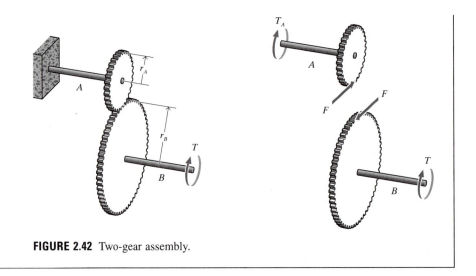

**FIGURE 2.42** Two-gear assembly.

## 2.4.2 Torsional Stresses

The stresses and deformations induced in a circular shaft by a twisting moment can be found by what is sometimes called the *direct method* of stress analysis. Here an expression of the geometrical form of displacement in the structure is proposed, after which the kinematic, constitutive, and equilibrium equations are applied sequentially to develop expressions for the strains and stresses. In the case of simple twisting of a circular shaft, the geometric statement is simply that the circular symmetry of the shaft is maintained, which implies in turn that plane cross sections remain plane, without warping. As depicted in Fig. 2.43, the deformation is like that of a stack of poker chips that rotate relative to one another while remaining flat. The sequence of direct analysis then takes the following form:

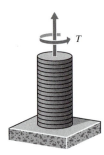

**FIGURE 2.43** Poker-chip visualization of torsional deformation.

1. *Geometrical statement.* To quantify the geometry of deformation, consider an increment of length $dz$ from the shaft, as seen in Fig. 2.44, in which the top rotates relative to the bottom by an increment of angle $d\theta$. The relative tangential displacement of the top of a vertical line drawn at a distance $r$ from the center is then

$$\delta = r\,d\theta \qquad (2.33)$$

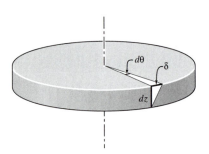

**FIGURE 2.44** Incremental deformation in torsion.

2. *Kinematic or strain-displacement equation.* The geometry of deformation fits our earlier description of shear strain exactly, so we can write

$$\gamma_{z\theta} = \frac{\delta}{dz} = r\frac{d\theta}{dz} \qquad (2.34)$$

The subscript indicates a shearing of the $z$ plane (the plane normal to the $z$ axis) in the $\theta$ direction. As with the shear stresses, $\gamma_{z\theta} = \gamma_{\theta z}$, so the order of subscripts is arbitrary.

3. *Constitutive equation.* The shear stress is given directly from Hooke's law as

$$\tau_{\theta z} = G\gamma_{\theta z} = Gr\frac{d\theta}{dz} \qquad (2.35)$$

The sign convention here is that positive twisting moments (moment vector along the $+z$ axis) produce positive shear stresses and strains. However, it is probably easier simply to intuit in which direction the applied moment will tend to slip adjacent horizontal planes. Here the upper $(+z)$ plane is clearly being twisted to the right relative to the lower $(-z)$ plane, so the upper arrow points to the right. The other three arrows are then determined as well.

4. *Equilibrium equation.* In order to maintain rotational equilibrium, the sum of the moments contributed by the shear stress acting on each differential area $dA$ on the cross section must balance the applied moment $T$ as shown in Fig. 2.45:

$$T = \int_A \tau_{\theta z} r \, dA = \int_A Gr \frac{d\theta}{dz} r = G \frac{d\theta}{dz} \int_A r^2 \, dA$$

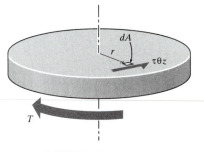

**FIGURE 2.45** Torque balance.

The quantity $\int r^2 dA$ is the *polar moment of inertia J*, which for a hollow circular cross section is calculated as

$$J = \int_{R_i}^{R_o} r^2 2\pi r \, dr = \frac{\pi(R_o^4 - R_i^4)}{2} \tag{2.36}$$

where $R_i$ and $R_o$ are the inside and outside radii. For solid shafts, $R_i = 0$. The quantity $d\theta/dz$ can now be found as

$$\frac{d\theta}{dz} = \frac{T}{GJ} \rightarrow \theta = \int_z \frac{T}{JG} dz$$

Since in the simple twisting case under consideration the quantities $T, J, G$ are constant along $z$, the angle of twist can be written as

$$\frac{d\theta}{dz} = \text{constant} = \frac{\theta}{L}$$

$$\boxed{\theta = \frac{TL}{GJ}} \tag{2.37}$$

This is analogous to the expression $\delta = PL/AE$ for the elongation of a uniaxial tensile specimen.

5. An explicit formula for the stress can be obtained by using this in Eq. 2.35:

$$\tau_{\theta z} = Gr \frac{d\theta}{dz} = Gr \frac{\theta}{L} = \frac{Gr}{L} \frac{TL}{GJ}$$

$$\boxed{\tau_{\theta z} = \frac{Tr}{J}} \tag{2.38}$$

Note that the material property $G$ has been canceled out of this final expression for stress, so that the stresses are independent of the choice of material. Earlier, we noted that stresses are independent of materials properties in certain pressure vessels and truss elements because those structures are statically determinate. The shaft in torsion isn't statically determinate, however; we had to use geometrical considerations and a statement of material linear elastic response as well as static equilibrium in obtaining the result. Since the material properties don't appear in the resulting equation for stress, it is easy to forget that the derivation depended on geometrical and material linearity. It is always important to keep in mind the assumptions used in derivations such as this and to be on guard against using the result in instances for which the assumptions are not justified.

For instance, we might twist a shaft until it breaks at a final torque of $T = T_f$ and then use Eq. 2.38 to compute an apparent ultimate shear strength: $\tau_f = T_f r/J$. However, the material may very well have been stressed beyond its elastic limit in this test, and the assumption of material linearity may not have been valid at failure. The resulting value of $\tau_f$ obtained from the elastic analysis is therefore fictitious unless proven otherwise, and it could be substantially different from the actual stress. The fictitious value might be used, however, to estimate failure torques in shafts of the same material but of different sizes, because the actual failure stress would scale with the fictitious stress in that case. The fictitious failure stress calculated using the elastic analysis is often called the *modulus of rupture in torsion*.

Equation 2.38 shows one reason why most drive shafts are hollow: There isn't much point in using material at the center, where the stresses are zero. Also, for a given quantity of material the designer will want to maximize the moment of inertia by placing the material as far from the center as possible. This is a powerful tool, because $J$ varies as the fourth power of the radius.

---

## EXAMPLE 2.15

An automobile engine is delivering 100 hp (horsepower) at 1800 rpm (revolutions per minute) to the drive shaft, and we wish to compute the shearing stress. From Eq. 2.32 the torque on the shaft is

$$T = \frac{W}{\omega} = \frac{100\text{hp}\left(\dfrac{1}{1.341 \times 10^{-3}}\right)\dfrac{\text{N} \cdot \text{m}}{\text{s} \cdot \text{hp}}}{1800\dfrac{\text{rev}}{\text{min}}2\pi\dfrac{\text{rad}}{\text{rev}}\left(\dfrac{1}{60}\right)\dfrac{\text{min}}{\text{s}}} = 396 \text{ N} \cdot \text{m}$$

The present drive shaft is a solid rod with a circular cross section and a diameter of $d = 10$ mm. Using Eq. 2.38, the maximum stress occurs at the outer surface of the rod as

$$\tau_{\theta z} = \frac{Tr}{J}, \qquad r = d/2, \qquad J = \pi(d/2)^4/2$$

$$\tau_{\theta z} = 252 \text{ MPa}$$

Now consider what the shear stress would be if the shaft were made annular rather than solid, keeping the amount of material the same. The outer-surface shear stress for an annular shaft with outer radius $r_o$ and inner radius $r_i$ is

$$\tau_{\theta z} = \frac{Tr_o}{J}, \qquad J = \frac{\pi}{2}\left(r_o^4 - r_i^4\right)$$

To keep the amount of material in the annular shaft the same as in the solid one, the cross-sectional areas must be the same. Since the cross-sectional area of the solid shaft is $A_0 = \pi r^2$, the inner radius $r_i$ of an annular shaft with outer radius $r_o$ and area $A_0$ is found as

$$A_0 = \pi \left( r_o^2 - r_i^2 \right) \rightarrow r_i = \sqrt{r_o^2 - (A_0/\pi)}$$

Evaluating these equations using the same torque and with $r_o = 30$ mm, we find $r_i = 28.2$ mm (a 1.8 mm wall thickness) and a stress of $\tau_{\theta z} = 44.5$ MPa. This is an 82 percent reduction in stress. The value of $r$ in the elastic shear stress formula went up when we went to the annular rather than solid shaft, but this was more than offset by the increase in moment of inertia $J$, which varies as $r^4$.

## EXAMPLE 2.16

Just as with trusses, the angular displacements in systems of torsion rods may be found from direct geometrical considerations. In the case of the two-rod geared system described earlier, the angle of twist of rod $A$ is

$$\theta_A = \left( \frac{L}{GJ} \right)_A T_A = \left( \frac{L}{GJ} \right)_A T \cdot \frac{r_A}{r_B}$$

This rotation will be experienced by gear $A$ as well, so a point on its periphery will sweep through an arc $S$ of

$$S = \theta_A r_A = \left( \frac{L}{GJ} \right)_A T \cdot \frac{r_A}{r_B} \cdot r_A$$

Gears $A$ and $B$ are connected at their peripheries, so gear $B$ will rotate through an angle of

$$\theta_{\text{gear } B} = \frac{S}{r_B} = \left( \frac{L}{GJ} \right)_A T \cdot \frac{r_A}{r_B} \cdot \frac{r_A}{r_B}$$

(see Fig. 2.46). Finally, the total angular displacement at the end of rod $B$ is the rotation of gear $B$ plus the twist of rod $B$ itself:

$$\theta = \theta_{\text{gear } B} + \theta_{\text{rod } B} = \left( \frac{L}{GJ} \right)_A T \left( \frac{r_A}{r_B} \right)^2 + \left( \frac{L}{GJ} \right)_B T$$

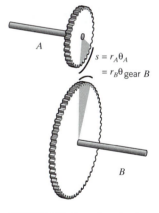

$$s = r_A \theta_A$$
$$= r_B \theta \text{ gear } B$$

**FIGURE 2.46** Rotations in the two-gear assembly.

## 2.4.3 Energy Method for Rotational Displacement

The angular deformation may also be found using Castigliano's theorem, and in some problems this approach may be easier. The strain energy per unit volume in a material subjected to elastic shearing stresses $\tau$ and strains $\gamma$ arising from simple torsion is

$$U^* = \int \tau \, d\gamma = \frac{1}{2} \tau \gamma = \frac{\tau^2}{2G} = \frac{1}{2G} \left( \frac{Tr}{J} \right)^2$$

This is then integrated over the specimen volume to obtain the total energy:

$$U = \int_V U^* dV = \int_L \int_A \frac{1}{2G} \left(\frac{Tr}{J}\right)^2 dA\,dz = \int_L \frac{T^2}{2GJ^2} \int_A r^2 dA$$

$$\boxed{U = \int_L \frac{T^2}{2GJ} dz} \tag{2.39}$$

If $T$, $G$, and $J$ are constant along the length $z$, this becomes simply

$$U = \frac{T^2 L}{2GJ} \tag{2.40}$$

which is analogous to the expression $U = P^2 L/2AE$ for tensile specimens.

In torsion the angle $\theta$ is the generalized displacement congruent to the applied moment $T$, so Castigliano's theorem is applied for a single torsion rod as

$$\theta = \frac{\partial U}{\partial T} = \frac{TL}{GJ}$$

as before.

---

## EXAMPLE 2.17

Consider the two shafts geared together discussed earlier (Fig. 2.46). The energy method requires no geometrical reasoning and follows immediately once the torques transmitted by the two shafts are known. Because the torques are constant along the lengths, we can write

$$U = \sum_i \left(\frac{T^2 L}{2GJ}\right)_i = \left(\frac{L}{2GJ}\right)_A \left(T\frac{r_A}{r_B}\right)^2 + \left(\frac{L}{2GJ}\right)_B T^2$$

$$\theta = \frac{\partial U}{\partial T} = \left(\frac{L}{GJ}\right)_A \left(T \cdot \frac{r_A}{r_B}\right)\left(\frac{r_A}{r_B}\right) + \left(\frac{L}{GJ}\right)_B T$$

---

### 2.4.4 Noncircular Sections: the Prandtl Membrane Analogy

Shafts with noncircular sections are not uncommon. Although a circular shape is optimal from a stress analysis point of view, square or prismatic shafts may be easier to produce. Also, round shafts often have keyways or other geometrical features that are needed for joining them to gears or other parts. All these considerations make it necessary to be able to cope with noncircular sections. We will outline one means of doing this here, partly for its inherent usefulness and partly to introduce a type of experimental stress analysis. Chapter 5 will expand on these methods and will present a more complete treatment of the underlying mathematical theory.

The lack of axial symmetry in noncircular sections renders the direct approach that led to Eq. 2.38 invalid, and a thorough treatment must attack the differential governing equations of the problem mathematically. These equations will be discussed in

Chapters 3 and 5, but suffice it to say that they can be difficult to solve in closed form for arbitrarily shaped cross sections. The advent of finite element and other computer methods to solve these equations numerically has removed this difficulty to some degree, but one important limitation of numerical solutions is that they usually fail to provide intuitive insight into why the stress distributions are the way they are and thus to provide hints as to how the stresses might be modified favorably by design changes. This intuition is one of the designer's most important tools.

In an elegant insight, Prandtl[7] pointed out that the stress distribution in torsion can be described by a "Poisson" differential equation, identical in form to that describing the deflection of a flexible membrane supported and pressurized from below.[8] This provides the basis of the *Prandtl membrane analogy,* which was used for many years to provide a form of experimental stress analysis for noncircular shafts in torsion. Although this experimental use has been supplanted by the more convenient computer methods, the analogy provides a visualization of torsionally induced stresses that can provide the sort of design insight we seek.

The analogy works such that the shear stresses in a torsionally loaded shaft of arbitrary cross section are proportional to the *slope* of a suitably inflated flexible membrane. The membrane is clamped so that its edges follow a shape similar to that of the noncircular section, and then it is displaced by air pressure. Visualize a horizontal sheet of metal with a circular hole in it, a sheet of rubber placed below the hole, and the rubber now made to bulge upward by pressure acting from beneath the plate (see Fig. 2.47). The bulge will be steepest at the edges and horizontal at its center; that is, its slope will be zero at the center and largest at the edges, just like the stresses in a twisted circular shaft.

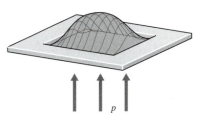

**FIGURE 2.47** Membrane inflated through a circular hole.

It is not difficult to visualize that if the hole were square, as in Fig. 2.48, rather than round, the membrane would be forced to lie flat (have zero slope) in the corners and would have the steepest slopes at the midpoints of the outside edges. This is just what the stresses do. One good reason for not using square sections for torsion rods, then, is that the corners carry no stress and are therefore wasted material. The designer could remove them without consequence, the decision just being whether the cost of making circular rather than square shafts is more or less than the cost of the wasted material. To generalize the lesson in stress analysis, a *protruding* angle is not dangerous in terms of stress—only wasteful of material.

An *entrant* angle, however, can be extremely dangerous. A sharp notch cut into the shaft is like a knife edge cutting into the rubber membrane, causing the rubber to be almost vertical. Such notches or keyways are notorious stress risers, very often acting as the origination sites for fatigue cracks. They may be necessary in some cases, but the designer must be painfully aware of their consequences.

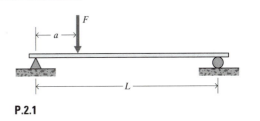

**FIGURE 2.48** Membrane inflated through a square hole.

## 2.5 PROBLEMS

**1.** A rigid beam of length $L$ rests on two supports, which resist vertical motion, and is loaded by a vertical force $F$ a distance $a$ from the left support (see Fig. P.2.1). Draw a free-body diagram for the beam, replacing the supports by the reaction forces $R_1$ and $R_2$ that they exert on the beam. Solve for the reaction forces in terms of $F$, $a$, and $L$.

**P.2.1**

---

[7]Ludwig Prandtl (1875–1953) is best known for his pioneering work in aerodynamics.

[8]J.P. Den Hartog, *Advanced Strength of Materials,* McGraw-Hill, New York, 1952.

**2.** A third support is added to the beam of Problem 1, as shown in Fig. P.2.2. Draw the free-body diagram for this case and write the equilibrium equations available to solve for the reaction forces at each support. Is it possible to solve for all the reaction forces?

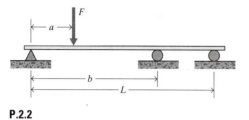

**P.2.2**

**3.** The handles of a pair of pliers shown in Fig. P.2.3 are squeezed with a force $F$. Draw a free-body diagram for one of the pliers' arms. What is the force exerted on an object gripped between the pliers' faces?

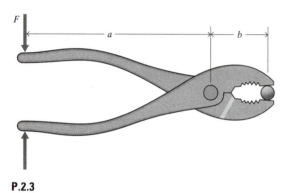

**P.2.3**

**4.** Two people are carrying a couch upstairs, one at each end, as shown in Fig. P.2.4. What is the force exerted by each person? Does the answer you calculate agree with your own experience as to which person has the easier chore?

**P.2.4**

**5.** An object of weight $W$ is suspended from a frame as shown in Fig. P.2.5. What is the tension in the restraining cable $AB$?

**6.** Coulomb's simple law of friction states that a body can develop a "friction force" $F_f$ that acts to resist sliding and that the maximum

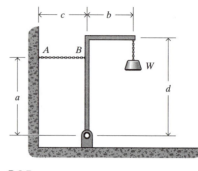

**P.2.5**

friction force is the "coefficient of friction" $\mu$ times the "normal force" $N$ that exists between the body and the surface: $F_f = \mu N$. Write an expression for the force $F$ required to induce sliding in a body of weight $W$ resting on a concave-upward circular surface (see Fig. P.2.6) as a function of the angle $\theta$.

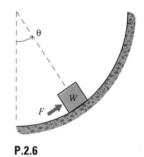

**P.2.6**

**7.** Determine the force in each element of the trusses drawn in Figs. P.2.7a–h.

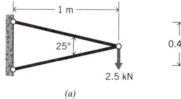

*(a)*

2.5 kN

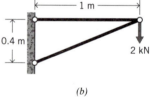

*(b)*

2 kN

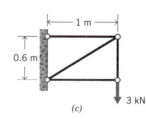

*(c)*

3 kN

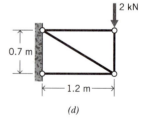

*(d)*

2 kN

*(e)*

1.5 kN

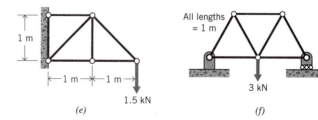

*(f)*

3 kN

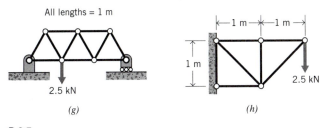

All lengths = 1 m

2.5 kN

(g)

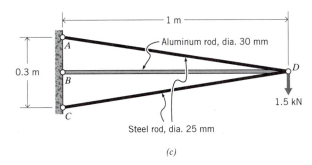

⊢ 1 m ─┼─ 1 m ─┤

1 m

2.5 kN

(h)

**P.2.7**

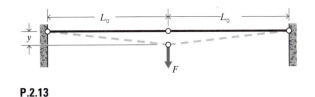

⊢─────────── 1 m ───────────┤

Aluminum rod, dia. 30 mm

0.3 m

A

B

D

C

1.5 kN

Steel rod, dia. 25 mm

(c)

**P.2.12**

**8.** Using geometrical considerations, determine the deflection of the loading point (the point at which the load is applied, in the direction of the load) for the trusses in Problem 7 (a) through (h). All elements are constructed of 20-mm-diameter round carbon steel rods.

**9.** Same as Problem 8, but using Castigliano's theorem.

**10.** Same as Problem 8, but using finite element analysis.

**11.** Find the element forces and deflection at the loading point for the truss shown in Fig. P.2.11, using the method of your own choice.

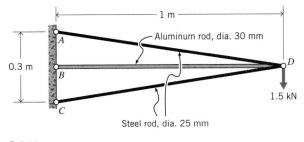

⊢──────────── 1 m ────────────┤

Aluminum rod, dia. 30 mm

0.3 m

A

B

D

1.5 kN

C

Steel rod, dia. 25 mm

**P.2.11**

**12.** Write out the global stiffness matrices for the trusses shown in Figs. P.2.12a–c, and solve for the unknown forces and displacements. For each element assume $E = 30$ Mpsi and $A = 0.1$ in$^2$.

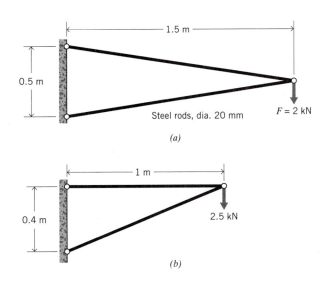

⊢─────────── 1.5 m ───────────┤

0.5 m

Steel rods, dia. 20 mm

$F = 2$ kN

(a)

⊢──────── 1 m ────────┤

0.4 m

2.5 kN

(b)

**13.** Two truss elements of equal initial length $L_0$ are connected horizontally as shown in Fig. P.2.13. Assuming the elements remain linearly elastic at all strains, determine the downward deflection $y$ as a function of a load $F$ applied transversely to the joint.

⊢────── $L_0$ ──────┼────── $L_0$ ──────┤

y

F

**P.2.13**

**14.** A closed-end cylindrical pressure vessel constructed of carbon steel has a wall thickness of 0.075″, a diameter of 6″, and a length of 30″. What are the hoop and axial stresses $\sigma_\theta$, $\sigma_z$ when the cylinder carries an internal pressure of 1500 psi? What is the radial displacement $\delta_r$?

**15.** What will be the safe pressure of the cylinder in the previous problem, using a safety factor of 2?

**16.** A compound pressure vessel with dimensions as shown in Fig. P.2.16 is constructed of an aluminum inner layer and a carbon-overwrapped outer layer. Determine the circumferential stresses ($\sigma_\theta$) in the two layers when the internal pressure is 15 MPa. The modulus of the graphite layer in the circumferential direction is 15.5 GPa.

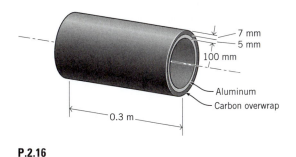

7 mm

5 mm

100 mm

Aluminum

Carbon overwrap

0.3 m

**P.2.16**

**17.** A copper cylinder is fitted snugly inside a steel one as shown in Fig. P.2.17. What is the contact pressure generated between the two cylinders if the temperature is increased by 10°C? What if the copper cylinder is on the outside?

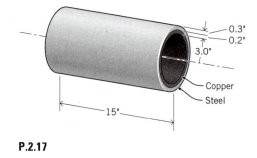

0.3"
0.2"
3.0"
Copper
Steel
15"

**P.2.17**

**18.** Three cylinders are fitted together to make a compound pressure vessel. The inner cylinder is of carbon steel with a thickness of 2 mm, the central cylinder is of copper alloy with a thickness of 4 mm, and the outer cylinder is of aluminum with a thickness of 2 mm. The inside radius of the inner cylinder is 300 mm, and the internal pressure is 1.4 MPa. Determine the radial displacement and circumferential stress in the inner cylinder.

**19.** A pressure vessel is constructed with an open-ended steel cylinder of diameters 6", length 8", and wall thickness 0.375". The ends are sealed with rigid end plates held by four $\frac{1}{4}$" diameter bolts. The bolts have 18 threads per inch, and the retaining nuts have been tightened $\frac{1}{4}$ turn beyond their just-snug point before pressure is applied. Find the internal pressure that will just cause incipient leakage from the vessel.

**20.** An aluminum cylinder, with 1.5" inside radius and thickness 0.1", is to be fitted inside a steel cylinder of thickness 0.25". The inner radius of the steel cylinder is 0.005" smaller than the outer radius of the aluminum cylinder; this is called an *interference fit*. In order to fit the two cylinders together initially, the inner cylinder is shrunk by cooling. By how much should the temperature of the aluminum cylinder be lowered in order to fit it inside the steel cylinder? Once the assembled compound cylinder has warmed to room temperature, how much contact pressure is developed between the aluminum and the steel?

**21.** A polyvinyl chloride (PVC) pipe, 0.2 m in diameter and with a 10 mm wall thickness, is carrying water at 100 kPa pressure and 75°C. At this temperature we can model the material as a Standard Linear Solid with a glassy modulus of $E_g = 2$ GPa, a rubbery modulus of $E_r = 5$ MPa, and a single relaxation time of $\tau = 100$ s. Plot the radial expansion $\delta_r(t)$ of the pipe over a time from 0 to 500 s.

**22.** Repeat Problem 21, but with an internal pressure that increases linearly from zero to 100 kPa in 200 s and remains constant at 100 kPa thereafter.

**23.** Assuming the material in a spherical rubber balloon can be modeled as linearly elastic with modulus $E$ and Poisson's ratio $\nu = 0.5$, show that the internal pressure $p$ needed to expand the balloon varies with the radial expansion ratio $\lambda_r = r/r_0$ as

$$\frac{pr_0}{4Eb_0} = \frac{1}{\lambda_r^2} - \frac{1}{\lambda_r^3}$$

where $b_0$ is the initial wall thickness. Plot this function and determine its critical values.

**24.** Repeat Problem 23, but using the constitutive relation obtained in Chapter 1, Problem 25:

$$_t\sigma_x = \frac{E}{3}\left(\lambda_x^2 - \frac{1}{\lambda_x^2\lambda_y^2}\right)$$

**25.** What pressure is needed to expand a balloon, initially 3" in diameter and with a wall thickness of 0.1", to a diameter of 30"? The balloon is constructed of a rubber with a specific gravity of 0.9 and a molecular weight between crosslinks of 3000 g/mol. The temperature is 20°C.

**26.** After the balloon of Problem 25 has been inflated, the temperature is increased by 25°C. How do the pressure and radius change?

**27.** It is possible to obtain an estimate for the theoretical stiffness of a polymer molecule by visualizing it as a sequence of truss-type elements that are connected so as to resist force-induced changes in the covalent angle between elements (approximately 112° for polyethylene) as well as in their length. The axial stiffness of the bond elements has been determined spectrographically to be $k_l = 435$ N/m, and the torsional stiffness of the bond angle is $k_\phi/l^2 = 35$ N/m, where here $l = 153$ pm is the bond length and $\phi = 112/2 = 56°$ is the half-angle. The torsional stiffness is defined as the change in torque $\Delta T$ required to induce a unit incremental change in angle $\Delta\phi$; that is, $k_\phi = \Delta T/\Delta\phi$.

(a) Show that the extension $\delta$ of a polymer chain having $n$ such links and subjected to a tensile force $F$ is

$$\delta = nF\left[\frac{\sin^2\phi}{k_l} + \frac{l^2\cos^2\phi}{4k_\phi}\right]$$

*Hint:* This problem can be done using Castigliano's method, with the strain energy of extension being $U_l = F_l^2/(2k_l)$ and that of torsion being $U_\phi = T^2/(2k_\phi)$. The force $F_l$ is the component of force along the element.

(b) Compute the theoretical Young's modulus, using

$$E = \frac{\sigma}{\epsilon} = \frac{F}{A}\frac{L}{\delta}$$

where the effective molecular area $A$ as determined by X-ray diffraction measurements of crystal packing is $A = 0.181$ nm$^2$.

**28.** A torsion bar 1.5 m in length and 30 mm in diameter is clamped at one end, and the free end is twisted through an angle of 10°, as shown in Fig. P.2.28. Find the maximum torsional shear stress induced in the bar.

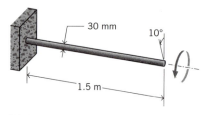

30 mm
10°
1.5 m

**P.2.28**

**29.** The torsion bar of Problem 28 fails when the applied torque is 1500 N-m. What is the modulus of rupture in torsion? Is this the same as the material's maximum shear stress?

**30.** A solid steel drive shaft is to be capable of transmitting 50 hp at 500 rpm. What should its diameter be if the maximum torsional shear stress is to be kept less than half the tensile yield strength?

**31.** How much power could the shaft referred to in Problem 30 transmit (at the same maximum torsional shear stress) if the same quantity of material were used in an annular rather than a solid shaft? Take the inside diameter to be half the outside diameter.

**32.** Two shafts, each 1 ft long and $1''$ in diameter, are connected by a 2 : 1 gearing, as shown in Fig. P.2.32, and the free end is loaded with a 100 ft-lb torque. Find the angle of twist at the loaded end.

**33.** A shaft of length $L$, diameter $d$, and shear modulus $G$ is loaded with a uniformly distributed twisting moment of $T_0$(N-m/m), as shown in Fig. P.2.33. (The twisting moment $T(x)$ at a distance $x$ from the free end is therefore $T_0 x$.) Find the angle of twist at the free end.

**34.** A composite shaft 3 ft in length is constructed assembling an aluminum rod, $2''$ in diameter, over which is bonded an annular steel cylinder of $0.5''$ wall thickness. Determine the maximum torsional shear stress when the composite cylinder is subjected to a torque of 10,000 in-lb.

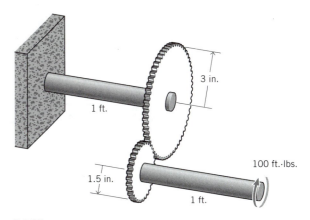

**P.2.32**

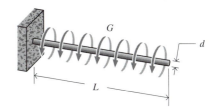

**P.2.33**

# 3 GENERAL CONCEPTS OF STRESS AND STRAIN

In extending the direct method of stress analysis presented in the last chapter to geometrically more complex structures, it will be convenient to have available somewhat more general mathematical statements of the kinematic, equilibrium, and constitutive equations; this is the objective of the present chapter. These equations also form the basis for more theoretical methods in stress analysis as well as for numerical approaches such as the finite element method. We will also seek to introduce some of the notational schemes used widely in the technical literature for such entities as stress and strain. Depending on the specific application, both index and matrix notations can be very convenient; these are described in Appendix D.

## 3.1 THE KINEMATIC EQUATIONS

### 3.1.1 Infinitesimal Strain

The *kinematic* or *strain-displacement* equations describe how the strains—the stretching and distortion—within a loaded body relate to the body's displacements. The displacement components in the $x$, $y$, and $z$ directions are denoted by the vector $\mathbf{u} \equiv u_i \equiv (u, v, w)$, and are functions of position within the body: $\mathbf{u} = \mathbf{u}(x, y, z)$. If all points within the material experience the same displacement ($\mathbf{u} = $ constant), the structure moves as a rigid body but does not stretch or deform internally. For strain to occur, points within the body must experience *different* displacements.

Consider two points $A$ and $B$ separated initially by a small distance $dx$ as shown in Fig. 3.1, and experiencing motion in the $x$ direction. If the displacement at point $A$ is $u_A$, the displacement at $B$ can be expressed by a Taylor's series expansion of $u(x)$ around the point $x = A$:

$$u_B = u_A + du = u_A + \frac{\partial u}{\partial x} dx$$

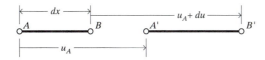

**FIGURE 3.1** Incremental deformation.

where here the expansion has been truncated after the second term. The *differential motion* δ between the two points is then

$$\delta = u_B - u_A = \left(u_A + \frac{\partial u}{\partial x}dx\right) - u_A = \frac{\partial u}{\partial x}dx$$

In our concept of stretching as being the differential displacement per unit length, the x component of strain is then

$$\epsilon_x = \frac{\delta}{dx} = \frac{\partial u}{\partial x} \tag{3.1}$$

Hence, the strain is a *displacement gradient*. Applying similar reasoning to differential motion in the y direction, the y-component of strain is the gradient of the vertical displacement v with respect to y:

$$\epsilon_y = \frac{\partial v}{\partial y} \tag{3.2}$$

The distortion of the material, which we have previously described as the change in originally right angles, is the sum of the tilts imparted to vertical and horizontal lines. As shown in Fig. 3.2, the tilt of an originally vertical line is the relative horizontal displacement of two nearby points along the line:

$$\delta = u_B - u_A = \left(u_A + \frac{\partial u}{\partial y}dy\right) - u_A = \frac{\partial u}{\partial y}dy$$

The change in angle is then

$$\gamma_1 \approx \tan\gamma_1 = \frac{\delta}{dy} = \frac{\partial u}{\partial y}$$

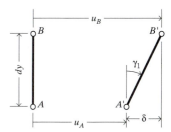

**FIGURE 3.2** Shearing distortion.

Similarly (see Fig. 3.3), the tilt $\gamma_2$ of an originally horizontal line is the gradient of v with respect to x. The shear strain in the xy plane is then

$$\gamma_{xy} = \gamma_1 + \gamma_2 = \frac{\partial v}{\partial x} + \frac{\partial u}{\partial y} \tag{3.3}$$

This notation, using ε for normal strain and γ for shearing strain, is sometimes known as the "classical" description of strain.

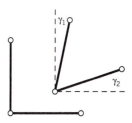

**FIGURE 3.3** Shearing strain.

### 3.1.2 Index and Matrix Notations

The "indicial notation" described in Appendix D provides a concise method of writing out all the components of three-dimensional states of strain:

$$\epsilon_{ij} = \frac{1}{2}\left(\frac{\partial u_i}{\partial x_j} + \frac{\partial u_j}{\partial x_i}\right) \equiv \frac{1}{2}\left(u_{i,j} + u_{j,i}\right) \tag{3.4}$$

This double-subscript index notation leads naturally to a matrix arrangement of the strain components, in which the $i$-$j$ component of the strain becomes the matrix element in the $i$th row and the $j$th column:

$$\epsilon_{ij} = \begin{bmatrix} \dfrac{\partial u}{\partial x} & \dfrac{1}{2}\left(\dfrac{\partial u}{\partial y} + \dfrac{\partial v}{\partial x}\right) & \dfrac{1}{2}\left(\dfrac{\partial u}{\partial z} + \dfrac{\partial w}{\partial x}\right) \\[2ex] \dfrac{1}{2}\left(\dfrac{\partial u}{\partial y} + \dfrac{\partial v}{\partial x}\right) & \dfrac{\partial v}{\partial y} & \dfrac{1}{2}\left(\dfrac{\partial v}{\partial z} + \dfrac{\partial w}{\partial y}\right) \\[2ex] \dfrac{1}{2}\left(\dfrac{\partial w}{\partial x} + \dfrac{\partial u}{\partial z}\right) & \dfrac{1}{2}\left(\dfrac{\partial v}{\partial z} + \dfrac{\partial w}{\partial y}\right) & \dfrac{\partial w}{\partial z} \end{bmatrix} \tag{3.5}$$

Note that the strain matrix is symmetric; that is, $\epsilon_{ij} = \epsilon_{ji}$. This symmetry means that there are six independent strains rather than nine, as might be expected in a $3 \times 3$ matrix. Also note that the indicial description of strain yields the same result for the normal components as in the classical description: $\epsilon_{11} = \epsilon_x$. However, the indicial components of shear strain are half their classical counterparts: $\epsilon_{12} = \gamma_{xy}/2$.

In still another useful notational scheme, the classical strain-displacement equations can be written out in a vertical list, similar to a vector:

$$\begin{Bmatrix} \epsilon_x \\ \epsilon_y \\ \epsilon_z \\ \gamma_{yz} \\ \gamma_{xz} \\ \gamma_{xy} \end{Bmatrix} = \begin{Bmatrix} \partial u/\partial x \\ \partial v/\partial y \\ \partial w/\partial z \\ \partial v/\partial z + \partial w/\partial y \\ \partial u/\partial z + \partial w/\partial x \\ \partial u/\partial y + \partial v/\partial x \end{Bmatrix}$$

This vectorlike arrangement of the strain components is for convenience only and is sometimes called a *pseudovector*. Strain is actually a *second-rank tensor,* like stress or moment of inertia, and has mathematical properties very different from those of vectors. The ordering of the elements in the pseudovector form is arbitrary, but it is conventional to list them as we have here by moving down the diagonal of the strain matrix of Eq. 3.5 from upper left to lower right, then move up the third column, and finally move one column to the left on the first row; this gives the ordering 1,1; 2,2; 3,3; 2,3; 1,3; 1,2.

Following the rules of matrix multiplication, the strain pseudovector can also be written in terms of the displacement vector as

$$\begin{Bmatrix} \epsilon_x \\ \epsilon_y \\ \epsilon_z \\ \gamma_{yz} \\ \gamma_{xz} \\ \gamma_{xy} \end{Bmatrix} = \begin{bmatrix} \partial/\partial x & 0 & 0 \\ 0 & \partial/\partial y & 0 \\ 0 & 0 & \partial/\partial z \\ 0 & \partial/\partial z & \partial/\partial y \\ \partial/\partial z & 0 & \partial/\partial x \\ \partial/\partial y & \partial/\partial x & 0 \end{bmatrix} \begin{Bmatrix} u \\ v \\ w \end{Bmatrix} \tag{3.6}$$

abbreviated as **L**:

$$
\mathbf{L} = \begin{bmatrix}
\partial/\partial x & 0 & 0 \\
0 & \partial/\partial y & 0 \\
0 & 0 & \partial/\partial y \\
0 & \partial/\partial z & \partial/\partial y \\
\partial/\partial z & 0 & \partial/\partial x \\
\partial/\partial y & \partial/\partial x & 0
\end{bmatrix}
\tag{3.7}
$$

The strain-displacement equations can then be written in the concise "pseudovector-matrix" form

$$
\boxed{\boldsymbol{\epsilon} = \mathbf{L}\mathbf{u}}
\tag{3.8}
$$

Equations such as this must be used in a well-defined context, because they apply only when the somewhat arbitrary pseudovector listing of the strain components is used.

### 3.1.3 Volumetric Strain

Since the normal strain is just the change in length per unit of original length, the new length $L'$ after straining is found as

$$
\epsilon = \frac{L' - L_0}{L_0} \Rightarrow L' = (1 + \epsilon)L_0
\tag{3.9}
$$

If a cubical volume element, originally of dimension $abc$, is subjected to normal strains in all three directions, the change in the element's volume is

$$
\frac{\Delta V}{V} = \frac{a'b'c' - abc}{abc} = \frac{a(1 + \epsilon_x)b(1 + \epsilon_y)c(1 + \epsilon_z) - abc}{abc}
\tag{3.10}
$$

$$
= (1 + \epsilon_x)(1 + \epsilon_y)(1 + \epsilon_z) - 1 \approx \epsilon_x + \epsilon_y + \epsilon_z
$$

where products of strains are neglected in comparison with individual values. The volumetric strain is therefore the sum of the normal strains—that is, the sum of the diagonal elements in the strain matrix (this is also called the *trace* of the matrix, denoted tr[$\boldsymbol{\epsilon}$]). In index notation, this can be written simply

$$
\frac{\Delta V}{V} = \epsilon_{kk}
$$

This is known as the volumetric, or "dilatational" component of the strain.

**EXAMPLE 3.1**

To illustrate how volumetric strain is calculated, consider a thin sheet of steel subjected to strains in its plane given by $\epsilon_x = 3$, $\epsilon_y = -4$, and $\gamma_{xy} = 6$ (all in $\mu\text{in/in}$). The sheet is not in plane strain, because it can undergo a Poisson strain in the $z$ direction given by $\epsilon_z = \nu(\epsilon_x + \epsilon_y) = 0.3(3 - 4) = -0.3$. The total state of strain can therefore be written as the matrix

$$[\epsilon] = \begin{bmatrix} 3 & 6 & 0 \\ 6 & -4 & 0 \\ 0 & 0 & -0.3 \end{bmatrix} \times 10^{-6}$$

where the brackets on the $[\epsilon]$ symbol emphasize that the matrix rather than pseudovector form of the strain is being used. The volumetric strain is

$$\frac{\Delta V}{V} = (3 - 4 - 0.3) \times 10^{-6} = -1.3 \times 10^{-6}$$

Engineers often refer to "microinches" of strain; they really mean microinches per inch. In the case of volumetric strain, the corresponding (but awkward) unit would be micro–cubic inches per cubic inch.

### 3.1.4 Finite Strain

The infinitesimal strain-displacement relations given by Eqs. 3.1 through 3.3 are used in the vast majority of mechanical analyses, but they do not describe stretching accurately when the displacement gradients become large. This often occurs when polymers (especially elastomers) are being considered. Large strains also occur during deformation processing operations, such as stamping of steel automotive body panels. The kinematics of large displacement or strain can be complicated and subtle, but the following section will outline a simple description of *Lagrangian* finite strain to illustrate some of the concepts involved.

Consider two orthogonal lines $OB$ and $OA$ as shown in Fig. 3.4, originally of length $dx$ and $dy$, along the x-y axes, where for convenience we set $dx = dy = 1$. After strain the endpoints of these lines move to new positions $A_1 O_1 B_1$ as shown. We will describe these new positions using the coordinate scheme of the original x-y axes, although we could also allow the new positions to define a new set of axes. In following the motion of the lines, we are using the so-called *Lagrangian* viewpoint. We could also have selected a fixed control volume and monitored material as it passes through it; this is the *Eulerian* view often used in fluid mechanics.

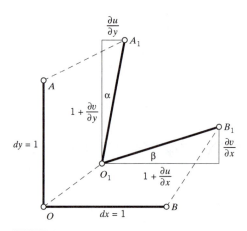

**FIGURE 3.4** Finite displacements.

After straining, the distance $dx$ becomes

$$(dx)' = \left(1 + \frac{\partial u}{\partial x}\right)dx$$

Using our earlier "small" thinking, the $x$-direction strain would be just $\partial u/\partial x$. However, when the strains become larger, we must also consider that the upward motion of point $B_1$ relative to $O_1$—that is, $\partial v/\partial x$—also helps stretch the line $OB$. Considering both these effects, the Pythagorean theorem gives the new length $O_1 B_1$ as

$$O_1 B_1 = \sqrt{\left(1 + \frac{\partial u}{\partial x}\right)^2 + \left(\frac{\partial v}{\partial x}\right)^2}$$

We now define our Lagrangian strain as

$$\epsilon_x = \frac{O_1 B_1 - OB}{OB} = O_1 B_1 - 1$$

$$= \sqrt{1 + 2\frac{\partial u}{\partial x} + \left(\frac{\partial u}{\partial x}\right)^2 + \left(\frac{\partial v}{\partial x}\right)^2} - 1$$

Using the series expansion $\sqrt{1 + x} = 1 + x/2 + x^2/8 + \cdots$ and neglecting terms beyond first order, this becomes

$$\epsilon_x \approx \left\{1 + \frac{1}{2}\left[2\frac{\partial u}{\partial x} + \left(\frac{\partial u}{\partial x}\right)^2 + \left(\frac{\partial v}{\partial x}\right)^2\right]\right\} - 1$$

$$= \frac{\partial u}{\partial x} + \frac{1}{2}\left[\left(\frac{\partial u}{\partial x}\right)^2 + \left(\frac{\partial v}{\partial x}\right)^2\right] \tag{3.11}$$

Similarly, we can show

$$\epsilon_y = \frac{\partial v}{\partial y} + \frac{1}{2}\left[\left(\frac{\partial v}{\partial y}\right)^2 + \left(\frac{\partial u}{\partial y}\right)^2\right] \tag{3.12}$$

$$\gamma_{xy} = \frac{\partial u}{\partial y} + \frac{\partial v}{\partial x} + \frac{\partial^2 u}{\partial x\,\partial y} + \frac{\partial^2 v}{\partial x\,\partial y} \tag{3.13}$$

When the strains are sufficiently small that the quadratic terms are negligible compared with the linear ones, these reduce to the infinitesimal-strain expressions shown earlier.

## EXAMPLE 3.2

The displacement function $u(x)$ for a tensile specimen of uniform cross section and length $L$, fixed at one end and subjected to a displacement $\delta$ at the other, is just the linear relation

$$u(x) = \left(\frac{x}{L}\right)\delta$$

The Lagrangian strain is then given by Eq. 3.11 as

$$\epsilon_x = \frac{\delta}{L} + \frac{1}{2}\left(\frac{\delta}{L}\right)^2$$

The first term is the familiar small-strain expression, with the second nonlinear term becoming more important as $\delta$ becomes larger. When $\delta = L$—that is, the conventional strain is 100 percent—there is a 50 percent difference between the conventional and Lagrangian strain measures.

The Lagrangian strain components can be generalized using index notation as

$$\epsilon_{ij} = \tfrac{1}{2}(u_{i,j} + u_{j,i} + u_{r,i}u_{r,j})$$

where, as indicated in Appendix D, the comma denotes differentiation. A pseudovector form is also convenient occasionally:

$$\begin{Bmatrix} \epsilon_x \\ \epsilon_y \\ \gamma_{xy} \end{Bmatrix} = \begin{Bmatrix} u_{,x} \\ v_{,y} \\ u_{,y} + v_{,x} \end{Bmatrix} + \frac{1}{2}\begin{bmatrix} u_{,x} & v_{,x} & 0 & 0 \\ 0 & 0 & u_{,y} & v_{,y} \\ u_{,y} & v_{,y} & u_{,x} & v_{,x} \end{bmatrix}\begin{Bmatrix} u_{,x} \\ v_{,x} \\ u_{,y} \\ v_{,y} \end{Bmatrix}$$

$$= \left(\begin{bmatrix} \partial/\partial x & 0 \\ 0 & \partial/\partial y \\ \partial/\partial y & \partial/\partial x \end{bmatrix} + \frac{1}{2}\begin{bmatrix} u_{,x} & v_{,x} & 0 & 0 \\ 0 & 0 & u_{,y} & v_{,y} \\ u_{,y} & v_{,y} & u_{,x} & v_{,x} \end{bmatrix}\begin{bmatrix} \partial/\partial x & 0 \\ 0 & \partial/\partial x \\ \partial/\partial y & 0 \\ 0 & \partial/\partial y \end{bmatrix}\right)\begin{Bmatrix} u \\ v \end{Bmatrix}$$

which can be abbreviated

$$\boldsymbol{\epsilon} = [\mathbf{L} + \mathbf{A(u)}]\mathbf{u} \tag{3.14}$$

The matrix $\mathbf{A(u)}$ contains the nonlinear effect of large strain and becomes negligible when strains are small.

## 3.2 THE EQUILIBRIUM EQUATIONS

### 3.2.1 Cauchy Stress

In Chapters 1 and 2 we expressed the normal stress as force per unit area acting perpendicularly to a selected area, and a shear stress was a force per unit area acting transversely to the area. To generalize this concept, consider the situation depicted in Fig. 3.5a, in which a *traction vector* **T** acts on an arbitrary plane within or on the external boundary of the body and at an arbitrary direction with respect to the orientation of the plane. The traction is a simple force vector having magnitude and direction, but its magnitude is expressed in terms of force per unit of area:

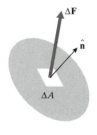

*(a)*

**FIGURE 3.5a**
Traction vector.

$$\mathbf{T} = \lim_{\Delta A \to 0}\left(\frac{\Delta \mathbf{F}}{\Delta A}\right) \tag{3.15}$$

where $\Delta A$ is the magnitude of the area on which $\Delta \mathbf{F}$ acts. The *Cauchy*[1] stresses, which are a generalization of our earlier definitions of stress, are the forces per unit area acting on the Cartesian $x$, $y$, and $z$ planes to balance the traction. In two dimensions this balance can be written by drawing a simple free-body diagram with the traction vector acting on an area of arbitrary size $A$ (Fig. 3.5b), remembering to obtain the forces by multiplying by the appropriate area:

$$\sigma_x(A\cos\theta) + \tau_{xy}(A\sin\theta) = T_x A$$

$$\tau_{xy}(A\cos\theta) + \sigma_y(A\sin\theta) = T_y A$$

Canceling the factor $A$, this can be written in matrix form as

$$\begin{bmatrix} \sigma_x & \tau_{xy} \\ \tau_{xy} & \sigma_y \end{bmatrix} \begin{Bmatrix} \cos\theta \\ \sin\theta \end{Bmatrix} = \begin{Bmatrix} T_x \\ T_y \end{Bmatrix} \tag{3.16}$$

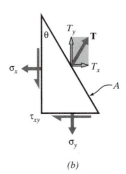

(b)

**FIGURE 3.5b** Cauchy stress.

## EXAMPLE 3.3

Consider a circular cavity containing an internal pressure $p$, as shown in Fig. 3.6. The components of the traction vector are then $T_x = -p\cos\theta$, $T_y = -p\sin\theta$. The Cartesian Cauchy stresses in the material at the boundary must then satisfy the relations

$$\sigma_x\cos\theta + \tau_{xy}\sin\theta = -p\cos\theta$$

$$\tau_{xy}\cos\theta + \sigma_y\sin\theta = -p\sin\theta$$

At $\theta = 0$, $\sigma_x = -p$, $\sigma_y = \tau_{xy} = 0$; at $\theta = \pi/2$, $\sigma_y = -p$, $\sigma_x = \tau_{xy} = 0$. The shear stress $\tau_{xy}$ vanishes for $\theta = 0$ or $\pi/2$; in a later section it will be seen that the normal stresses $\sigma_x$ and $\sigma_y$ are therefore *principal* stresses at those points.

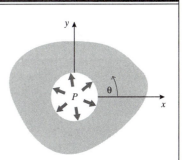

**FIGURE 3.6** Constant pressure on internal circular boundary.

The vector $(\cos\theta, \sin\theta)$ on the left-hand side of Eq. 3.16 is also the vector $\hat{\mathbf{n}}$ of direction cosines of the normal to the plane on which the traction acts, and serves to define the orientation of this plane. This matrix equation, which is sometimes called *Cauchy's relation,* can be abbreviated as

$$\boxed{[\sigma]\,\hat{\mathbf{n}} = \mathbf{T}} \tag{3.17}$$

The brackets here serve as a reminder that the stress is being written as the square matrix of Eq. 3.16 rather than in pseudovector form. This relation serves to define the stress concept as an entity that relates the traction (a vector) acting on an arbitrary surface to the orientation of the surface (another vector). The stress is therefore of a higher degree of abstraction than a vector and is technically a *second-rank tensor.* The difference between vectors (first-rank tensors) and second-rank tensors shows up in how they *transform* with respect to coordinate rotations, which is treated in a later section. As illustrated by the previous example, Cauchy's relation serves both to define the stress and to compute its magnitude at boundaries where the tractions are known.

[1]Baron Augustin-Louis Cauchy (1789–1857) was a prolific French engineer and mathematician.

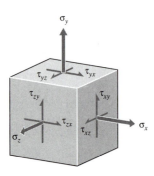

**FIGURE 3.7** Cartesian Cauchy stress components in three dimensions.

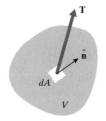

**FIGURE 3.8** Traction vector **T** acting on differential area $dA$ with direction cosines $\hat{n}$.

In three dimensions, the matrix form of the stress state shown in Fig. 3.7 is the symmetric $3 \times 3$ array obtained by an obvious extension of the one in Eq. 3.16:

$$[\sigma] = \sigma_{ij} = \begin{bmatrix} \sigma_x & \tau_{xy} & \tau_{xz} \\ \tau_{xy} & \sigma_y & \tau_{yz} \\ \tau_{xz} & \tau_{yz} & \sigma_z \end{bmatrix} \tag{3.18}$$

The element in the $i$th row and the $j$th column of this matrix is the stress on the $i$th face in the $j$th direction. Moment equilibrium requires that the stress matrix be symmetric (see Problem 8 in this chapter), so the order of subscripts of the off-diagonal shearing stresses is immaterial.

### 3.2.2 Differential Governing Equations

Determining the variation of the stress components as functions of position within the interior of a body is obviously a principal goal in stress analysis. This is a type of *boundary value problem* often encountered in the theory of differential equations, in which the *gradients* of the variables, rather than the explicit variables themselves, are specified. In the case of stress, the gradients are governed by conditions of static equilibrium: The stresses cannot change arbitrarily between two points $A$ and $B$, or the material between those two points may not be in equilibrium.

To develop this idea formally, we require that the integrated value of the surface traction **T** over the surface $A$ of an arbitrary volume element $dV$ within the material (see Fig. 3.8) must sum to zero in order to maintain static equilibrium:

$$0 = \int_A \mathbf{T} dA = \int_A [\sigma] \hat{n} dA$$

Here we assume the lack of gravitational, centripetal, or other "body" forces acting on material within the volume. The surface integral in this relation can be converted to a volume integral by Gauss' divergence theorem:[2]

$$\int_V \nabla [\sigma] dV = 0$$

Since the volume $V$ is arbitrary, this requires that the integrand be zero:

$$\nabla [\sigma] = 0 \tag{3.19}$$

For Cartesian problems in three dimensions this expands to

$$\begin{aligned} \frac{\partial \sigma_x}{\partial x} + \frac{\partial \tau_{xy}}{\partial y} + \frac{\partial \tau_{xz}}{\partial z} &= 0 \\ \frac{\partial \tau_{xy}}{\partial x} + \frac{\partial \sigma_y}{\partial y} + \frac{\partial \tau_{yz}}{\partial z} &= 0 \\ \frac{\partial \tau_{xz}}{\partial x} + \frac{\partial \tau_{yz}}{\partial y} + \frac{\partial \sigma_z}{\partial x} &= 0 \end{aligned} \tag{3.20}$$

---

[2]Gauss' theorem states that $\int_A X\hat{n}\, dA = \int_S \nabla X dV$, where $X$ is a scalar, vector, or tensor quantity.

$$\sigma_{ij,j} = 0 \tag{3.21}$$

Or in pseudovector-matrix form, we can write

$$
\begin{bmatrix}
\frac{\partial}{\partial x} & 0 & 0 & 0 & \frac{\partial}{\partial z} & \frac{\partial}{\partial y} \\
0 & \frac{\partial}{\partial y} & 0 & \frac{\partial}{\partial z} & 0 & \frac{\partial}{\partial x} \\
0 & 0 & \frac{\partial}{\partial z} & \frac{\partial}{\partial y} & \frac{\partial}{\partial x} & 0
\end{bmatrix}
\begin{Bmatrix}
\sigma_x \\ \sigma_y \\ \sigma_z \\ \tau_{yz} \\ \tau_{xz} \\ \tau_{xy}
\end{Bmatrix}
=
\begin{Bmatrix} 0 \\ 0 \\ 0 \end{Bmatrix}
\tag{3.22}
$$

Noting that the differential operator matrix in the brackets is just the transform of the one that appeared in Eq. 3.7, we can write this as

$$\mathbf{L}^T \boldsymbol{\sigma} = 0 \tag{3.23}$$

## EXAMPLE 3.4

It isn't hard to come up with functions of stress that satisfy the equilibrium equations; any constant will do, because the stress gradients will then be identically zero. The catch is that they must satisfy the boundary conditions as well, and this complicates things considerably. Chapter 5 will outline several approaches to solving the equations directly, but in some simple cases a solution can be seen by inspection.

Consider a tensile specimen subjected to a load $P$, as shown in Fig. 3.9. A trial solution that certainly satisfies the equilibrium equations is

$$
[\sigma] =
\begin{bmatrix}
c & 0 & 0 \\
0 & 0 & 0 \\
0 & 0 & 0
\end{bmatrix}
$$

where $c$ is a constant we must choose so as to satisfy the boundary conditions. To maintain horizontal equilibrium in the free-body diagram of Fig. 3.9b, it is immediately obvious that $cA = P$, or $\sigma_x = c = P/A$. This familiar relation was used in Chapter 1 to define the stress, but we see here that it can be viewed as a consequence of equilibrium considerations rather than a basic definition.

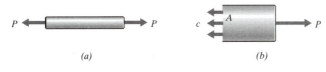

(a)                    (b)

**FIGURE 3.9**  A tensile specimen.

# 3.3 TENSOR TRANSFORMATIONS

## 3.3.1 Direct Approach

One of the most common problems in mechanics of materials involves *transformation of axes*. For instance, we may know the stresses acting on $xy$ planes but be really more interested in the stresses acting on planes oriented at, say, 30° to the $x$ axis as seen in

Fig. 3.10, perhaps because these are close-packed atomic planes on which sliding is prone to occur, or because this angle is the angle at which two pieces of lumber are glued together in a "scarf" joint. We seek a means to transform the stresses to these new $x'y'$ planes.

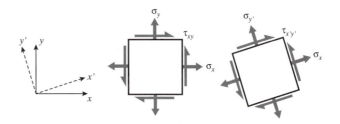

FIGURE 3.10 Rotation of axes in two dimensions.

The rules for stress transformations can be developed directly from considerations of static equilibrium. For illustration, consider the case of uniaxial tension shown in Fig. 3.11 in which all stresses other than $\sigma_y$ are zero. A free-body diagram is then constructed in which the specimen is "cut" along the inclined plane on which the stresses, labeled $\sigma_{y'}$ and $\tau_{x'y'}$, are desired. The key here is to note that the area on which these transformed stresses act is different from the area normal to the $y$ axis, so that both the areas and the forces acting on them need to be "transformed." Balancing forces in the $y'$ direction (the direction normal to the inclined plane):

$$(\sigma_y A) \cos \theta = \sigma_{y'} \left( \frac{A}{\cos \theta} \right)$$

$$\sigma_{y'} = \sigma_y \cos^2 \theta \tag{3.24}$$

Similarly, a force balance in the tangential direction gives

$$\tau_{x'y'} = \sigma_y \sin \theta \cos \theta \tag{3.25}$$

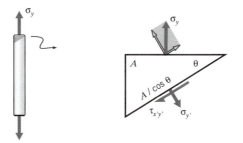

FIGURE 3.11 An inclined plane in a tensile specimen.

## EXAMPLE 3.5

Consider a unidirectionally reinforced composite ply with strengths $\hat{\sigma}_1$ in the fiber direction, $\hat{\sigma}_2$ in the transverse direction, and $\hat{\tau}_{12}$ in shear. As the angle $\theta$ between the fiber direction and an applied tensile stress $\sigma_y$ is increased, the stress in the fiber direction will decrease according to Eq. 3.24. If the ply were to fail by fiber fracture alone, the stress $\sigma_{y,b}$ needed to cause failure would *increase* with misalignment according to $\sigma_{y,b} = \hat{\sigma}_1 / \cos^2 \theta$.

However, the shear stresses as given by Eq. 3.25 *increase* with $\theta$, so the $\sigma_y$ stress needed for shear failure drops. The strength $\sigma_{y,b}$ is the smaller of the stresses needed to cause fiber-direction or shear failure, so the strength becomes limited by shear after only a few degrees of misalignment. In fact, a 15° off-axis tensile specimen has been proposed as a means of measuring intralaminar shear strength. When the orientation angle approaches 90°, failure is dominated by the transverse strength. The experimental data shown in Fig. 3.12 are for glass-epoxy composites,[3] which show good but not exact agreement with these simple expressions.

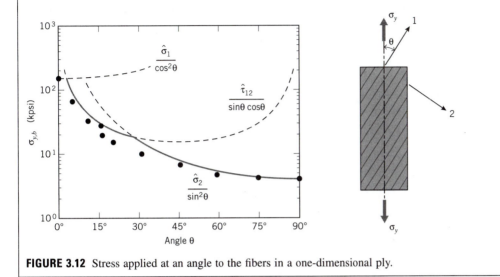

**FIGURE 3.12** Stress applied at an angle to the fibers in a one-dimensional ply.

A similar approach, but generalized to include stresses $\sigma_x$ and $\tau_{xy}$ on the original $xy$ planes as shown in Fig. 3.13 (see Problem 3 in this chapter) gives

$$
\begin{aligned}
\sigma_{x'} &= \sigma_x \cos^2 \theta + \sigma_y \sin^2 \theta + 2\tau_{xy} \sin \theta \cos \theta \\
\sigma_{y'} &= \sigma_x \sin^2 \theta + \sigma_y \cos^2 \theta - 2\tau_{xy} \sin \theta \cos \theta \\
\tau_{x'y'} &= (\sigma_y - \sigma_x) \sin \theta \cos \theta + \tau_{xy}(\cos^2 \theta - \sin^2 \theta)
\end{aligned}
\tag{3.26}
$$

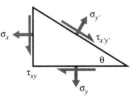

**FIGURE 3.13** Stresses on inclined plane.

These relations can be written in pseudovector-matrix form as

$$
\left\{ \begin{array}{c} \sigma_{x'} \\ \sigma_{y'} \\ \tau_{x'y'} \end{array} \right\} =
\left[ \begin{array}{ccc} c^2 & s^2 & 2sc \\ s^2 & c^2 & -2sc \\ -sc & sc & c^2 - s^2 \end{array} \right]
\left\{ \begin{array}{c} \sigma_x \\ \sigma_y \\ \tau_{xy} \end{array} \right\}
\tag{3.27}
$$

where $c = \cos \theta$ and $s = \sin \theta$. This can be abbreviated as

$$
\boldsymbol{\sigma'} = \mathbf{A}\boldsymbol{\sigma} \tag{3.28}
$$

where $\mathbf{A}$ is the transformation matrix in brackets in Eq. 3.27. This expression in Eq. 3.28 would be valid for three-dimensional as well as two-dimensional stress states, although the particular form of $\mathbf{A}$ given in Eq. 3.27 is valid in two dimensions only (plane stress) and for Cartesian coordinates.

---

[3]R.M. Jones, *Mechanics of Composite Materials,* McGraw-Hill, 1975.

Using either mathematical or geometric arguments (see Problems 17 and 18 in this chapter), it can be shown that the components of infinitesimal strain transform by *almost* the same relations:

$$\left\{ \begin{array}{c} \epsilon_{x'} \\ \epsilon_{y'} \\ \frac{1}{2}\gamma_{x'y'} \end{array} \right\} = \mathbf{A} \left\{ \begin{array}{c} \epsilon_{x} \\ \epsilon_{y} \\ \frac{1}{2}\gamma_{xy} \end{array} \right\} \tag{3.29}$$

The factor of $\frac{1}{2}$ on the shear components arises from the classical definition of shear strain, which is twice the tensorial shear strain. This introduces some awkwardness into the transformation relations, some of which can be reduced by defining *Reuter's matrix* as

$$[\mathbf{R}] = \begin{bmatrix} 1 & 0 & 0 \\ 0 & 1 & 0 \\ 0 & 0 & 2 \end{bmatrix} \quad \text{or} \quad [\mathbf{R}]^{-1} = \begin{bmatrix} 1 & 0 & 0 \\ 0 & 1 & 0 \\ 0 & 0 & \frac{1}{2} \end{bmatrix} \tag{3.30}$$

We can now write

$$\left\{ \begin{array}{c} \epsilon_{x'} \\ \epsilon_{y'} \\ \gamma_{x'y'} \end{array} \right\} = \mathbf{R} \left\{ \begin{array}{c} \epsilon_{x'} \\ \epsilon_{y'} \\ \frac{1}{2}\gamma_{x'y'} \end{array} \right\} = \mathbf{RA} \left\{ \begin{array}{c} \epsilon_{x} \\ \epsilon_{y} \\ \frac{1}{2}\gamma_{xy} \end{array} \right\} = \mathbf{RAR}^{-1} \left\{ \begin{array}{c} \epsilon_{x} \\ \epsilon_{y} \\ \gamma_{xy} \end{array} \right\}$$

or

$$\boldsymbol{\epsilon}' = \mathbf{RAR}^{-1}\boldsymbol{\epsilon} \tag{3.31}$$

As can be verified by expanding this relation, the transformation equations for strain can also be obtained from the stress transformation equations (such as Eq. 3.26) by replacing $\sigma$ with $\epsilon$ and $\tau$ with $\gamma/2$:

$$\boxed{\begin{array}{l} \epsilon_{x'} = \epsilon_x \cos^2\theta + \epsilon_y \sin^2\theta + \gamma_{xy}\sin\theta\cos\theta \\ \epsilon_{y'} = \epsilon_x \sin^2\theta + \epsilon_y \cos^2\theta - \gamma_{xy}\sin\theta\cos\theta \\ \gamma_{x'y'} = 2(\epsilon_y - \epsilon_x)\sin\theta\cos\theta + \gamma_{xy}(\cos^2\theta - \sin^2\theta) \end{array}} \tag{3.32}$$

## EXAMPLE 3.6

Consider the biaxial strain state

$$\boldsymbol{\epsilon} = \left\{ \begin{array}{c} \epsilon_{x'} \\ \epsilon_{y'} \\ \gamma_{x'y'} \end{array} \right\} = \left\{ \begin{array}{c} 0.01 \\ -0.01 \\ 0 \end{array} \right\}$$

The state of strain $\epsilon'$ referred to axes rotated by $\theta = 45°$ from the *x-y* axes can be computed by matrix multiplication as

$$\mathbf{A} = \begin{bmatrix} c^2 & s^2 & 2sc \\ s^2 & c^2 & -2sc \\ -sc & sc & c^2 - s^2 \end{bmatrix} = \begin{bmatrix} 0.5 & 0.5 & 1.0 \\ 0.5 & 0.5 & -1.0 \\ -0.5 & 0.5 & 0.0 \end{bmatrix}$$

Then

$$\boldsymbol{\epsilon}' = \mathbf{RAR}^{-1}\boldsymbol{\epsilon} = \begin{bmatrix} 1.0 & 0.0 & 0.0 \\ 0.0 & 1.0 & 0.0 \\ 0.0 & 0.0 & 2.0 \end{bmatrix} \begin{bmatrix} 0.5 & 0.5 & 1.0 \\ 0.5 & 0.5 & -1.0 \\ -0.5 & 0.5 & 0.0 \end{bmatrix} \begin{bmatrix} 1.0 & 0.0 & 0.0 \\ 0.0 & 1.0 & 0.0 \\ 0.0 & 0.0 & 0.5 \end{bmatrix} \begin{Bmatrix} 0.01 \\ -0.01 \\ 0 \end{Bmatrix}$$

$$= \begin{Bmatrix} 0.00 \\ 0.00 \\ -0.02 \end{Bmatrix}$$

Obviously, the matrix multiplication method is tedious unless matrix-handling software is available, in which case it becomes very convenient.

## 3.3.2 Mohr's Circle

Everyday experience with such commonplace occurrences as pushing objects at an angle gives us all a certain intuitive sense of how vector transformations work. Second-rank tensor transformations seem more abstract at first, and a device to help visualize them is of great value. As it happens, the transformation equations have a famous (among engineers) graphical interpretation known as *Mohr's circle*.[4] The Mohr procedure is justified mathematically by using the trigonometric double-angle relations to show that Eqs. 3.26 have a circular representation (see Problem 19 in this chapter), but it can probably best be learned simply by memorizing the following recipe:

1. Draw the stress square, noting the values on the $x$ and $y$ faces; Fig. 3.14$a$ shows a hypothetical case for illustration. *For the purpose of Mohr's circle only,* regard a shear stress acting in a clockwise rotation sense as being positive and counterclockwise as negative. The shear stresses on the $x$ and $y$ faces must then have opposite signs. The normal stresses are positive in tension and negative in compression as usual.

2. Construct a graph with $\tau$ as the ordinate ($y$ axis) and $\sigma$ as abscissa, and plot the stresses on the $x$ and $y$ faces of the stress square as two points on this graph. Since the shear stresses on these two faces are the negatives of each other, one of these points will be above the $\sigma$-axis exactly as far as the other is below. It is helpful to label the two points as $x$ and $y$.

3. Connect these two points with a straight line. It will cross the $\sigma$ axis at the line's midpoint. This point will be at $(\sigma_x + \sigma_y)/2$, which in our illustration is $[5 + (-3)]/2 = 1$.

4. Place the point of a compass at the line's midpoint and set the pencil at the end of the line. Draw a circle with the line as a diameter. The completed circle for our illustrative stress state is shown in Fig. 3.14$b$.

5. To determine the stresses on a stress square that has been rotated through an angle $\theta$ with respect to the original square, rotate the diametral line *in the same direction* through *twice* this angle, or $2\theta$. The new end points of the line can now be labeled $x'$ and $y'$, and their $\sigma$-$\tau$ values are the stresses on the rotated $x'$-$y'$ axes as shown in Fig. 3.14$c$.

---

[4]Presented in 1900 by the German engineer Otto Mohr (1835–1918).

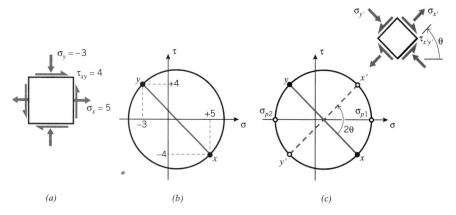

(a)  (b)  (c)

**FIGURE 3.14** Stress square (a) and Mohr's circle (b) for $\sigma_x = +5$, $\sigma_y = -3$, $\gamma_{xy} = -4$. (c) Stress state on inclined plane.

There is nothing mysterious or magical about the Mohr's circle; it is simply a device to help visualize how stresses and other second-rank tensors change when the axes are rotated.

It is clear in looking at the Mohr's circle in Fig. 3.14c that there is something special about axis rotations that cause the diametral line to become either horizontal or vertical. In the first case, the normal stresses assume maximal values and the shear stresses are zero. These normal stresses are known as the *principal* stresses, $\sigma_{p1}$ and $\sigma_{p2}$, and the planes on which they act are the *principal* planes. If the material is prone to fail by tensile cracking, it will do so by cracking along the principal planes when the value of $\sigma_{p1}$ exceeds the tensile strength.

## EXAMPLE 3.7

It is instructive to use a Mohr's circle construction to predict how a piece of black-board chalk will break in torsion, and then verify it in practice. The torsion produces a state of pure shear, as shown in Fig. 3.15, which causes the principal planes to appear at $\pm 45°$ to the chalk's long axis. The crack will appear transverse to the

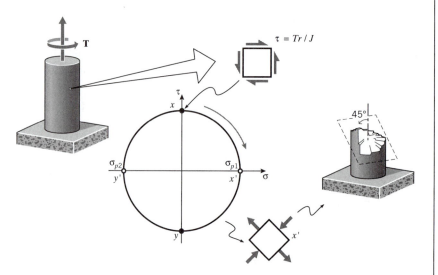

**FIGURE 3.15** Mohr's circle for simple torsion.

principal tensile stress, producing a spiral-like failure surface. (As the crack progresses into the chalk, the state of pure shear is replaced by a more complicated stress distribution, so the last part of the failure surface deviates from this ideal path to one running along the axial direction.) This is the same type of fracture that occurred all too often in skiers' thighbones (femurs) before the advent of modern safety bindings.

By direct Pythagorean construction, as shown in Fig. 3.16, the Mohr's circle shows that the angle from the $x$-$y$ axes to the principal planes is

$$\tan 2\theta_p = \frac{\tau_{xy}}{(\sigma_x - \sigma_y)/2} \qquad (3.33)$$

and the values of the principal stresses are

$$\sigma_{p1,p1} = \frac{\sigma_x + \sigma_y}{2} \pm \sqrt{\left(\frac{\sigma_x - \sigma_y}{2}\right)^2 + \tau_{xy}^2} \qquad (3.34)$$

where the first term is the $\sigma$-coordinate of the circle's center and the second is its radius.

When the Mohr's circle diametral line is vertical, the shear stresses become maximum, equal in magnitude to the radius of the circle:

$$\tau_{max} = \sqrt{\left(\frac{\sigma_x - \sigma_y}{2}\right)^2 + \tau_{xy}^2} = \frac{\sigma_{p1} - \sigma_{p2}}{2} \qquad (3.35)$$

The points of maximum shear are 90° away from the principal stress points on the Mohr's circle, so on the actual specimen the planes of maximum shear are 45° from the principal planes. The molecular sliding associated with yield is driven by shear and usually takes place on the planes of maximum shear. A tensile specimen has principal planes along and transverse to its loading direction, so shear slippage will occur on planes ±45° from the loading direction. These slip planes can often be observed as "shear bands" on the specimen.

Note that normal stresses may appear on the planes of maximum shear, so the situation is not quite the converse of that of the principal planes, on which the shear stresses vanish while the normal stresses are maximum. If the normal stresses happen to vanish on the planes of maximum shear, the stress state is said to be one of "pure shear," such as is induced by simple torsion. A state of pure shear is therefore one for which a rotation of axes exists such that the normal stresses vanish, which is possible only if the center of Mohr's circle is at the origin; that is, if $(\sigma_x + \sigma_y)/2 = 0$. More generally, a state of pure shear is one in which the trace of the stress (and strain) matrix vanishes.

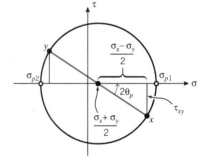

**FIGURE 3.16** Principal stresses on Mohr's circle.

# EXAMPLE 3.8

Mohr's circles can be drawn for strains as well as stresses, with shear strain plotted on the ordinate and normal strain on the abscissa. However, the ordinate must be $\gamma/2$ rather than just $\gamma$, because of the way classical infinitesimal strains are defined. Consider a state of pure shear with strain $\gamma$ and stress $\tau$ as shown in Fig. 3.17, such as might be produced by placing a circular shaft in torsion. A Mohr's circle for strain

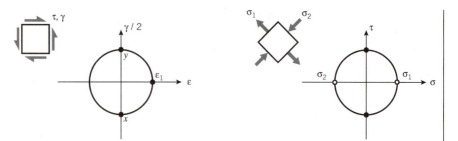

**FIGURE 3.17** Strain and stress Mohr's circles for simple shear.

quickly shows that the principal strain, on a plane 45° away, is given by $\epsilon_1 = \gamma/2$. Hooke's law for shear gives $\tau = G\gamma$, so $\epsilon_1 = \tau/2G$. The principal strain is also related to the principal stresses by

$$\epsilon_1 = \frac{1}{E}(\sigma_1 - \nu\sigma_2)$$

The Mohr's circle for stress gives $\sigma_1 = -\sigma_2 = \tau$, so this can be written

$$\frac{\tau}{2G} = \frac{1}{E}[\tau - \nu(-\tau)]$$

Canceling $\tau$ and rearranging, we have the relation among elastic constants stated earlier without proof:

$$G = \frac{E}{2(1 + \nu)}$$

### 3.3.3 General Approach

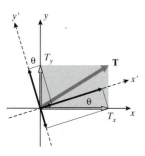

**FIGURE 3.18** Transformation of vectors.

An approach to the stress transformation equations, capable of easy extension to three dimensions, starts with the familiar relations by which *vectors* are transformed in two dimensions (see Fig. 3.18):

$$T_{x'} = T_x \cos\theta + T_y \sin\theta$$
$$T_{y'} = -T_x \sin\theta + T_y \cos\theta$$

In matrix form this is

$$\begin{Bmatrix} T_{x'} \\ T_{y'} \end{Bmatrix} = \begin{bmatrix} \cos\theta & \sin\theta \\ -\sin\theta & \cos\theta \end{bmatrix} \begin{Bmatrix} T_x \\ T_y \end{Bmatrix}$$

or

$$\mathbf{T'} = \mathbf{aT} \qquad (3.36)$$

where $\mathbf{a}$ is another transformation matrix that serves to transform the vector components in the original coordinate system to those in the primed system. In index-notation terms, this could also be denoted $a_{ij}$, so that

$$T_i' = a_{ij}T_j$$

The individual elements of $a_{ij}$ are the cosines of the angles between the $i$th *unprimed* axis and the $j$th *primed* axis.

It can be shown by direct examination that the **a** matrix has the useful property that its inverse equals its transpose; that is, $\mathbf{a}^{-1} = \mathbf{a}^T$. We can multiply Eq. 3.36 by $\mathbf{a}^T$ to give

$$\mathbf{a}^T\mathbf{T}' = (\mathbf{a}^T\mathbf{a})\mathbf{T} = \mathbf{T} \tag{3.37}$$

so the transformation can go from primed to unprimed, or the reverse.

These relations can be extended to yield an expression for transformation of stresses (or strains, or moments of inertia, or other similar quantities). Recall Cauchy's relation in matrix form:

$$[\sigma]\hat{\mathbf{n}} = \mathbf{T}$$

Using Eq. 3.37 to transform the $\hat{\mathbf{n}}$ and $\mathbf{T}$ vectors into their primed counterparts, we have

$$[\sigma]\mathbf{a}^T\hat{\mathbf{n}}' = \mathbf{a}^T\mathbf{T}'$$

Multiplying through by **a**,

$$(\mathbf{a}[\sigma]\mathbf{a}^T)\hat{\mathbf{n}}' = (\mathbf{a}\mathbf{a}^T)\mathbf{T}' = \mathbf{T}'$$

This is just Cauchy's relation again, but in the primed coordinate frame. The quantity in parentheses must therefore be $[\sigma']$:

$$[\sigma'] = \mathbf{a}[\sigma]\mathbf{a}^T \tag{3.38}$$

Therefore, transformation of stresses can be done by pre- and postmultiplying by the same transformation matrix applicable to vector transformation. This can also be written out using index notation, which provides another illustration of the transformation differences between scalars (zero-rank tensors), vectors (first-rank tensors), and second-rank tensors:

$$\text{Rank 0:} \quad b' = b$$
$$\text{Rank 1:} \quad T_i' = a_{ij}T_j \tag{3.39}$$
$$\text{Rank 2:} \quad \sigma_{ij}' = a_{ij}a_{kl}\sigma_{kl}$$

In practical work it is not always a simple matter to write down the nine elements of the **a** matrix needed in Eq. 3.38. The squares of the components of $\hat{\mathbf{n}}$ for any given plane must sum to unity, and for the three planes of the transformed stress cube to be mutually perpendicular the dot product between any two plane normals must vanish, so not just any nine numbers will make sense. Obtaining **a** is made much easier by using *Euler angles* to describe axis transformations in three dimensions.

As shown in Fig. 3.19, the final transformed axes are visualized as being achieved in three steps:

**1.** Rotate the original $x$-$y$-$z$ axes by an angle $\psi$ (psi) around the $z$-axis to obtain a new frame we may call $x'$-$y'$-$z$.

**2.** Rotate this new frame by an angle $\theta$ about the $x'$ axis to obtain another frame we can call $x'$-$y''$-$z'$.

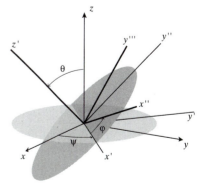

**FIGURE 3.19** Transformation in terms of Euler angles.

**3.** Rotate this frame by an angle $\phi$ (phi) around the $z'$ axis to obtain the final frame $x''\text{-}y'''\text{-}z'$.

These three transformations correspond to the transformation matrix

$$
\mathbf{a} = \begin{bmatrix} \cos\psi & \sin\psi & 0 \\ -\sin\psi & \cos\psi & 0 \\ 0 & 0 & 1 \end{bmatrix} \begin{bmatrix} 1 & 0 & 0 \\ 0 & \cos\theta & \sin\theta \\ 0 & -\sin\theta & \cos\theta \end{bmatrix} \begin{bmatrix} \cos\phi & \sin\phi & 0 \\ -\sin\phi & \cos\phi & 0 \\ 0 & 0 & 1 \end{bmatrix}
$$

This multiplication would certainly be a pain if done manually, but it is a natural for a computational approach. The `strs3d` code described in Appendix F uses this method for three-dimensional transformations.

### 3.3.4 Principal Stresses and Planes in Three Dimensions

The Mohr's circle procedure is not capable of finding principal stresses for three-dimensional stress states, and a more general method is needed. In three dimensions we seek orientations of axes such that no shear stresses appear, leaving only normal stresses in three orthogonal directions. The vanishing of shear stresses on a plane means that the stress vector $\mathbf{T}$ is normal to the plane, illustrated in two dimensions in Fig. 3.20. The traction vector can therefore be written as

$$\mathbf{T} = \sigma_p \hat{\mathbf{n}}$$

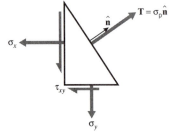

**FIGURE 3.20** Traction vector normal to principal plane.

where $\sigma_p$ is a simple scalar quantity, the *magnitude* of the stress vector. Using this in Cauchy's relation,

$$\boldsymbol{\sigma}\hat{\mathbf{n}} = \mathbf{T} = \sigma_p\hat{\mathbf{n}}$$

$$(\boldsymbol{\sigma} - \sigma_p \mathbf{I})\hat{\mathbf{n}} = \mathbf{0} \tag{3.40}$$

Here $\mathbf{I}$ is the unit matrix. This system will have a nontrivial solution ($\hat{\mathbf{n}} \neq 0$) only if its determinant is zero:

$$
|\boldsymbol{\sigma} - \sigma_p \mathbf{I}| = \begin{vmatrix} \sigma_x - \sigma_p & \tau_{xy} & \tau_{xz} \\ \tau_{xy} & \sigma_y - \sigma_p & \tau_{yz} \\ \tau_{xz} & \tau_{yz} & \sigma_z - \sigma_p \end{vmatrix} = 0
$$

Expanding the determinant yields a cubic polynomial equation in $\sigma_p$:

$$f(\sigma_p) = \sigma_p^3 - I_1\sigma_p^2 + I_2\sigma_p - I_3 = 0 \tag{3.41}$$

This is the *characteristic equation* for stress, where the coefficients are

$$I_1 = \sigma_x + \sigma_y + \sigma_z = \sigma_{kk} \tag{3.42}$$

$$I_2 = \sigma_x\sigma_y + \sigma_x\sigma_z + \sigma_y\sigma_z - \tau_{xy}^2 - \tau_{yz}^2 - \tau_{xz}^2 = \tfrac{1}{2}\sigma_{ij}\sigma_{ji} \tag{3.43}$$

$$I_3 = \det|\sigma| = \tfrac{1}{3}\sigma_{ij}\sigma_{jk}\sigma_{ki} \tag{3.44}$$

These $I$ parameters are known as the *invariants* of the stress state; they do not change with transformation of the coordinates and can be used to characterize the overall nature of the stress. For instance, $I_1$, which has been identified earlier as the trace of the stress matrix, will be seen in a later section to be a measure of the tendency of the stress state to induce hydrostatic dilation or compression. We have already noted that the stress state is one of pure shear if its trace vanishes.

Since the characteristic equation is cubic in $\sigma_p$, it will have three roots, and it can be shown that all three roots must be real. These roots are just the principal stresses $\sigma_{p1}$, $\sigma_{p2}$, and $\sigma_{p3}$.

---

**EXAMPLE 3.9**

Consider a state of simple shear with $\tau_{xy} = 1$ and all other stresses zero:

$$[\sigma] = \begin{bmatrix} 0 & 1 & 0 \\ 1 & 0 & 0 \\ 0 & 0 & 0 \end{bmatrix}$$

The invariants are

$$I_1 = 0, \qquad I_2 = -1, \qquad I_3 = 0$$

and the characteristic equation is

$$\sigma_p^3 - \sigma_p = 0$$

This equation has roots of $(-1, 0, 1)$, corresponding to principal stresses $\sigma_{p1} = 1$, $\sigma_{p2} = 0$, $\sigma_{p3} = -1$, and is plotted in Fig. 3.21. This is the same stress state considered in Example 3.6, and the roots of the characteristic equation agree with the principal values shown by the Mohr's circle.

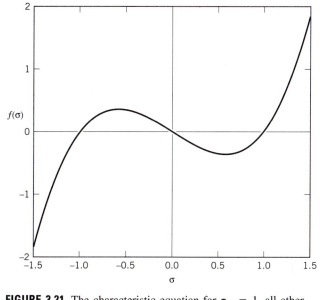

**FIGURE 3.21** The characteristic equation for $\tau_{xy} = 1$, all other stresses zero.

## 3.4 CONSTITUTIVE EQUATIONS

### 3.4.1 Isotropic Elastic Materials

In the general case of a linear relation between components of the strain and stress tensors, we might propose a statement of the form

$$\epsilon_{ij} = S_{ijkl}\sigma_{kl}$$

where the $S_{ijkl}$ is a *fourth-rank* tensor. This constitutes a sequence of nine equations, because each component of $\epsilon_{ij}$ is a linear combination of all the components of $\sigma_{ij}$. For instance:

$$\epsilon_{23} = S_{2311}\epsilon_{11} + S_{2312}\epsilon_{12} + \cdots + S_{2333}\epsilon_{33}$$

Based on each of the indices of $S_{ijkl}$ taking on values from 1 to 3, we might expect a total of 81 independent components in $S$. However, both $\epsilon_{ij}$ and $\sigma_{ij}$ are symmetric, with six rather than nine independent components each. This reduces the number of $S$ components to 36, as can be seen from a linear relation between the pseudovector forms of the strain and stress:

$$
\begin{Bmatrix} \epsilon_x \\ \epsilon_y \\ \epsilon_z \\ \gamma_{yz} \\ \gamma_{xz} \\ \gamma_{xy} \end{Bmatrix}
=
\begin{bmatrix}
S_{11} & S_{12} & \cdots & S_{16} \\
S_{21} & S_{22} & \cdots & S_{26} \\
\vdots & \vdots & \ddots & \vdots \\
S_{61} & S_{26} & \cdots & S_{66}
\end{bmatrix}
\begin{Bmatrix} \sigma_x \\ \sigma_y \\ \sigma_z \\ \tau_{yz} \\ \tau_{xz} \\ \tau_{xy} \end{Bmatrix}
\tag{3.45}
$$

It can be shown that the **S** matrix in this form is also symmetric. It therefore contains only 21 independent elements, as can be seen by counting the elements in the upper right triangle of the matrix, including the diagonal elements ($1 + 2 + 3 + 4 + 5 + 6 = 21$).

If the material exhibits symmetry in its elastic response, the number of independent elements in the **S** matrix can be reduced still further. In the simplest case of an *isotropic* material, whose stiffnesses are the same in all directions, only *two* elements are independent. We have earlier shown that in two dimensions the relations between strains and stresses in isotropic materials can be written as

$$
\begin{aligned}
\epsilon_x &= \frac{1}{E}\left(\sigma_x - \nu\sigma_y\right) \\
\epsilon_y &= \frac{1}{E}\left(\sigma_y - \nu\sigma_x\right) \\
\gamma_{xy} &= \frac{1}{G}\tau_{xy}
\end{aligned}
\tag{3.46}
$$

along with the relation

$$G = \frac{E}{2(1 + \nu)}$$

$$
\begin{Bmatrix} \epsilon_x \\ \epsilon_y \\ \epsilon_z \\ \gamma_{yz} \\ \gamma_{xz} \\ \gamma_{xy} \end{Bmatrix} = \begin{bmatrix} \frac{1}{E} & \frac{-\nu}{E} & \frac{-\nu}{E} & 0 & 0 & 0 \\ \frac{-\nu}{E} & \frac{1}{E} & \frac{-\nu}{E} & 0 & 0 & 0 \\ \frac{-\nu}{E} & \frac{-\nu}{E} & \frac{1}{E} & 0 & 0 & 0 \\ 0 & 0 & 0 & \frac{1}{G} & 0 & 0 \\ 0 & 0 & 0 & 0 & \frac{1}{G} & 0 \\ 0 & 0 & 0 & 0 & 0 & \frac{1}{G} \end{bmatrix} \begin{Bmatrix} \sigma_x \\ \sigma_y \\ \sigma_z \\ \tau_{yz} \\ \tau_{xz} \\ \tau_{xy} \end{Bmatrix}
\tag{3.47}
$$

The quantity in brackets is called the *compliance matrix* of the material, denoted $\mathbf{S}$ or $S_{ij}$. It is important to grasp the physical significance of its various terms. Directly from the rules of matrix multiplication, the element in the $i$th row and $j$th column of $S_{ij}$ is the contribution of the $i$th stress to the $j$th strain. For instance, the component in the 1,2 position is the contribution of the $y$-direction stress to the $x$-direction strain. Multiplying $\sigma_y$ by $1/E$ gives the $y$-direction strain generated by $\sigma_y$, and then multiplying this by $-\nu$ gives the Poisson strain induced in the $x$ direction. The zero elements show the lack of coupling between the normal and shearing components.

The isotropic constitutive law can also be written using index notation as (see Problem 26 in this chapter)

$$
\epsilon_{ij} = \frac{1+\nu}{E}\sigma_{ij} - \frac{\nu}{E}\delta_{ij}\sigma_{kk}
\tag{3.48}
$$

where here the indicial form of strain is used and $G$ has been eliminated using $G = E/2(1+\nu)$. The symbol $\delta_{ij}$ is the *Kronecker delta*, described in Appendix D.

If we wish to write the stresses in terms of the strains, Eqs. 3.47 can be inverted. In cases of plane strain ($\epsilon_z = \gamma_{xz} = \gamma_{yz} = 0$), this yields

$$
\begin{Bmatrix} \sigma_x \\ \sigma_y \\ \tau_{xy} \end{Bmatrix} = \frac{E}{1-\nu^2} \begin{bmatrix} 1 & \nu & 0 \\ \nu & 1 & 0 \\ 0 & 0 & (1-\nu)/2 \end{bmatrix} \begin{Bmatrix} \epsilon_x \\ \epsilon_y \\ \gamma_{xy} \end{Bmatrix}
\tag{3.49}
$$

where again $G$ has been replaced by $E/2(1+\nu)$. Or, in abbreviated notation,

$$
\boldsymbol{\sigma} = \mathbf{D}\boldsymbol{\epsilon}
\tag{3.50}
$$

where $\mathbf{D} = \mathbf{S}^{-1}$ is the *stiffness matrix*.

## 3.4.2 Hydrostatic and Distortional Components

A state of hydrostatic compression, depicted in Fig. 3.22, is one in which no shear stresses exist and where all the normal stresses are equal to the hydrostatic pressure:

$$
\sigma_x = \sigma_y = \sigma_z = -p
$$

where the minus sign indicates that compression is conventionally positive for pressure but negative for stress. From this it is obviously true that

$$
\tfrac{1}{3}\left(\sigma_x + \sigma_y + \sigma_z\right) = \tfrac{1}{3}\sigma_{kk} = -p
$$

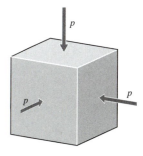

**FIGURE 3.22** Hydrostatic compression.

so that the hydrostatic pressure is the negative *mean normal stress*. This quantity is just one-third of the stress invariant $I_1$, which follows because hydrostatic pressure is the same in all directions and does not vary with axis rotations.

In many cases other than direct hydrostatic compression, it is still convenient to "dissociate" the hydrostatic (or "dilatational") component from the stress tensor:

$$\sigma_{ij} = \tfrac{1}{3}\sigma_{kk}\delta_{ij} + \Sigma_{ij} \tag{3.51}$$

Here $\Sigma_{ij}$ is what is left over from $\sigma_{ij}$ after the hydrostatic component is subtracted. The $\Sigma_{ij}$ tensor can be shown to represent a state of pure shear; in other words, there exists an axis transformation such that all normal stresses vanish (see Problem 30 in this chapter). The $\Sigma_{ij}$ is called the distortional, or deviatoric, component of the stress. Hence, all stress states can be thought of as having two components, as shown in Fig. 3.23, one purely extensional and one purely distortional. This concept is convenient because the material responds to these stress components in very different ways. For instance, plastic and viscous flow is driven dominantly by distortional components, with the hydrostatic component causing only elastic deformation.

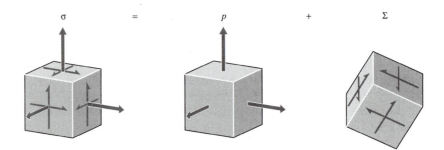

**FIGURE 3.23** Dilatational and deviatoric components of the stress tensor.

## EXAMPLE 3.10

Consider the stress state

$$\sigma = \begin{bmatrix} 5 & 6 & 7 \\ 6 & 8 & 9 \\ 7 & 9 & 2 \end{bmatrix}, \qquad \text{GPa}$$

The mean normal stress is $\sigma_{kk}/3 = (5 + 8 + 2)/3 = 5$, so the stress decomposition is

$$\sigma = \frac{1}{3}\sigma_{kk}\delta_{ij} + \Sigma_{ij} = \begin{bmatrix} 5 & 0 & 0 \\ 0 & 5 & 0 \\ 0 & 0 & 5 \end{bmatrix} + \begin{bmatrix} 0 & 6 & 7 \\ 6 & 3 & 9 \\ 7 & 9 & -3 \end{bmatrix}$$

It is not obvious that the deviatoric component given in the second matrix represents pure shear, because there are nonzero components on its diagonal. However, a stress transformation using Euler angles $\psi = \phi = 0, \theta = -9.22°$ gives the stress state

$$\Sigma' = \begin{bmatrix} 0.00 & 4.80 & 7.87 \\ 4.80 & 0.00 & 9.49 \\ 7.87 & 9.49 & 0.00 \end{bmatrix}$$

The hydrostatic component of stress is related to the volumetric strain through the modulus of compressibility ($-p = K\Delta V/V$), so

$$\tfrac{1}{3}\sigma_{kk} = K\epsilon_{kk} \tag{3.52}$$

Like the stress, the strain can also be dissociated as

$$\epsilon_{ij} = \tfrac{1}{3}\epsilon_{kk}\delta_{ij} + e_{ij}$$

where $e_{ij}$ is the deviatoric component of strain. The deviatoric components of stress and strain are related by the material's shear modulus:

$$\Sigma_{ij} = 2Ge_{ij} \tag{3.53}$$

where the factor 2 is needed because tensor descriptions of strain are half the classical strains for which values of $G$ have been tabulated. Writing the constitutive equations in the form of Eqs. 3.52 and 3.53 produces a simple form without the coupling terms in the conventional $E$-$\nu$ form.

## EXAMPLE 3.11

Using the stress state of the previous example along with the elastic constants for steel ($E = 207$ GPa, $\nu = 0.3$, $K = E/3(1 - 2\nu) = 173$ GPa, $G = E/2(1 + \nu) = 79.6$ GPa), the dilatational and distortional components of strain are

$$\delta_{ij}\epsilon_{kk} = \frac{\delta_{ij}\sigma_{kk}}{3K} = \begin{bmatrix} 0.0289 & 0 & 0 \\ 0 & 0.0289 & 0 \\ 0 & 0 & 0.0289 \end{bmatrix}$$

$$e_{ij} = \frac{\Sigma_{ij}}{2G} = \begin{bmatrix} 0 & 0.0378 & 0.0441 \\ 0.0378 & 0.0189 & 0.0567 \\ 0.0441 & 0.0567 & -0.0189 \end{bmatrix}$$

The total strain is then

$$\epsilon_{ij} = \frac{1}{3}\epsilon_{kk}\delta_{ij} + e_{ij} = \begin{bmatrix} 0.00960 & 0.0378 & 0.0441 \\ 0.0378 & 0.0285 & 0.0567 \\ 0.0441 & 0.0567 & -0.00930 \end{bmatrix}$$

If we evaluate the total strain using Eq. 3.48, we have

$$\epsilon_{ij} = \frac{1 + \nu}{E}\sigma_{ij} - \frac{\nu}{E}\delta_{ij}\sigma_{kk} = \begin{bmatrix} 0.00965 & 0.0377 & 0.0440 \\ 0.0377 & 0.0285 & 0.0565 \\ 0.0440 & 0.0565 & -0.00915 \end{bmatrix}$$

These results are the same, differing only by roundoff error.

### 3.4.3 Finite Strain Model

When deformations become large, geometrical as well as material nonlinearities can arise that are important in many practical problems. In such cases the analyst must employ not only a different strain measure, such as the Lagrangian strain described earlier, but also different stress measures (the "second Piola-Kirchoff stress" replaces the Cauchy stress when Lagrangian strain is used) and different stress-strain constitutive laws as well. A treatment of these formulations is beyond the scope of this text, but a simple nonlinear stress-strain model for rubbery materials will be outlined here to illustrate some aspects of finite strain analysis. The text by Bathe[5] provides a more extensive discussion of this area, including finite element implementations.

In the case of small displacements, the strain $\epsilon_x$ is given by the expression

$$\epsilon_x = \frac{1}{E}\left[\sigma_x - \nu(\sigma_y + \sigma_z)\right]$$

For the case of elastomers with $\nu = 0.5$, this can be rewritten in terms of the mean stress $\sigma_m = (\sigma_x + \sigma_y + \sigma_z)/3$ as

$$2\epsilon_x = \frac{3}{E}(\sigma_x - \sigma_m)$$

For the large-strain case the following analogous stress-strain relation has been proposed:

$$\lambda_x^2 = 1 + 2\epsilon_x = \frac{3}{E}(\sigma_x - \sigma_m^*) \tag{3.54}$$

where here $\epsilon_x$ is the Lagrangian strain and $\sigma_m^*$ is a parameter not necessarily equal to $\sigma_m$. The $\sigma_m^*$ parameter can be found for the case of uniaxial tension by considering the transverse contractions $\lambda_y = \lambda_z$:

$$\lambda_y^2 = \frac{3}{E}(\sigma_y - \sigma_m^*)$$

For rubber $\lambda_x\lambda_y\lambda_z = 1$, so $\lambda_y^2 = 1/\lambda_x$. Making this substitution and solving for $\sigma_m^*$,

$$\sigma_m^* = \frac{-E\lambda_y^2}{3} = \frac{-E}{3\lambda_x}$$

Substituting this back into Eq. 3.54,

$$\lambda_x^2 = \frac{3}{E}\left[\sigma_x - \frac{E}{3\lambda_x}\right]$$

Solving for $\sigma_x$,

$$\sigma_x = \frac{E}{3}\left(\lambda_x^2 - \frac{1}{\lambda_x}\right)$$

---

[5]K.-J. Bathe, *Finite Element Procedures in Engineering Analysis,* Prentice-Hall, 1982.

Here the stress $\sigma_x = F/A$ is the "true" stress based on the actual (contracted) cross-sectional area. The "engineering" stress $\sigma_e = F/A_0$ based on the original area $A_0 = A\lambda_x$ is

$$\sigma_e = \frac{\sigma_x}{\lambda_x} = G\left(\lambda_x - \frac{1}{\lambda_x^2}\right)$$

where $G = E/2(1 + \nu) = E/3$ for $\nu = \frac{1}{2}$. This result is the same as that obtained earlier by considering the force arising from the reduced entropy as molecular segments spanning crosslink sites are extended. It appears here from a simple hypothesis of stress-strain response, using a suitable measure of finite strain.

### 3.4.4 Anisotropic Materials

If the material has a texture like wood or unidirectionally reinforced fiber composites, as shown in Fig. 3.24, the modulus $E_1$ in the fiber direction will typically be larger than those in the transverse directions ($E_2$ and $E_3$). When $E_1 \neq E_2 \neq E_3$, the material is said to be *orthotropic*. It is common, however, for the properties in the plane transverse to the fiber direction to be isotropic to a good approximation ($E_2 = E_3$); such a material is called *transversely isotropic*. The elastic constitutive laws must be modified to account for this anisotropy, and the following form is an extension of Eq. 3.47 for transversely isotropic materials:

$$\left\{\begin{array}{c} \epsilon_1 \\ \epsilon_2 \\ \gamma_{12} \end{array}\right\} = \left[\begin{array}{ccc} 1/E_1 & -\nu_{21}/E_2 & 0 \\ -\nu_{12}/E_1 & 1/E_2 & 0 \\ 0 & 0 & 1/G_{12} \end{array}\right]\left\{\begin{array}{c} \sigma_1 \\ \sigma_2 \\ \tau_{12} \end{array}\right\} \tag{3.55}$$

The parameter $\nu_{12}$ is the *principal Poisson ratio;* it is the ratio of the strain induced in the 2-direction to a strain applied in the 1-direction. This parameter is not limited to values less than 0.5 as in isotropic materials. Conversely, $\nu_{21}$ gives the strain induced in the 1-direction by a strain applied in the 2-direction. Since the 2-direction (transverse to the fibers) usually has much less stiffness than the 1-direction, it should be clear that a given strain in the 1-direction will usually develop a much larger strain in the 2-direction than the same strain in the 2-direction will induce in the 1-direction. Hence, we will usually have $\nu_{12} > \nu_{21}$. There are five constants in Eq. 3.55: $E_1, E_2, \nu_{12}, \nu_{21}$, and $G_{12}$. However, only four of them are independent; the **S** matrix is symmetric, so $\nu_{21}/E_2 = \nu_{12}/E_1$.

A table of elastic constants and other properties for widely used anisotropic materials can be found in Appendix B.

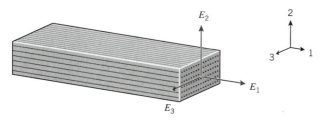

**FIGURE 3.24** An orthotropic material.

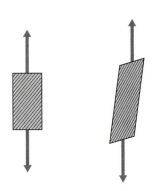

**FIGURE 3.25** Shear-normal
coupling.

The simple form of Eq. 3.55, with zeroes in the terms representing coupling be-tween normal and shearing components, is obtained only when the axes are aligned along the *principal material directions;* that is, along and transverse to the fiber axes. If the axes are oriented along some other direction, all terms of the compliance ma-trix will be populated, and the symmetry of the material will not be evident. If, for instance, the fiber direction is off-axis from the loading direction, the material will develop shear strain as the fibers try to orient along the loading direction, as shown in Fig. 3.25. There will therefore be a coupling between a normal stress and a shearing strain, which never occurs in an isotropic material.

The transformation law for compliance can be developed from the transformation laws for strains and stresses, using the procedures described in Section 3.3.1. By suc-cessive transformations, the pseudovector form for strain in an arbitrary *x-y* direction, shown in Fig. 3.26, is related to strain in the 1-2 (principal material) directions, then to the stresses in the 1-2 directions, and finally to the stresses in the *x-y* directions. The final grouping of transformation matrices relating the *x-y* strains to the *x-y* stresses is then the transformed compliance matrix in the *x-y* direction:

$$\left\{\begin{array}{c} \epsilon_x \\ \epsilon_y \\ \gamma_{xy} \end{array}\right\} = \mathbf{R}\left\{\begin{array}{c} \epsilon_x \\ \epsilon_y \\ \frac{1}{2}\gamma_{xy} \end{array}\right\} = \mathbf{R}\mathbf{A}^{-1}\left\{\begin{array}{c} \epsilon_1 \\ \epsilon_2 \\ \frac{1}{2}\gamma_{12} \end{array}\right\} = \mathbf{R}\mathbf{A}^{-1}\mathbf{R}^{-1}\left\{\begin{array}{c} \epsilon_1 \\ \epsilon_2 \\ \gamma_{12} \end{array}\right\}$$

$$= \mathbf{R}\mathbf{A}^{-1}\mathbf{R}^{-1}\mathbf{S}\left\{\begin{array}{c} \sigma_1 \\ \sigma_2 \\ \tau_{12} \end{array}\right\} = \mathbf{R}\mathbf{A}^{-1}\mathbf{R}^{-1}\mathbf{S}\mathbf{A}\left\{\begin{array}{c} \sigma_x \\ \sigma_y \\ \tau_{xy} \end{array}\right\} \equiv \overline{\mathbf{S}}\left\{\begin{array}{c} \sigma_x \\ \sigma_y \\ \tau_{xy} \end{array}\right\}$$

where $\overline{\mathbf{S}}$ is the *transformed compliance matrix* relative to *x-y* axes. Here $\mathbf{A}$ is the transformation matrix defined in Eq. 3.27, and $\mathbf{R}$ is the Reuter's matrix defined in Eq. 3.30. The inverse of $\overline{\mathbf{S}}$ is $\overline{\mathbf{D}}$, the stiffness matrix relative to *x-y* axes:

$$\boxed{\overline{\mathbf{S}} = \mathbf{R}\mathbf{A}^{-1}\mathbf{R}^{-1}\mathbf{S}\mathbf{A}, \qquad \overline{\mathbf{D}} = \overline{\mathbf{S}}^{-1}} \qquad (3.56)$$

## EXAMPLE 3.12

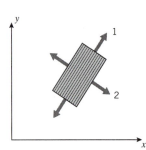

**FIGURE 3.26** Axis trans-formation for constitutive
equations.

Consider a ply of Kevlar-epoxy composite with stiffnesses $E_1 = 82$, $E_2 = 4$, $G_{12} = 2.8$ (all GPa) and $\nu_{12} = 0.25$. The compliance matrix $\mathbf{S}$ in the 1-2 (mate-rial) direction is

$$\mathbf{S} = \begin{bmatrix} 1/E_1 & -\nu_{21}/E_2 & 0 \\ -\nu_{12}/E_1 & 1/E_2 & 0 \\ 0 & 0 & 1/G_{12} \end{bmatrix}$$

$$= \begin{bmatrix} .1220 \times 10^{-10} & -.3050 \times 10^{-11} & 0 \\ -.3050 \times 10^{-11} & .2500 \times 10^{-9} & 0 \\ 0 & 0 & .3571 \times 10^{-9} \end{bmatrix}$$

If the ply is oriented with the fiber direction (the "1" direction) at $\theta = 30°$ from the *x-y* axes, the appropriate transformation matrix is

$$\mathbf{A} = \begin{bmatrix} c^2 & s^2 & 2sc \\ s^2 & c^2 & -2sc \\ -sc & sc & c^2 - s^2 \end{bmatrix} = \begin{bmatrix} .7500 & .2500 & .8660 \\ .2500 & .7500 & -.8660 \\ -.4330 & .4330 & .5000 \end{bmatrix}$$

The compliance matrix relative to the $x$-$y$ axes is then

$$\bar{\mathbf{S}} = \mathbf{R}\mathbf{A}^{-1}\mathbf{R}^{-1}\mathbf{S}\mathbf{A} = \begin{bmatrix} .8830 \times 10^{-10} & -.1970 \times 10^{-10} & -.1222 \times 10^{-9} \\ -.1971 \times 10^{-10} & .2072 \times 10^{-9} & -.8371 \times 10^{-10} \\ -.1222 \times 10^{-9} & -.8369 \times 10^{-10} & -.2905 \times 10^{-9} \end{bmatrix}$$

Note that this matrix is symmetric (to within numerical roundoff error) but that nonzero coupling values exist. A user not aware of the internal composition of the material would consider it completely anisotropic.

The apparent engineering constants that would be observed if the ply were tested in the $x$-$y$ rather than 1-2 directions can be found directly from the transformed $\bar{\mathbf{S}}$ matrix. For instance, the apparent elastic modulus in the $x$ direction is $E_x = 1/\bar{S}_{1,1} = 1/(.8830 \times 10^{-10}) = 11.33$ GPa.

### 3.4.5 Associated Viscoelastic Formulation

The constitutive laws for linearly viscoelastic materials presented in Chapters 1 and 2 can be generalized to appear as an extension of the elastic laws. As an example that does not restrict the generality of the discussion, consider the *Wiechert* model shown in Fig. 3.27; this model is an extension of the Standard Linear Solid presented earlier in Section 1.5.2. A real polymer does not relax with a single relaxation time as was presented in Chapter 1. Segments of varying length contribute to the relaxation, with the simpler and shorter segments relaxing much more quickly than the long ones. This leads to a distribution of relaxation times, which in turn produces a relaxation spread over a much longer time than can be modeled accurately with a single relaxation time. When the engineer considers it necessary to incorporate this effect, the Wiechert model can have as many spring-dashpot Maxwell elements as are needed to approximate the distribution satisfactorily.

The total stress $\sigma$ transmitted by the model is the stress in the isolated spring (of stiffness $k_e$) plus that in each of the Maxwell spring-dashpot arms:

$$\sigma = \sigma_e + \sum_j \sigma_j$$

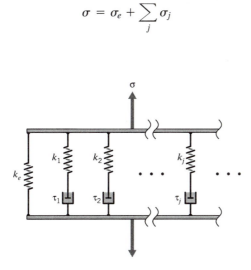

**FIGURE 3.27** The Wiechert viscoelastic model.

As was shown in the discussion leading to Eq. 1.25, the stress $\sigma_j$ in the $j$th Maxwell arm is stated implicitly by the relation

$$\dot{\epsilon} = \dot{\epsilon}_s + \dot{\epsilon}_d = \frac{\dot{\sigma}_j}{k_j} + \frac{\sigma_j}{\eta_j}$$

where the $k_j$ and $\eta_j$ are the spring stiffness and the dashpot viscosity, respectively. The difficulty in extracting $\sigma_j$ from this expression is much reduced by taking Laplace transforms (see Appendix C):

$$s\bar{\epsilon} = \frac{s\bar{\sigma}_j}{k_j} + \frac{\bar{\sigma}_j}{\eta_j}$$

Here $s$ is the Laplace transform parameter, and the overbar indicates the transformed variable. Multiplying through by $k_j$ and using the expression for relaxation time $\tau_j = \eta_j/k_j$:

$$k_j s\bar{\epsilon} = \left(s + \frac{1}{\tau_j}\right)\bar{\sigma}_j$$

$$\bar{\sigma}_j = \frac{k_j s\bar{\epsilon}}{\left(s + \dfrac{1}{\tau_j}\right)}$$

Then

$$\bar{\sigma} = \bar{\sigma}_e + \sum_j \bar{\sigma}_j = \left\{ k_e + \sum_j \frac{k_j s}{\left(s + \frac{1}{\tau_j}\right)} \right\}\bar{\epsilon} \tag{3.57}$$

The quantity in braces is an operator we shall denote by $\mathscr{C}$. The expression can then be written

$$\bar{\sigma} = \mathscr{C}\bar{\epsilon} \tag{3.58}$$

This form, which is clearly reminiscent of Hooke's law $\sigma = E\epsilon$ but in the Laplace plane, is called the *associated viscoelastic constitutive equation*. Here the specific expression for $\mathscr{C}$ is that of the Wiechert model, but other models could have been used as well.

For a given strain input function $\epsilon(t)$, we obtain the resulting stress function in three steps:

1. Obtain an expression for the transform of the strain function, $\bar{\epsilon}(s)$.
2. Form the algebraic product $\bar{\sigma}(s) = \mathscr{C}\bar{\epsilon}(s)$.
3. Obtain the inverse transform of the result to yield the stress function in the time plane.

EXAMPLE 3.13

In stress relaxation tests, we have

$$\epsilon(t) = \epsilon_0 \Rightarrow \bar{\epsilon}(s) = \epsilon_0/s$$

$$\bar{\sigma}(s) = \mathscr{E}(s)\bar{\epsilon}(s) = \left\{ k_e + \sum_j \frac{k_j s}{s + \frac{1}{\tau_j}} \right\} \frac{\epsilon_0}{s} = \left\{ \frac{k_e}{s} + \sum_j \frac{k_j}{s + \frac{1}{\tau_j}} \right\} \epsilon_0$$

$$\sigma(t) = \mathscr{L}^{-1}[\bar{\sigma}(s)] = \left\{ k_e + \sum_j k_j \exp\left(\frac{-t}{\tau_j}\right) \right\} \epsilon_0 \qquad (3.59)$$

The quantity in brackets is clearly just the time-dependent relaxation modulus, $E_{\text{rel}} = \sigma(t)/\epsilon_0$, for the Wiechert model.

The material constants that appear in the model formulations (the $k$'s and $\tau$'s) must be chosen by reference to appropriate experimental data, and several means for accomplishing this are available. In most cases polymers are observed to exhibit not just one but several major relaxations during stress relaxation or other viscoelastic response modes; these are usually labeled with Greek letters beginning with $\alpha$ for the major glass-rubber transition and working through $\beta$, $\gamma$, and so on for the shorter-time relaxations that are generated by increasingly simple molecular motions (see Fig. 3.28). If we are satisfied by modeling each of the material's major transitions as a single-relaxation-time process, then model fitting is straightforward: $k_e$ is set equal to the long-time equilibrium (rubbery) modulus, and each successive $k_j$ is chosen to raise the modulus to the next higher plateau. The $\tau_j$ are chosen to position the relaxation properly along the time axis. Such a model fit may be acceptable for illustrative purposes or for rough design work.

However, a single Maxwell arm generally predicts too sharp a change in a given transition, and we can improve the fit by assigning several elements to each major relaxation. A convenient means of doing this is by a "collocation" process, in which the model relaxation times are located arbitrarily through the transition and the corresponding spring stiffnesses then chosen to make the model equation match the experimental data.

To illustrate a common form of this method, consider a hypothetical experimental plot of relaxation modulus through a transition spanning seven decades[6] of time, as

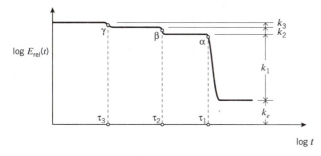

**FIGURE 3.28** Stress relaxation showing multiple transitions.

---

[6]Here a "decade" does not mean ten years but a *tenfold increase* in time. Thus the interval between $10^{-3}$ s and $10^{-2}$ s is one decade, and so is the interval from $10^3$ s to $10^4$ s.

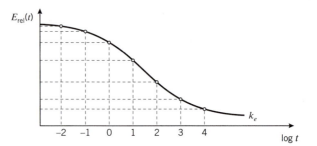

**FIGURE 3.29**  A hypothetical relaxation.

seen in Fig. 3.29. We seek to determine the constants $k_e$, $k_j$, and $\tau_j$ in the Wiechert model prediction for the relaxation modulus:

$$E_{\text{rel}}(t) = k_e + \sum_{j=1}^{n} k_j \exp(-t/\tau_j) \tag{3.60}$$

First, we locate the $\tau_j$ at the integer values of log time; in the example of Fig. 3.29 this gives $\tau_1 = 10^{-2}$, $\tau_2 = 10^{-1}$, $\tau_3 = 10, \cdots, \tau_7 = 10^4$. Next, we evaluate Eq. 3.60 at these seven values of $\tau$ to obtain the system of seven equations:

$$k_1 e^{-(10^{-2}/10^{-2})} + k_2 e^{-(10^{-2}/10^{-1})} + \cdots + k_7 e^{-(10^{-2}/10^4)} = E_{\text{rel}}(10^{-2}) - k_e$$

$$k_1 e^{-(10^{-1}/10^{-2})} + k_2 e^{-(10^{-1}/10^{-1})} + \cdots + k_7 e^{-(10^{-1}/10^4)} = E_{\text{rel}}(10^{-1}) - k_e$$

$$\vdots$$

$$k_1 e^{-(10^4/10^{-2})} + k_2 e^{-(10^4/10^{-1})} + \cdots + k_7 e^{-(10^4/10^4)} = E_{\text{rel}}(10^4) - k_e$$

Noting that terms like $e^{-(10^{-1}/10^{-2})} = e^{-10} \approx 0$ and terms like $e^{-(10^{-2}/10^{-1})} = e^{-0.1} \approx 1$, the equations can be written as the triangular system

$$\begin{bmatrix} 0.368 & 0.905 & 1 & \cdots & 1 \\ 0 & 0.368 & 0.905 & \cdots & 1 \\ 0 & 0 & 0.368 & \cdots & 1 \\ & & \vdots & & \vdots \\ 0 & 0 & 0 & \cdots & 0.368 \end{bmatrix} \begin{Bmatrix} k_1 \\ k_2 \\ k_3 \\ \vdots \\ k_7 \end{Bmatrix} = \begin{Bmatrix} E_{10^{-2}} - k_e \\ E_{10^{-1}} - k_e \\ E_{10^0} - k_e \\ \vdots \\ E_{10^4} - k_e \end{Bmatrix} \tag{3.61}$$

On setting $k_e$ to the equilibrium rubbery modulus, we can easily solve this system by back-substitution. This collocation approach was of considerable interest to viscoelasticians before the advent of easy access to computers, since the back-substitution process could be performed with a desk calculator.

## EXAMPLE 3.14

The strain function resulting from a constant-strain-rate test, such as might be performed with a tensile testing instrument, is

$$\epsilon = Rt \Rightarrow \bar{\epsilon}(s) = R/s^2$$

where $R$ is the strain rate. Then

$$\bar{\sigma}(s) = \mathcal{E}(s)\bar{\epsilon}(s) = \left\{ k_e + \sum_j \frac{k_j s}{s + \frac{1}{\tau_j}} \right\} \frac{R}{s^2} = \frac{k_e R}{s^2} + \sum_j \frac{k_j R}{s\left(s + \frac{1}{\tau_j}\right)}$$

(3.62)

$$\sigma(t) = k_e R t + \sum_j k_j R \tau_j \left[ 1 - \exp(-t/\tau_j) \right]$$

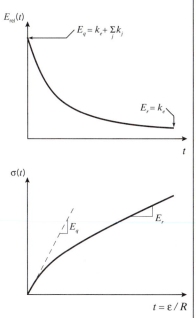

The equations used here are linear, and the material is "linearly viscoelastic." However, note that the stress-time function, and hence the stress-strain curve obtained in a constant-strain-rate test, is not linear. It is not true, therefore, that a curved stress-strain diagram implies that the material response is nonlinear. This result is due to the constantly changing input strain, not to any nonlinearity of the material. It is also interesting to note that the slope of the constant-strain-rate stress-strain curve is related to the value of the relaxation modulus evaluated at the same time (see Fig. 3.30):

$$\frac{d\sigma}{d\epsilon} = \frac{d\sigma}{dt} \cdot \frac{dt}{d\epsilon} = \frac{d\sigma}{dt} \cdot \frac{1}{R} = \left[ k_e R + \sum_j k_j R \exp(-t/\tau_j) \right] \frac{1}{R}$$

$$= \left[ k_e + \sum_j k_j \exp(-t/\tau_j) \right] \equiv E_{\mathrm{rel}}(t) \Big|_{t = \epsilon/R}$$

(3.63)

**FIGURE 3.30** Relation between relaxation and constant-strain-rate tests.

In problems for which the strain resulting from a specified stress input is desired, it is usually awkward mathematically to put a modulus operator in the denominator, as in

$$\bar{\epsilon} = \frac{\bar{\sigma}}{\mathcal{E}}$$

In most cases it is more convenient to develop a *compliance* operator $\mathcal{C}$ from scratch and work with an equation of the form

$$\bar{\epsilon} = \mathcal{C}\bar{\sigma}$$

This is done in a manner exactly analogous to the foregoing developments using $\mathcal{E}$, and in Section 1.5.2 we used this approach to develop an expression for the creep compliance of the standard linear solid. Problem 33 in this chapter illustrates the procedure for a more generalized spring-dashpot model.

The Laplace transformation method is entirely equivalent to the formulation using superposition integrals presented in Chapter 2. This can be seen by writing Eq. 3.58 for the case of stress relaxation, with $\bar{\epsilon} = \epsilon_0/s$:

$$\bar{\sigma} = \mathcal{E} \frac{\epsilon_0}{s}$$

Then

$$\frac{\bar{\sigma}}{\epsilon_0} \equiv \bar{E}_{\mathrm{rel}} = \frac{\mathcal{E}}{s}$$

(It is important to understand the difference between $\mathscr{E}$, which is the general operator relating the time-dependent stress to an arbitrary strain-time function in the Laplace plane, and $\overline{E}_{rel}$, which is the transform of the very specific modulus in which strain is held constant and only the stress varies with time.) Now replacing $\mathscr{E}$ in Eq. 3.58 with $s\overline{E}_{rel}$ and invoking the rule for transforms of time derivatives, we have

$$\overline{\sigma} = s\overline{E}_{rel} \cdot \overline{\epsilon} = \overline{E}_{rel}\overline{\dot{\epsilon}}$$

The inverse of a product of two functions is a convolution integral (see Appendix C), so this gives

$$\sigma(t) = \int_{-\infty}^{t} E_{rel}(t - \xi)\dot{\epsilon}(\xi)\, d\xi \qquad (3.64)$$

This is of the same form developed by physical reasoning earlier for the creep response of a pressure vessel in Chapter 2. Three other equivalent convolution integrals could have been written as well, because it is arbitrary which function in the product has the overdot or the $t - \xi$ argument:

$$\sigma(t) = \begin{cases} \int_{-\infty}^{t} \dot{E}_{rel}(t - \xi)\epsilon(\xi)\, d\xi \\ \int_{-\infty}^{t} E_{rel}(t)\dot{\epsilon}(t - \xi)\, d\xi \\ \int_{-\infty}^{t} \dot{E}_{rel}(t)\epsilon(t - \xi)\, d\xi \end{cases}$$

## EXAMPLE 3.15

In working with viscoelastic convolution integrals it is often convenient to employ the *Heaviside function*, defined as

$$u(t) = \begin{cases} 0, & t < 0 \\ 1, & t \geq 0 \end{cases}$$

If for instance a material is subjected to a strain ramp applied at time $t = t_1$, we can write

$$\epsilon(t) = R \cdot (t - t_1)\, u(t - t_1) = Rtu(t - t_1) - Rt_1 u(t - t_1)$$

The time derivative of this is

$$\dot{\epsilon}(t) = Ru(t - t_1) + Rt\dot{u}(t - t_1) - Rt_1\dot{u}(t - t_1)$$

The time derivative of the Heaviside function $\dot{u}$ is the closely related *Dirac function* $\delta(t)$. The Dirac function is an infinitely high, infinitely narrow spike at time $t = 0$ formally defined as

$$\delta(t) = \dot{u}(t) = \begin{cases} 0, & t \neq 0 \\ \infty, & t = 0 \end{cases}$$

Further, the Dirac function has the property that

$$\int_{-\infty}^{+\infty} \delta(t)\, dt = \int_{0^-}^{0^+} \delta(t)\, dt = 1$$

Using this, the strain derivative is

$$\dot{\epsilon}(t) = Ru(t - t_1) + Rt\delta(t - t_1) - Rt_1\delta(t - t_1)$$

The last two terms here cancel, since $\delta$ is zero unless $t = t_1$, in which case they are identical. Taking the Standard Linear Solid form of the relaxation modulus, the convolution integral can then be written

$$\sigma(t) = \int_0^t \left[ k_e + k_1 e^{-(t-\xi)/\tau} \right] Ru(\xi - t_1)\, d\xi$$

It can be shown that for any continuous function of time $f(t)$

$$\int_{-\infty}^t f(\xi)u(\xi - t_1)d\xi = u(t - t_1) \int_{t_1}^t f(\xi)d\xi$$

Using this, the stress is

$$\sigma(t) = u(t - t_1) \int_{t_1}^t \left[ k_e + k_1 e^{-(t-\xi)/\tau} \right] R\, d\xi$$

$$= u(t - t_1)\left\{ Rk_e(t - t_1) + Rk_1\tau \left[ 1 - e^{-(t-t_1)/\tau} \right] \right\}$$

This is the same expression developed as Eq. 3.62, with only one arm in the model and the times shifted to the right by an amount $t_1$.

The foregoing discussion should make it clear that a spring-dashpot formulation of material viscoelasticity is not the only one possible but that it is equivalent to alternative approaches such as superposition integrals. Another useful and often used formulation employs a generalized ordinary differential equation of arbitrary order to relate stress and strain, and this approach can be shown to be equivalent to the spring-dashpot formulation and thus the integral formulation as well.

Recall the Laplace form of the Wiechert constitutive model:

$$\bar{\sigma} = \left[ k_e + \sum_j \frac{k_j s}{s + \frac{1}{\tau_j}} \right] \bar{\epsilon}$$

If the expression in brackets is expanded and recast with a common denominator, it takes the form

$$\bar{\sigma} = \frac{\left[ b_m s^m + b_{m-1}s^{m-1} + \cdots + b_1 s + b_0 \right]}{\left[ a_n s^n + a_{n-1}s^{n-1} + \cdots + a_1 s + a_0 \right]}\bar{\epsilon}$$

Remembering that a term such as $s^2\bar{\sigma}$, which would be created if this relation were multiplied out, would in turn lead to a second derivative $\ddot{\sigma}$ on inversion, this relation can be written in operational form in the time plane as

$$\sigma = \frac{\left[ b_m \dfrac{\partial^m}{\partial t^m} + \cdots + b_1 \dfrac{\partial}{\partial t} + b_0 \right]}{\left[ a_n \dfrac{\partial^n}{\partial t^n} + \cdots + a_1 \dfrac{\partial}{\partial t} + a_0 \right]} \epsilon$$

This higher-order differential equation relating strain to stress is sometimes taken as a starting point for linear viscoelasticity. (The $\partial$ symbol is used here even though the equation is ordinary, since in the more general case the stresses and strains will depend on position as well as time.)

## EXAMPLE 3.16

Letting the Wiechert model consist of two Maxwell elements plus the equilibrium spring, the constitutive equation is

$$\mathscr{E} = k_e + \frac{k_1 s}{s + \dfrac{1}{\tau_1}} + \frac{k_2 s}{s + \dfrac{1}{\tau_2}}$$

This has the common denominator

$$\left( s + \frac{1}{\tau_1} \right)\left( s + \frac{1}{\tau_2} \right) = s^2 + \left( \frac{1}{\tau_1} + \frac{1}{\tau_2} \right) s + \frac{1}{\tau_1 \tau_2}$$

so the viscoelastic constitutive law may be written as

$$\overline{\sigma} = \frac{\left\{ k_g s^2 + \left[ \dfrac{1}{\tau_1}(k_e + k_1) + \dfrac{1}{\tau_2}(k_e + k_2) \right] s + \dfrac{k_e}{\tau_1 \tau_2} \right\} \epsilon}{s^2 + \left( \dfrac{1}{\tau_1} + \dfrac{1}{\tau_2} \right) s + \dfrac{1}{\tau_1 \tau_2}}$$

where $k_g = k_e + k_1 + k_2$. By multiplying the denominator across the equal sign and inverting term by term, this is seen to be equivalent to a second-order differential equation of the form

$$a_2 \frac{\partial^2 \sigma}{\partial t^2} + a_1 \frac{\partial \sigma}{\partial t} + a_0 \sigma = b_2 \frac{\partial^2 \epsilon}{\partial t^2} + b_1 \frac{\partial \epsilon}{\partial t} + b_0 \epsilon$$

where

$$a_2 = 1, \qquad a_1 = \left( \frac{1}{\tau_1} + \frac{1}{\tau_2} \right), \qquad a_0 = \frac{k_e}{\tau_1 \tau_2}$$

$$b_2 = k_g, \qquad b_1 = \left[ \frac{1}{\tau_1}(k_e + k_1) + \frac{1}{\tau_2}(k_e + k_2) \right], \qquad b_0 = \frac{k_e}{\tau_1 \tau_2}$$

It should be apparent that a Wiechert model with three Maxwell arms is equivalent to a third-order equation, a model with four arms gives a fourth-order equation, and so on.

Many authors employ the differential equation formulation written in the operator form $P\sigma = Q\epsilon$, where $P$ and $Q$ are the differential operators

$$P = a_n \frac{\partial^n}{\partial t^n} + \cdots + a_1 \frac{\partial}{\partial t} + a_0$$

$$Q = b_m \frac{\partial^m}{\partial t^m} + \cdots + b_1 \frac{\partial}{\partial t} + b_0$$

In terms of our earlier nomenclature,

$$\mathscr{E}(s) = \mathscr{L}(Q/P)$$

The foregoing viscoelastic expressions have assumed a simple stress state in which a specimen is subjected to uniaxial tension. This loading is germane to laboratory characterization tests, but the information obtained from these tests must be cast in a form that allows application to the multiaxial stress states that are encountered in actual design.

For isotropic materials the shear modulus $G$ and the bulk modulus $K$ are related to Young's modulus $E$ and Poisson's ratio $\nu$ as

$$E = \frac{9GK}{3K + G} \tag{3.65}$$

$$\nu = \frac{3K - 2G}{6K + 2G} \tag{3.66}$$

The viscoelastic analogs of Eqs. 3.52 and 3.53 can be written as

$$\Sigma_{ij} = 2\mathscr{G}(s)\overline{e}_{ij} \tag{3.67}$$

$$\overline{\sigma}_{ij} = 3\mathscr{K}(s)\overline{\epsilon}_{ij} \tag{3.68}$$

and these viscoelastic operators may be related to the tensile operators by

$$\mathscr{E}(s) = \frac{9\mathscr{G}(s)\mathscr{K}(s)}{3\mathscr{K}(s) + \mathscr{G}(s)}$$

$$\mathscr{N}(s) = \frac{3\mathscr{K}(s) - 2\mathscr{G}(s)}{6\mathscr{K}(s) + 2\mathscr{G}(s)}$$

In uniaxial testing, full characterization would require observation of both longitudinal and transverse response, so that both $E(t)$ and $\nu(t)$ could be determined. Models could then be fitted to both deformation modes to find $\mathscr{E}$ and $\mathscr{N}$. Other approaches could be used as well; for instance, we might find $\mathscr{E}$ from uniaxial testing and $\mathscr{K}$ from dilatometry. $\mathscr{K}$ could be used directly in Eq. 3.68, and $\mathscr{G}$ is

$$\mathscr{G} = \frac{3\mathscr{K}\mathscr{E}}{9\mathscr{K} - \mathscr{E}}$$

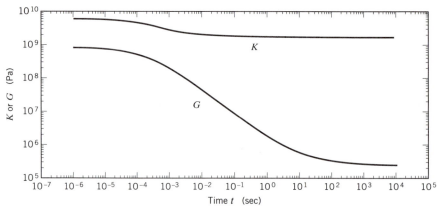

**FIGURE 3.31** Relaxation moduli of polyisobutylene in dilation ($K$) and shear ($G$). From Huang, M.G., Lee, E.H., and Rogers, T.G., "On the Influence of Viscoelastic Compressibilty in Stress Analysis," *Stanford University Technical Report No. 140* (1963).

In practice $K(t)$ is usually much larger than $G(t)$, and $K(t)$ usually experiences much smaller relaxations than $G(t)$ (see Fig. 3.31). These observations lead to idealizations of compressibility that greatly simplify analysis. First, if one takes $K_{\text{rel}} = K_e$ to be finite but constant (that is, only the shear response is viscoelastic), then

$$\mathcal{K} = s\overline{K}_{\text{rel}} = s\frac{K_e}{s} = K_e$$

$$\mathcal{G} = \frac{3K_e\mathcal{E}}{9K_e - \mathcal{E}}$$

Secondly, if $K$ is assumed not only constant but infinite (that is, that the material is incompressible, no hydrostatic deformation), then

$$\mathcal{G} = \frac{\mathcal{E}}{3}$$

$$\mathcal{N} = \nu = \frac{1}{2}$$

## EXAMPLE 3.17

The shear modulus of polyvinyl chloride (PVC) is observed to relax from a glassy value of $G_g = 800$ MPa to a rubbery value of $G_r = 1.67$ MPa. The relaxation time at 75°C is approximately $\tau = 100$ s, although the transition is much broader than would be predicted by a single-relaxation-time model. Assuming a Standard Linear Solid model as an approximation, the shear operator is

$$\mathcal{G} = G_r + \frac{(G_g - G_r)s}{s + \frac{1}{\tau}}$$

The bulk modulus is constant to a good approximation at $K_e = 1.33$ GPa. These data can be used to predict the time dependence of Poisson's ratio, using the expression

$$\mathcal{N} = \frac{3K_e - 2\mathcal{G}}{6K_e + 2\mathcal{G}}$$

On substituting the numerical values and simplifying, this becomes

$$\mathcal{N} = 0.25 + \frac{9.97 \times 10^8}{4.79 \times 10^{11}s + 3.99 \times 10^9}$$

The "relaxation" Poisson's ratio—the time-dependent strain in one direction induced by a constant strain in a transverse direction—is then

$$\overline{\nu}_{\text{rel}} = \frac{\mathcal{N}}{s} = \frac{0.25}{s} + \frac{1}{s}\left(\frac{9.97 \times 10^8}{4.79 \times 10^{11}s + 3.99 \times 10^9}\right)$$

Inverting, this gives

$$\nu_{\text{rel}} = 0.5 - 0.25e^{-t/120}$$

This function is plotted in Fig. 3.32. Poisson's ratio is seen to *rise* from a glassy value of 0.25 to a rubbery value of 0.5 as the material moves from the glassy to the rubbery regime over time. Note that the time constant of 120 s in this expression is not the same as the relaxation time $\tau$ for the pure shear response.

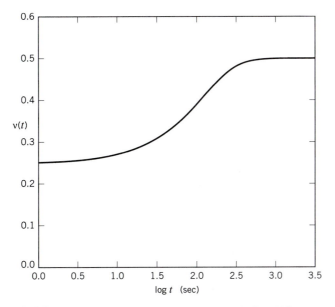

**FIGURE 3.32** Time dependence of Poisson's ratio for PVC at 75°C, assuming viscoelastic shear response and elastic hydrostatic response.

In the case of material isotropy (properties not dependent on direction of measurement), at most two viscoelastic operators—say $\mathcal{G}$ and $\mathcal{H}$—will be necessary for a full characterization of the material. For materials exhibiting lower orders of symmetry, more descriptors will be necessary: a transversely isotropic material requires four constitutive descriptors, an orthotropic material requires nine, and a triclinic material 21. If the material is both viscoelastic and anisotropic, these are the numbers of viscoelastic operators that will be required. Clearly, the analyst must be discerning in finding the proper balance between realism and practicality in choosing models.

## 3.5 PROBLEMS

**1.** Write out the abbreviated strain-displacement equation $\epsilon = \mathbf{Lu}$ (Eq. 3.8) for two dimensions.

**2.** Write out the components of the Langrangian strain tensor in three dimensions:

$$\epsilon_{ij} = \tfrac{1}{2}(u_{i,j} + u_{j,i} + u_{\tau,i}u_{\tau,j})$$

**3.** Show that for small strains the fractional volume change is the trace of the infinitesimal strain tensor:

$$\frac{\Delta V}{V} \equiv \epsilon_{kk} = \epsilon_x + \epsilon_y + \epsilon_z$$

**4.** When the material is incompressible, show that the extension ratios are related by

$$\lambda_x\lambda_y\lambda_z = 1$$

**5.** Show that the fractional volume change can be written in terms of the stresses as

$$\frac{\Delta V}{V} = \frac{(\sigma_x + \sigma_y + \sigma_z)(1 - 2\nu)}{E}$$

**6.** Show that for an isotropic elastic material the modulus of compressibility is related to Young's modulus and Poisson's ratio as

$$K \equiv \frac{-p}{(\Delta V/V)} = \frac{E}{3(1 - 2\nu)}$$

**7.** Develop the two-dimensional form of the Cartesian equilibrium equations by drawing a free-body diagram of an infinitesimal section, as shown in Fig. P.3.7.

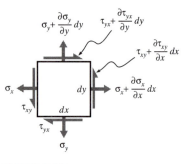

**FIGURE P.3.7**

**8.** Use the free-body diagram of Problem 7 to show that $\tau_{xy} = \tau_{yx}$.

**9.** Use a free-body diagram approach to show that in polar coordinates the equilibrium equations are

$$\frac{\partial \sigma_r}{\partial r} + \frac{1}{r}\frac{\partial \tau_r\theta}{\partial \theta} + \frac{\sigma_r - \sigma_\theta}{r} = 0$$

$$\frac{\partial r_{\tau\theta}}{\partial r} + \frac{1}{r}\frac{\partial \sigma_\theta}{\partial \theta} + 2\frac{\tau_{r\theta}}{r} = 0$$

**10.** Develop the equations for equilibrium in polar coordinates, already stated in Problem 9, by transforming the Cartesian equations using

$$x = r\cos\theta$$

$$y = r\sin\theta$$

**11.** If a long, slender bar is subjected to impact by a high-speed projectile, the stress may not be uniform along the bar but may propagate as a *stress wave*.

(a) Considering a small increment of length $dx$ along the bar over which the stress may vary by $d\sigma = (\partial\sigma/\partial x)\,dx$, use the full form of Newton's second law to develop the one-dimensional governing equation for wave propagation in a linear elastic solid:

$$\frac{\partial^2 u}{\partial t^2} = c^2\frac{\partial^2 u}{\partial x^2}$$

where $u$ is the $x$ component of displacement. $c = \sqrt{E/\rho}$ is the *wavespeed*, $E$ is Young's modulus and $\rho$ is the mass density.

(b) Show that this equation is satisfied by *traveling waves* of the form $u(x, t) = u(x \pm ct)$.

**12.** Consider two identical rods traveling at velocity $V$ and impacting one another.

(a) At a time $t$ after impact, compressive stress waves will have propagated a distance $ct$ into each bar. (In this region the practical velocity will be zero and the stress will be a constant value $\sigma$.) Show that

$$\sigma = \sqrt{\rho EV}$$

(b) When the stress waves have reached the outer edges of the rods, all particle velocities will be zero; both bars are temporarily at rest. Compare the strain energy in the bars with the initial kinetic energies before impact.

(c) The outer edge of each bar must be a stress-free surface, so when the compressive wave reaches the edge, the bar experiences what is equivalent to a *tensile* impact at velocity $V$ at the free edge to bring the stress there to zero. This sets up another wave, actually a reflection of the first, which propagates back toward the interface between the two bars. Describe the state of stress and velocity behind this reflected wave, (that is, between the wave and the free edge), and describe what happens when the reflected wave reaches the interface.

**13.** Show that the kinematic (strain-displacement) relations in for polar coordinates can be written

$$\epsilon_r = \frac{\partial u_r}{\partial r}$$

$$\epsilon_\theta = \frac{1}{r}\frac{\partial u_\theta}{\partial \theta} + \frac{u_r}{r}$$

$$\gamma_{r\theta} = \frac{1}{r}\frac{\partial u_r}{\partial \theta} + \frac{\partial u_\theta}{\partial r} - \frac{u_\theta}{r}$$

**14.** Show that the Langrangian strain $\epsilon_x$ in uniaxial extension is related to the stretch ratio $\lambda_x = L/L_0$ by $\lambda_x^2 = 1 + 2\epsilon_x$.

**15.** Develop an expression for the stress needed to cause transverse failure in a unidirectionally oriented composite as a function of the angle between the load direction and the fiber direction, and show this function in a plot of strength versus $\theta$.

**16.** Use a free-body force balance, as shown in Fig. P.3.16, to derive the two-dimensional Cartesian stress transformation equations as

$$\sigma_{x'} = \sigma_x \cos^2 \theta + \sigma_y \sin^2 \theta + 2\tau_{xy} \sin \theta \cos \theta$$

$$\sigma_{y'} = \sigma_x \sin^2 \theta + \sigma_y \cos^2 \theta - 2\tau_{xy} \sin \theta \cos \theta$$

$$\tau_{x'y'} = (\sigma_y - \sigma_x) \sin \theta \cos \theta + \tau_{xy}(\cos^2 \theta - \sin^2 \theta)$$

Or

$$\begin{Bmatrix} \sigma_{x'} \\ \sigma_{y'} \\ \tau_{x'y'} \end{Bmatrix} = \begin{bmatrix} c^2 & s^2 & 2sc \\ s^2 & c^2 & -2sc \\ -sc & sc & c^2 - s^2 \end{bmatrix} \begin{Bmatrix} \sigma_x \\ \sigma_y \\ \tau_{xy} \end{Bmatrix}$$

where $c = \cos \theta$ and $s = \sin \theta$.

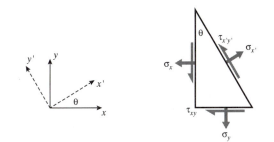

**FIGURE P.3.16**

**17.** Develop mathematical relations for displacements and gradients along transformed axes of the form

$$u' = u \cos \theta + v \sin \theta$$

$$\frac{\partial}{\partial x'} = \frac{\partial}{\partial x} \cdot \frac{\partial x}{\partial x'} + \frac{\partial}{\partial y} \cdot \frac{\partial y}{\partial x'} = \frac{\partial}{\partial x} \cdot \cos \theta + \frac{\partial}{\partial y} \cdot \sin \theta$$

with analogous expressions for $v'$ and $\partial/\partial y'$. Use these to obtain the strain transformation equations (Eq. 3.29).

**18.** Consider a line segment $AB$ of length $ds^2 = dx^2 + dy^2$, oriented at an angle $\theta$ from the Cartesian $x$-$y$ axes as shown in Fig. P.3.18. Let the differential displacement of end $B$ relative to end $A$ be

$$du = \frac{\partial u}{\partial x}dx + \frac{\partial u}{\partial y}dy$$

$$dv = \frac{\partial v}{\partial x}dx + \frac{\partial v}{\partial y}dy$$

Use this geometry to derive the strain transformation equations (Eq. 3.29), where the $x'$ axis is along line $AB$.

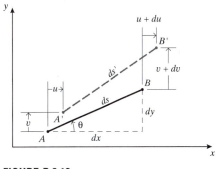

**FIGURE P.3.18**

**19.** Employ double-angle trigonometric relations to show that the two-dimensional Cartesian stress transformation equations can be written in the form

$$\sigma_{x'} = \frac{\sigma_x + \sigma_y}{2} + \frac{\sigma_x - \sigma_y}{2} \cos 2\theta + \tau_{xy} \sin 2\theta$$

$$\tau_{x'y'} = -\frac{\sigma_x - \sigma_y}{2} \sin 2\theta + \tau_{xy} \cos 2\theta$$

$$\sigma_{y'} = \frac{\sigma_x + \sigma_y}{2} + \frac{\sigma_x - \sigma_y}{2} \cos 2\theta - \tau_{xy} \sin 2\theta$$

Use these relations to justify the Mohr's circle construction.

**20.** Use matrix multiplication (Eqs. 3.28 or 3.31) to transform the following stress and strain states to axes rotated by $\theta = 30°$ from the original $x$-$y$ axes.

(a)

$$\sigma = \begin{Bmatrix} 1.0 \\ -2.0 \\ 3.0 \end{Bmatrix}$$

(b)

$$\epsilon = \begin{Bmatrix} 0.01 \\ -0.02 \\ 0.03 \end{Bmatrix}$$

**21.** Sketch the Mohr's circles for each of the stress states shown in Fig. P.3.21.

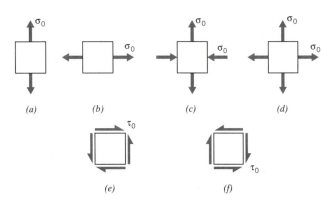

**FIGURE P.3.21**

**22.** Construct Mohr's circle solutions for the transformations of Problem 20.

**23.** Draw the Mohr's circles and determine the magnitudes of the principal stresses for the following stress states. Denote the principal stress state on a suitably rotated stress square.

(a) $\sigma_x = 30$ MPa, $\sigma_y = -10$ MPa, $\tau_{xy} = 25$ MPa

(b) $\sigma_x = -30$ MPa, $\sigma_y = -90$ MPa, $\tau_{xy} = -40$ MPa

(c) $\sigma_x = -10$ MPa, $\sigma_y = 210$ MPa, $\tau_{xy} = -15$ MPa

**24.** Show that the values of principal stresses given by Mohr's circle agree with those obtained mathematically by setting to zero the derivatives of the stress with respect to the transformation angle.

**25.** For the 3-dimensional stress state $\sigma_x = 25$, $\sigma_y = -15$, $\sigma_z = -30$, $\tau_{yz} = 20$, $\tau_{xz} = 10$, $\tau_{xy} = 30$ (all in MPa):

(a) Determine the stress state for Euler angles $\psi = 20°$, $\theta = 30°$, $\phi = 25°$.

(b) Plot the characteristic equation.

(c) Determine the principal stresses.

**26.** Expand the indicial forms of the governing equations for solid elasticity in three dimensions:

Equilibrium : $\sigma_{ij} = 0$

Kinematic : $\epsilon_{ij} = (u_{i,j} + u_{j,i})/2$

Constitutive : $\epsilon_{ij} = \dfrac{1+\nu}{E}\sigma_{ij} - \dfrac{\nu}{E}\delta_{ij}\sigma_{kk} + \alpha\delta_{ij}\Delta T$

where $\alpha$ is the coefficient of linear thermal expansion and $\Delta T$ is a temperature change.

**27.**

(a) Write out the compliance matrix **S** of Eq. 3.46 for polycarbonate using data in Appendix A.

(b) Use matrix inversion to obtain the stiffness matrix **D**.

(c) Use matrix multiplication to obtain the stresses needed to induce the strains

$$\epsilon = \begin{Bmatrix} \epsilon_x \\ \epsilon_y \\ \epsilon_z \\ \gamma_{yz} \\ \gamma_{xz} \\ \gamma_{xy} \end{Bmatrix} = \begin{Bmatrix} 0.02 \\ 0.0 \\ 0.03 \\ 0.01 \\ 0.025 \\ 0.0 \end{Bmatrix}$$

**28.**

(a) Write out the compliance matrix **S** of Eq. 3.46 for an aluminum alloy as described in Appendix A.

(b) Use matrix inversion to obtain the stiffness matrix **D**.

(c) Use matrix multiplication to obtain the stresses needed to induce the strains

$$\epsilon = \begin{Bmatrix} \epsilon_x \\ \epsilon_y \\ \epsilon_z \\ \gamma_{yz} \\ \gamma_{xz} \\ \gamma_{xy} \end{Bmatrix} = \begin{Bmatrix} 0.01 \\ 0.02 \\ 0.0 \\ 0.0 \\ 0.15 \\ 0.0 \end{Bmatrix}$$

**29.** Given the stress tensor

$$\sigma_{ij} = \begin{bmatrix} 1 & 2 & 3 \\ 2 & 4 & 5 \\ 3 & 5 & 7 \end{bmatrix} \quad \text{(MPa)}$$

(a) Dissociate $\sigma_{ij}$ into deviatoric and dilatational parts $\Sigma_{ij}$ and $\frac{1}{3}\sigma_{kk}\delta_{ij}$.

(b) Given $G = 357$ MPa and $K = 1.67$ GPa, obtain the deviatoric and dilatational strain tensors $e_{ij}$ and $\frac{1}{3}\epsilon_{kk}\delta_{ij}$.

(c) Add the deviatoric and dilatational strain components obtained above to obtain the total strain tensor $\epsilon_{ij}$.

(d) Compute the strain tensor $\epsilon_{ij}$ using the alternative form of the elastic constitutive law for isotropic elastic solids:

$$\epsilon_{ij} = \frac{1+\nu}{E}\sigma_{ij} - \frac{\nu}{E}\delta_{ij}\sigma_{kk}$$

Compare the result with that obtain in (c).

**30.** Provide an argument that any stress matrix having a zero trace can be transformed to one having only zeroes on its diagonal; in other words, that the deviatoric stress tensor $\Sigma_{ij}$ represents a state of pure shear.

**31.** Write out the $x$-$y$ two-dimensional compliance matrix $\overline{S}$ and stiffness matrix $\overline{D}$ (Eq. 3.56) for a single ply of graphite/epoxy composite with its fibers aligned along the $x$-$y$ axes.

**32.** Write out the $x$-$y$ two-dimensional compliance matrix $\overline{S}$ and stiffness matrix $\overline{D}$ (Eq. 3.56) for a single ply of graphite/epoxy composite with its fibers aligned 30° from the $x$ axis.

**33.** In cases where the stress rather than the strain is prescribed, the Kelvin model—a series arrangement of Voigt elements, as shown in Fig. P.3.33—is preferable to the Wiechert model. In Fig. P.3.33, where $\phi_j = 1/n_j = \dot{\epsilon}_j/\sigma_{dj}$ and $m_j = 1/k_j = \epsilon_j/\sigma_{sj}$. Using the relations $\epsilon = \epsilon_g + \Sigma\epsilon_j$, $\sigma = \sigma_{sj} + \sigma_{dj}$, show the viscoelastic operator equation to be

$$\epsilon(t) = \left[ m_g + \sum_j \frac{m_j}{\tau_j\left(D + \frac{1}{\tau_j}\right)} \right]\sigma_j$$

and for this model show the creep compliance to be

$$C_{\text{crp}}(t) = \frac{\epsilon(t)}{\sigma_0} = m_g + \sum_j k_j\left(1 - e^{-t/\tau_j}\right)$$

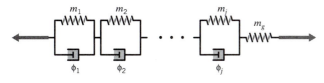

**FIGURE P.3.33**

**34.** Solithane 113 is a toluene diisocyanate–castor oil urethane rubber that has been used extensively as a model viscoelastic material. It exhibits a strong viscoelastic relation and is thermorheologically simple to a good approximation. In addition, it is strongly birefringent and thus useful in photoelasticity analyses, as will be discussed in Chapter 5. The "master" relaxation curve, obtained by time-temperature shifting methods to be discussed in Chapter 6, is shown in Fig. P.3.34. For this material, use collocation to obtain appropriate $k$'s and $\tau$'s in the *Prony series*

$$E_{\text{rel}} = k_e + \sum_{j=1}^{6} k_j \exp(-t/\tau_j)$$

The following approach might be used:

· Choose six values of $\tau_j$ through the relaxation, spaced at decade values from $10^{-6}$ to $10^{-1}$.
· Evaluate the Prony series at these six times, obtaining six simultaneous equations in $k_1, k_2, \ldots, k_6$:

$$k_1 e^{-(10^{-6}/10^{-6})} + k_2 e^{-(10^{-6}/10^{-5})} + \cdots + k_i e^{-(10^{-6}/10^{-1})}$$
$$= E_{\text{rel}}(10^{-6}) - k_e$$
$$k_1 e^{-(10^{-1}/10^{-6})} + k_2 e^{-(10^{-1}/10^{-5})} + \cdots + k_i e^{-(10^{-1}/10^{-1})}$$
$$= E_{\text{rel}}(10^{-1}) - k_e$$

· Neglect coefficients smaller than 0.01; solve the resulting system beginning with the last equation.
· Finally, we can improve the model a bit by recomputing $k_1$ so as to predict the glassy modulus correctly; otherwise $E_g$ will be overestimated.

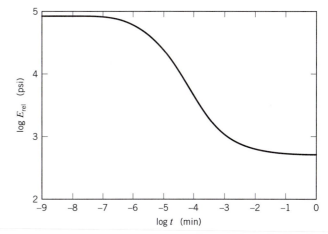

**FIGURE P.3.34**

**35.** Show that in a state of simple shear strain ($\gamma_{xy} = \gamma, \gamma_{xz} = \gamma_{yz} = \epsilon_x = \epsilon_y = \epsilon_z = 0$) the two principal extension ratios are $\lambda$ and $1/\lambda$ in the plane of shear, and, furthermore,

$$\gamma = \lambda - \frac{1}{\lambda}$$

**36.** In the case of an elastomeric material, use the relation for strain energy

$$W = \frac{NkT}{2}(\lambda_x + \lambda_y + \lambda_z - 3)$$

along with $dW = \tau d\gamma$ to show

$$G = NkT$$

**37.** Assuming a Standard Linear Solid viscoelastic material ($E_{\text{rel}} = k_e + k_1 \exp(-t/\tau)$), use the Boltzman superposition integral (Eq. 3.64) to evaluate the stress resulting from the following strain inputs:

(*a*) A constant strain applied at time $t = t_1$.

(*b*) A ramp-and-hold strain input. (*Hint:* This input can be viewed as two ramps, one beginning at $t_1$ and the other of opposite sign beginning at $t_2$.)

(*c*) A sinusoidal input, $\epsilon(t) = \epsilon_0 \sin(\omega t)$, where $\omega$ is the angular frequency.

**38.** Given a viscoelastic constitutive equation in differential form:

$$\ddot{\sigma} + 6\dot{\sigma} + 8\sigma = 9\ddot{\epsilon} + 34\dot{\epsilon} + 24\epsilon$$

(*a*) Determine the relaxation modulus

$$E_{\text{rel}}(t) = \mathcal{L}^{-1}[\mathcal{E}(s)/s] = ?$$

(*b*) Determine the $k$'s and $\tau$'s in the associated two-arm Wiechert model.

(c) Write the operator equation for the Wiechert model in part (b), then clear it of fractions so as to recover the original differential equation.

**39.** Show that for linearly viscoelastic materials the interconversion between relaxation moduli and creep compliances can be done by means of the integral relation

$$\int_0^t E_{rel}(\lambda)C_{crp}(t - \lambda)\,d\lambda = t$$

*Hint:* This is equivalent to the expression in the Laplace plane

$$\overline{E}_{rel}\overline{C}_{crp} = \frac{1}{s^2}$$

**40.** Verify the integral expression in Problem 39 for the simple Maxwell element.

**41.** For the simple Maxwell element with $k = 100$ and $\eta = 1000$:

(a) Plot $E_{rel}(t)$ and $E_{crp}(t) = 1/C_{crp}$ versus time.

(b) How good an approximation is the simple relation $C_{crp}(t) \approx 1/E_{rel}(t)$? Show this by plotting $C_{crp}(t) \times E_{rel}(t)$ versus log time through the transition for the Maxwell element of part (a).

# BENDING OF BEAMS AND PLATES

CHAPTER 4

*Beams* are long and slender structural elements, differing from truss elements in that they are called on to support transverse in addition to axial loads. Their attachment points can also be more complicated than those of truss elements: They may be bolted or welded together, so the attachments can transmit bending moments or transverse forces into the beam. Studied by Galileo (see Fig. 4.1), beams are among the most common of all structural elements, being the supporting frames of airplanes, buildings, cars, people, and much else.

The nomenclature of beams is rather standard. As shown in Fig. 4.2, $L$ is the length, or *span*; $b$ is the width; and $h$ is the height. The cross-sectional shape need not be rectangular; often it consists of a vertical *web* that separates horizontal *flanges* at the top and bottom of the beam.

This chapter will develop relations between stresses, deflections, and applied loads for beams and flat plates subjected to bending loads. This will be done using the direct method employed in Chapter 2 for circular shafts in torsion. However, bending problems have a higher order of dimensionality than twisted shafts, and it will be

**FIGURE 4.1** Galileo's illustration of a beam bending test. (*Polymers: An Encyclopedic Sourcebook of Engineering*/John Wiley & Sons Inc., 1987.)

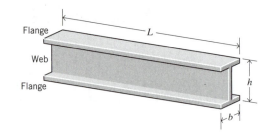

**FIGURE 4.2** Beam nomenclature.

convenient to use the more general formulations developed in Chapter 3. In particular, pseudovector-matrix notation will allow easy extension of beam concepts to flat plates.

# 4.1 STATICS CONSIDERATIONS

## 4.1.1 Free-Body Diagrams

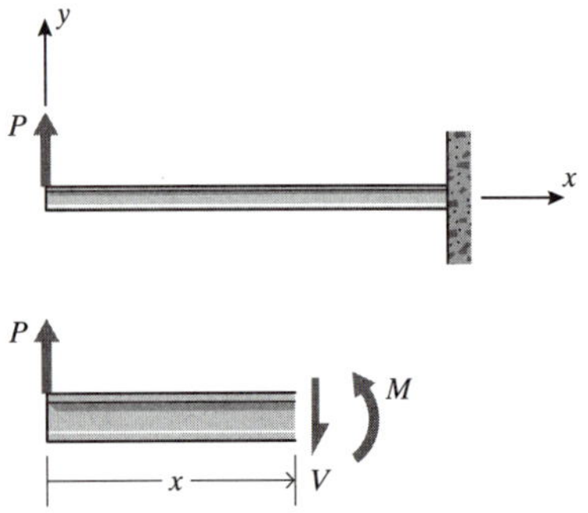

As a simple starting example, consider a beam clamped at one end and subjected to a load $P$ at the free end (*cantilevered*) as shown in Fig. 4.3. A free-body diagram of a section cut transversely at position $x$ shows that a shear force $V$ and a moment $M$ must exist on the cut section to maintain equilibrium. We will show in a later section that these are the resultants of shear and normal stresses that are set up on internal planes by the bending loads. As usual, we will consider section areas whose normals point in the $+x$ direction to be positive; then shear forces pointing in the $+y$ direction on $+x$ faces will be considered positive. Moments whose vector direction as given by the right-hand rule are in the $+z$ direction (vector out of the plane of the paper, or tending to cause counterclockwise rotation in the plane of the paper) will be positive when acting on $+x$ faces. Another way to recognize positive bending moments is that they cause the bending shape to be concave upward. For this example beam the statics equations give

$$\sum F_y = 0 = V + P \Rightarrow V = \text{constant} = -P \tag{4.1}$$

$$\sum M_0 = 0 = -M + Px \Rightarrow M = M(x) = Px \tag{4.2}$$

**FIGURE 4.3** A cantilevered beam.

Note that the moment increases with distance from the loaded end, so the magnitude of the maximum value of $M$ compared with $V$ increases as the beam becomes longer. This is true of most beams, so shear effects are usually more important in beams with small span-to-height ratios.

The stresses and deflections will be shown to be functions of $V$ and $M$, so it is important to be able to compute how these quantities vary along the beam's span. Plots of $V(x)$ and $M(x)$ are known as *shear* and *bending moment diagrams,* and it is necessary to obtain them before the stresses can be determined. For the end-loaded cantilever, the diagrams shown in Fig. 4.4 are obvious from Eqs. 4.1 and 4.2.

It was easiest to analyze the cantilevered beam by beginning at the free end, but the choice of origin is arbitrary. It is not always possible to guess the easiest way to proceed, so consider what would have happened if the origin were placed at the wall as in Fig. 4.5. Now when a free-body diagram is constructed, forces must be placed at

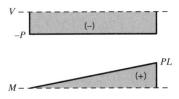

**FIGURE 4.4** Shear and bending moment diagrams.

the origin to replace the reactions that were imposed by the wall to keep the beam in equilibrium with the applied load. These reactions can be determined from free-body diagrams of the beam as a whole (if the beam is statically determinate) and must be found before the problem can proceed. For the beam of Fig. 4.5,

$$\sum F_y = 0 = -V_R + P \Rightarrow V_R = -P$$

$$\sum M_0 = 0 = M_R - PL \Rightarrow M_R = PL$$

The shear and bending moment at $x$ are then

$$V(x) = V_R = P = \text{constant}$$

$$M(x) = M_R - V_R x = PL - Px$$

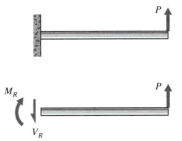

**FIGURE 4.5** Wall reactions for the cantilevered beam.

This choice of origin produces some extra algebra, but the $V(x)$ and $M(x)$ diagrams shown in Fig. 4.6 are the same as before (except for changes of sign): $V$ is constant and equal to $P$, and $M$ varies linearly from zero at the free end to $PL$ at the wall.

## 4.1.2 Distributed Loads

Transverse loads may be applied to beams in a distributed manner rather than at a point, as depicted in Fig. 4.7, which might be visualized as sand piled on the beam. It is convenient to describe these distributed loads in terms of *force per unit length,* so that $q(x)\,dx$ would be the load applied to a small section of length $dx$ by a distributed load $q(x)$. The shear force $V(x)$ set up in reaction to such a load is

**FIGURE 4.6** Alternative shear and bending moment diagrams for the cantilevered beam.

$$V(x) = -\int_{x_0}^{x} q(\xi)\,d\xi \tag{4.3}$$

where $x_0$ is the value of $x$ at which $q(x)$ begins, and $\xi$ is a dummy length variable that looks backward from $x$. Hence $V(x)$ is the area under the $q(x)$ diagram up to position $x$. The moment balance is obtained considering the increment of load $q(\xi)\,d\xi$ applied to a small width $d\xi$ of beam, a distance $\xi$ from point $x$. The incremental moment of this load around point $x$ is $q(\xi)\xi\,d\xi$, so the moment $M(x)$ is

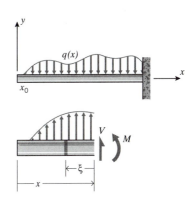

$$M = \int_{x_0}^{x} q(\xi)\,\xi\,d\xi \tag{4.4}$$

This can be related to the centroid of the area under the $q(x)$ curve up to $x$, whose distance from $x$ is

**FIGURE 4.7** A distributed load and a free-body section.

$$\bar{\xi} = \frac{\int q(\xi)\xi\,d\xi}{\int q(\xi)\,d\xi}$$

Hence Eq. 4.4 can be written

$$M = Q\bar{\xi} \tag{4.5}$$

where $Q = \int q(\xi)\,d\xi$ is the area. Therefore, the distributed load $q(x)$ is statically equivalent to a concentrated load of magnitude $Q$ placed at the centroid of the area under the $q(x)$ diagram.

EXAMPLE 4.1

Consider a simply supported beam carrying a triangular and a concentrated load as shown in Fig. 4.8. For the purpose of determining the support reaction forces $R_1$ and $R_2$, the distributed triangular load can be replaced by its static equivalent. The magnitude of this equivalent force is

$$Q = \int_0^2 (-600x)\,dx = -1200$$

The equivalent force acts through the centroid of the triangular area, which is 2/3 of the distance from its narrow end (see Problem 1 in this chapter). The reaction $R_2$ can now be found by taking moments around the left end:

$$\sum M_A = 0 = -500(1) - (1200)(2/3)(2) + R_2(2) \rightarrow R_2 = 1050$$

The other reaction can then be found from vertical equilibrium:

$$\sum F_y = 0 = R_1 - 500 - 1200 + 1050 \rightarrow R_1 = 650$$

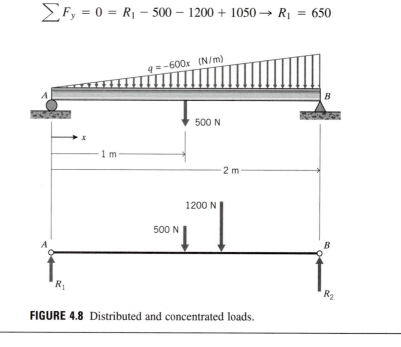

**FIGURE 4.8** Distributed and concentrated loads.

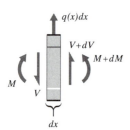

**FIGURE 4.9** Relations between distributed loads and internal shear forces and bending moments.

### 4.1.3 Successive Integration Method

We have already noted in Eq. 4.3 that the shear curve is the negative integral of the loading curve. Another way of developing this result is to consider a free-body balance on a small increment of length $dx$, over which the shear and moment change from $V$ and $M$ to $V + dV$ and $M + dM$ (see Fig. 4.9). The distributed load $q(x)$ can be taken as constant over the small interval, so the force balance is

$$\sum F_y = 0 = (V + dV) + q\,dx - V = 0$$

$$\frac{dV}{dx} = -q \tag{4.6}$$

or

$$V(x) = -\int q(x)\,dx \tag{4.7}$$

which is equivalent to Eq. 4.3. A moment balance around the center of the increment gives

$$\sum M_0 = (M + dM) + (V + dV)\frac{dx}{2} + V\frac{dx}{2} - M$$

As the increment $dx$ is reduced to the limit, the term containing the higher-order differential $dV\,dx$ vanishes in comparison with the others, leaving

$$\frac{dM}{dx} = -V \tag{4.8}$$

or

$$M(x) = -\int V(x)\,dx \tag{4.9}$$

Hence the value of the shear curve at any axial location along the beam is equal to the negative of the slope of the moment curve at that point, and the value of the moment curve at any point is equal to the negative of the area under the shear curve up to that point.

The shear and moment curves can be obtained by successive integration of the $q(x)$ distribution, as illustrated in the following example.

## EXAMPLE 4.2

Consider a cantilevered beam subjected to a negative distributed load $q(x) = -q_0 =$ constant as shown in Fig. 4.10; then

$$V(x) = -\int q(x)\,dx = q_0 x + c_1$$

where $c_1$ is a constant of integration. A free-body diagram of a small sliver of length near $x = 0$ shows that $V(0) = 0$, so the $c_1$ must be zero as well. The moment function is obtained by integrating again:

$$M(x) = -\int V(x)\, dx = -\frac{1}{2}q_0 x^2 + c_2$$

where $c_2$ is another constant of integration, which is also zero because $M(0) = 0$.

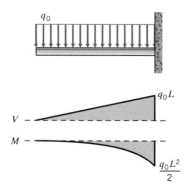

**FIGURE 4.10** Shear and moment distributions in a cantilevered beam.

Admittedly, this problem was easy, because we picked one with null boundary conditions and with only one loading segment. When concentrated or distributed loads are found at different positions along the beam, it is necessary to integrate over each section between loads separately. Each integration will produce an unknown constant, and these must be determined by invoking the continuity of forces and moments from section to section. This is a laborious process, but it can be made much easier using *singularity functions,* which will be introduced shortly.

It is often possible to sketch $V$ and $M$ diagrams without actually drawing free-body diagrams or writing equilibrium equations. This is made easier because the curves are integrals or derivatives of one another, so graphical sketching can take advantage of relations among slopes and areas.

These rules can be used to work gradually from the $q(x)$ curve to $V(x)$ and then to $M(x)$. Wherever a concentrated load appears on the beam, the $V(x)$ curve must jump by that value, but in the opposite direction; similarly, the $M(x)$ curve must jump discontinuously wherever a couple is applied to the beam.

## EXAMPLE 4.3

To illustrate this process, consider a simply-supported beam of length $L$, as shown in Fig. 4.11a, loaded over half its length by a negative distributed load $q = -q_0$. The solution for $V(x)$ and $M(x)$ takes the following steps:

**1.** The reactions at the supports are found from static equilibrium. Replacing the distributed load by a concentrated load $Q = -q_0(L/2)$ at the midpoint of the $q$ distribution (Fig. 4.11b) and taking moments around $A$:

$$R_B L = \left(\frac{q_0 L}{2}\right)\left(\frac{3L}{4}\right) \Rightarrow R_B = \frac{3q_0 L}{8}$$

The reaction at the right end is then found from a vertical force balance:

$$R_A = \frac{q_0 L}{2} - R_B = \frac{q_0 L}{8}$$

Note that only two equilibrium equations were available, because a horizontal force balance would provide no relevant information. Hence the beam will be statically indeterminate if more than two supports are present. The $q(x)$ diagram is then just the beam with the end reactions shown in Fig. 4.11c.

**2.** Beginning the shear diagram at the left, $V$ immediately jumps down to a value of $-q_0 L/8$ in opposition to the discontinuously applied reaction force at $A$; it remains at this value until $x = L/2$ as shown in Fig. 4.11d.

**3.** At $x = L/2$, the $V(x)$ curve starts to rise with a constant slope of $+q_0$ as the area under the $q(x)$ distribution begins to accumulate. When $x = L$, the shear curve will have risen by an amount $q_0 L/2$, the total area under the $q(x)$

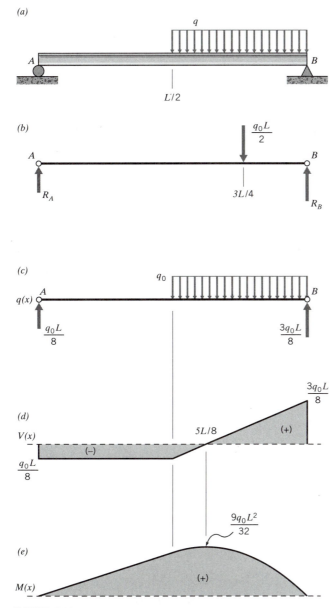

**FIGURE 4.11** (*a*) Simply supported beam; (*b*) equivalent concentrated load; (*c*) loading diagram; (*d*) shear diagram; (*e*) bending moment diagram.

curve; its value is then $(-q_0L/8) + (q_0L/2) = (3q_0L/8)$. The shear curve then drops to zero in opposition to the reaction force $R_B = (3q_0L/8)$. (The $V$ and $M$ diagrams should always close, and this provides a check on the work.)

**4.** The moment diagram starts from zero as shown in Fig. 4.11*e*, because there is no discontinuously applied moment at the left end. It moves upward at a constant slope of $+q_0L/2$, the value of the shear diagram in the first half of the beam. When $x = L/2$, the moment curve will have risen to a value of $q_0L^2/4$.

**5.** After $x = L/2$, the slope of the moment curve starts to fall as the shear curve rises. The moment curve is now parabolic, always being one order higher than the shear curve. The shear curve crosses the $V = 0$ axis at $x = 5L/8$, and at this point the slope of the moment curve will have dropped to zero. The maximum value of $M$ is $9q_0L^2/32$, the total area under the $V$ curve up to this point.

6. To the right of $x = 5L/8$ the moment curve falls parabolically, reaching zero at $x = L$.

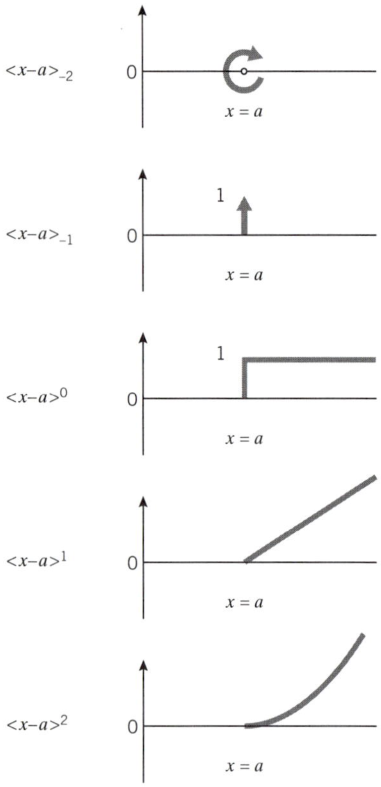

### 4.1.4 Singularity Functions

**FIGURE 4.12** Singularity functions.

This special family of functions provides an automatic way of handling the irregularities of loading that usually occur in beam problems. They are much like conventional polynomial factors, but they have the property of being zero until "activated" at desired points along the beam. The formal definition is

$$f_n(x) = \langle x - a \rangle^n = \begin{cases} 0, & x < a \\ (x - a)^n, & x > a \end{cases} \quad (4.10)$$

where $n = -2, -1, 0, 1, 2, \ldots$. The function $\langle x - a \rangle^0$ is a unit step function, $\langle x - a \rangle_{-1}$ is a concentrated load, and $\langle x - a \rangle_{-2}$ is a concentrated couple. The first five of these functions are sketched in Fig. 4.12.

The singularity functions are integrated much like conventional polynomials:

$$\int_{-\infty}^{x} \langle x - a \rangle^n \, dx = \frac{\langle x - a \rangle^{n+1}}{n + 1} \quad n \geq 0 \quad (4.11)$$

However, there are special integration rules for the $n = -1$ and $n = -2$ members, and this special handling is emphasized by using subscripts for the $n$ index:

$$\int_{-\infty}^{x} \langle x - a \rangle_{-2} \, dx = \langle x - a \rangle_{-1} \quad (4.12)$$

$$\int_{-\infty}^{x} \langle x - a \rangle_{-1} \, dx = \langle x - a \rangle^0 \quad (4.13)$$

---

## EXAMPLE 4.4

Applying singularity functions to the beam of Example 4.3, the loading function would be written

$$q(x) = +\frac{q_0 L}{8} \langle x - 0 \rangle_{-1} - q_0 \left\langle x - \frac{L}{2} \right\rangle^0$$

The reaction force at the right end could also be included, but it becomes activated only as the problem is over. Integrating once:

$$V(x) = -\int q(x) \, dx = -\frac{q_0 L}{8} \langle x \rangle^0 + q_0 \left\langle x - \frac{L}{2} \right\rangle^1$$

The constant of integration is included automatically here, because the influence of the reaction at $A$ has been included explicitly. Integrating again:

$$M(x) = -\int V(x)\,dx = \frac{q_0 L}{8}\langle x\rangle^1 - \frac{q_0}{2}\left\langle x - \frac{L}{2}\right\rangle^2$$

Examination of this result will show that it is the same as that developed previously.

## 4.2 NORMAL STRESSES IN BEAMS

A beam subjected to a positive bending moment will tend to develop a concave-upward curvature. Intuitively, this means that the material near the top of the beam is placed in compression along the $x$ direction and the lower region is in tension. At the transition between the compressive and tensile regions, the stress becomes zero; this is the *neutral axis* of the beam. If the material tends to fail in tension, like chalk or glass, it will do so by crack initiation and growth from the lower, tensile surface. If the material is strong in tension but weak in compression, it will fail at the top, compressive surface; this might be observed in a piece of wood by a compressive buckling of the outer fibers.

We seek an expression relating the magnitudes of these axial normal stresses to the shear and bending moment within the beam, analogously to the shear stresses induced in a circular shaft by torsion. In fact, the development of the needed relations follows exactly the same direct approach as that used for torsion:

1. *Geometrical statement:* We begin by stating that originally transverse planes within the beam remain planar under bending but rotate through an angle $\theta$ about points on the neutral axis as shown in Fig. 4.13. For small rotations this angle is given approximately by the $x$-derivative of the beam's vertical deflection function $v(x)$[1]:

$$u = -y v_{,x} \tag{4.14}$$

where the comma indicates differentiation with respect to the indicated variable ($v_{,x} \equiv dv/dx$). Here $y$ is measured positive upward from the neutral axis, whose location within the beam has not yet been determined.

2. *Kinematic equation:* The $x$-direction normal strain $\epsilon_x$ is then the gradient of the displacement:

$$\epsilon_x = \frac{du}{dx} = -y v_{,xx} \tag{4.15}$$

Note that the strains are zero at the neutral axis where $y = 0$, negative (compressive) above the axis, and positive (tensile) below. They increase in magnitude linearly with $y$, much as the shear strains increased linearly with $r$ in a torsionally loaded circular shaft. The quantity $v_{,xx} \equiv d^2 v/dx^2$ is the spatial

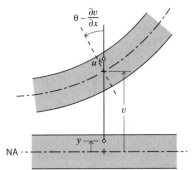

**FIGURE 4.13** Geometry of beam bending.

---

[1]The exact expression for curvature is

$$\frac{d\theta}{ds} = \frac{d^2 v/dx^2}{[1 + (dv/dx)^2]^{3/2}}$$

This gives $\theta \approx dv/dx$ when the squared derivative in the denominator is small compared to 1.

rate of change of the slope of the beam deflection curve, the "slope of the slope." This is called the *curvature* of the beam.

**3.** *Constitutive equation:* The stresses are obtained directly from Hooke's law as

$$\sigma_x = E\epsilon_x = -yEv_{,xx} \tag{4.16}$$

This restricts the applicability of this derivation to linear elastic materials. Hence the axial normal stress, like the strain, increases linearly from zero at the neutral axis to a maximum at the outer surfaces of the beam.

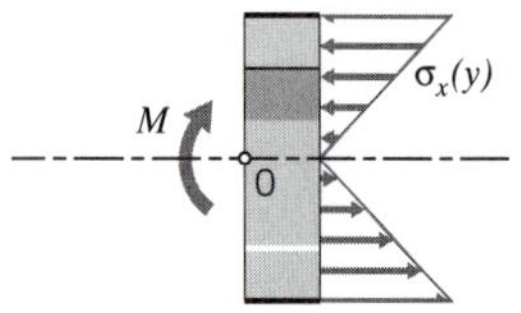

**4.** *Equilibrium relations:* Since there are no axial (*x*-direction) loads applied externally to the beam, the total axial force generated by the normal $\sigma_x$ stresses (shown in Fig. 4.14) must be zero. This can be expressed as

$$\sum F_x = 0 = \int_A \sigma_x\, dA = \int_A -yEv_{,xx}\, dA$$

which requires that

$$\int_A y\, dA = 0$$

The distance $\bar{y}$ from the neutral axis to the centroid of the cross-sectional area is

$$\bar{y} = \frac{\int_A y\, dA}{\int_A dA}$$

Hence $\bar{y} = 0$; that is, the neutral axis is coincident with the centroid of the beam cross-sectional area. This result is obvious on reflection because the stresses increase at the same linear rate, above the axis in compression and below the axis in tension. Only if the axis is exactly at the centroidal position will these stresses balance to give zero net horizontal force and keep the beam in horizontal equilibrium.

The normal stresses in compression and tension are balanced to give a zero net horizontal force, but they also produce a net clockwise moment. This moment must equal the value of $M(x)$ at that value of *x*, as seen by taking a moment balance around point $O$:

$$\sum M_O = 0 = M - \int_A \sigma_x \cdot y\, dA$$

$$M = \int_A (yEv_{,xx}) \cdot y\, dA = Ev_{,xx} \int_A y^2\, dA \tag{4.17}$$

The quantity $\int y^2\, dA$ is the rectangular moment of inertia with respect to the centroidal axis, denoted $I$. For a rectangular cross section of height *h* and width *b*, as shown in Fig. 4.15, this is

$$I = \int_{-h/2}^{h/2} y^2 b\, dy = \frac{bh^3}{12} \tag{4.18}$$

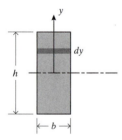

**FIGURE 4.14** Moment and force equilibrium in the beam.

**FIGURE 4.15** Moment of inertia for a rectangular section.

Solving Eq. 4.17 for $v_{,xx}$, the beam curvature is

$$v_{,xx} = \frac{M}{EI}$$  (4.19)

**5.** An explicit formula for the stress can be obtained by using this in Eq. 4.16:

$$\sigma_x = -yE\frac{M}{EI} = \frac{-My}{I}$$  (4.20)

The final expression for stress, Eq. 4.20, is similar to $\tau_{\theta z} = Tr/J$ for twisted circular shafts: The stress varies linearly from zero at neutral axis to a maximum at the outer surface, it varies inversely with the moment of inertia of the cross section, and it is independent of the material's properties. Just as a designer will favor annular drive shafts to maximize the polar moment of inertia $J$, beams are often made with wide flanges at the upper and lower surfaces to increase $I$.

## EXAMPLE 4.5

Consider a cantilevered T beam with dimensions as shown in Fig. 4.16, carrying a uniform loading of $w$ N/m. The maximum bending moment occurs at the wall and is easily found to be $M_{\max} = (wL)(L/2)$. The stress is then given by Eq. 4.20, which requires that we know the location of the neutral axis (since $y$ and $I$ are measured from there).

The distance $\bar{y}$ from the bottom of the beam to the centroidal neutral axis can be found using the *composite area theorem* (see Problem 8 in this chapter). This theorem states that the distance from an arbitrary axis to the centroid of an area made up of several subareas is the sum of the subareas times the distance to their individual centroids, divided by the sum of the subareas (that is, by the total area):

$$\bar{y} = \frac{\sum_i A_i \bar{y}_i}{\sum_i A_i}$$

For our example this is

$$\bar{y} = \frac{(d/2)(cd) + (d + b/2)(ab)}{cd + ab}$$

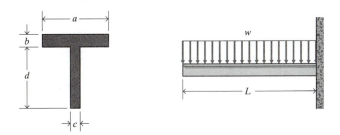

**FIGURE 4.16** A cantilevered T beam.

The moments of inertia of the individual parts of the compound area with respect to their own centroids are just $ab^3/12$ and $cd^3/12$. These moments can be referenced to the horizontal axis through the centroid of the compound area using the *parallel axis theorem* (see Problem 10 in this chapter). This theorem states that the moment of inertia $I_{z'}$ of an area $A$, relative to any arbitrary axis $z'$ parallel to an axis through the centroid but a distance $d$ from it, is the moment of inertia relative to the centroidal axis $I_z$ plus the product of the area $A$ and the square of the distance $d$:

$$I_{z'} = I_z + Ad^2$$

For our example this is

$$I^{(1)} = \frac{ab^3}{12} + (ab)\left(d + \frac{b}{2} - \bar{y}\right)^2$$

$$I^{(2)} = \frac{cd^3}{12} + (cd)\left(\frac{d}{2} - \bar{y}\right)^2$$

The moment of inertia of the entire compound area, relative to its centroid, is then the sum of these two contributions:

$$I = I^{(1)} + I^{(2)}$$

The maximum stress is then given by Eq. 4.20 using this value of $I$ and $y = \bar{y}$ (the distance from the neutral axis to the bottom fibers), along with the maximum bending moment $M_{max}$. The result of these substitutions is

$$\sigma_x = \frac{\left(cd^2 + 2abd + ab^2\right)wL^2}{c^2d^4 + 4abcd^3 + 6ab^2cd^2 + 4ab^3cd + a^2b^4}$$

In practice each step would likely be reduced to a numerical value rather than working toward an algebraic solution.

In pure bending (only bending moments applied, no transverse or longitudinal forces), the only stress is $\sigma_x$ as given by Eq. 4.20. All other stresses are zero ($\sigma_y = \sigma_z = \tau_{xy} = \tau_{xz} = \tau_{yz} = 0$). However, strains other than $\epsilon_x$ are present because of the Poisson effect. This does not generate shear strain ($\gamma_{xy} = \gamma_{xz} = \gamma_{yz} = 0$), but the normal strains are

$$\epsilon_x = \frac{1}{E}[\sigma_x - \nu(\sigma_y + \sigma_z)] = \frac{\sigma_x}{E}$$

$$\epsilon_y = \frac{1}{E}[\sigma_y - \nu(\sigma_x + \sigma_z)] = -\nu\frac{\sigma_x}{E}$$

$$\epsilon_z = \frac{1}{E}[\sigma_z - \nu(\sigma_x + \sigma_y)] = -\nu\frac{\sigma_x}{E}$$

The strains can also be written in terms of curvatures. From Eq. 4.15, the curvature along the beam is

$$v_{,xx} = -\frac{\epsilon_x}{y}$$

This is accompanied by a curvature transverse to the beam axis given by

$$v_{,zz} = -\frac{\epsilon_z}{y} = \frac{\nu\epsilon_x}{y} = -\nu v_{,xx}$$

This transverse curvature, shown in Fig. 4.17, is known as *anticlastic curvature;* it can be seen by bending a "Pink Pearl"–type eraser in the fingers.

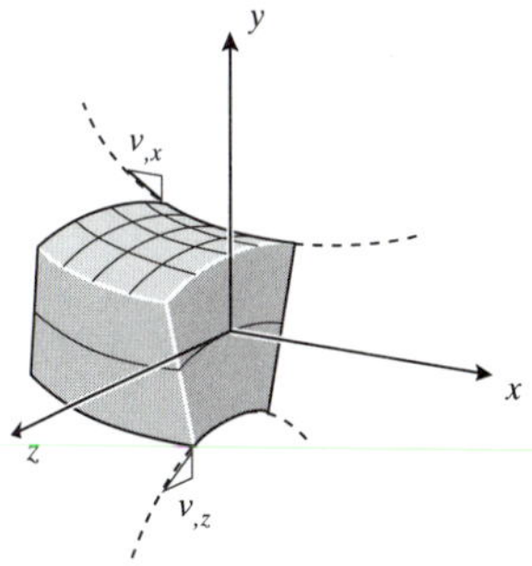

As with tension and torsion structures, bending problems can often be done more easily with energy methods. Knowing the stress from Eq. 4.20, the strain energy due to bending stress $U_b$ can be found by integrating the strain energy per unit volume $U^* = \sigma^2/2E$ over the specimen volume:

$$U_b = \int_V U^* dV = \int_L \int_A \frac{\sigma_x^2}{2E}\, dA\, dL$$

$$= \int_L \int_A \frac{1}{2E}\left(\frac{-My}{I}\right)^2 dA\, dL = \int_L \frac{M^2}{2EI^2}\int_A y^2\, dA\, dL$$

**FIGURE 4.17** Anticlastic curvature.

Since $\int_A y^2\, dA = I$, this becomes

$$\boxed{U_b = \int_L \frac{M^2\, dL}{2EI}} \qquad (4.21)$$

If the bending moment is constant along the beam (definitely *not* the usual case), this becomes

$$U = \frac{M^2 L}{2EI}$$

This is another analog to the expression for uniaxial tension, $U = P^2 L/2AE$.

## 4.3  BUCKLING

It was noted in Chapter 1 that long, slender columns placed in compression are prone to fail by buckling. This is actually a bending phenomenon, driven by the bending moment that develops if and when the beam undergoes a transverse deflection. Consider a beam loaded in axial compression and pinned at both ends as shown in Fig. 4.18. Now let the beam be made to deflect transversely by an amount $v$, perhaps by an adventitious sideward load or even an irregularity in the beam's cross section. Positions along the beam will experience a moment given by

$$M(x) = Pv(x) \qquad (4.22)$$

The beam's own stiffness will act to restore the deflection and recover a straight shape, but the effect of the bending moment is to deflect the beam *more*. The question is which influence will win out. If the tendency of the bending moment to increase the deflection dominates over the ability of the beam's elastic stiffness to resist bending, the beam will become unstable, continuing to bend at an accelerating rate until the structure fails.

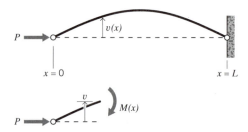

**FIGURE 4.18** Imminent buckling in a beam.

The bending moment is related to the beam curvature by Eq. 4.19, so combining this with Eq. 4.22 gives

$$v_{,xx} = \frac{P}{EI}v \tag{4.23}$$

Of course, this governing equation is satisfied identically if $v = 0$—that is, if the beam is straight. We wish to look beyond this trivial solution and ask whether the beam could adopt a *bent* shape that would also satisfy the governing equation; this would imply that the stiffness is insufficient to restore the unbent shape and therefore that the beam is beginning to buckle. Equation 4.23 will be satisfied by functions that are proportional to their own second derivatives. Trigonometric functions have this property, so candidate solutions will be of the form

$$v = c_1 \sin\sqrt{\frac{P}{EI}}x + c_2 \cos\sqrt{\frac{P}{EI}}x$$

It is obvious that $c_2$ must be zero, because the deflection must go to zero at $x = 0$ and $L$. The sine term must go to zero at these two positions as well, which requires that the length $L$ be exactly equal to a multiple of the half-wavelength of the sine function:

$$\sqrt{\frac{P}{EI}}L = n\pi, \qquad n = 1, 2, 3, \ldots$$

The lowest value of $P$ leading to the deformed shape corresponds to $n = 1$; then we have

$$\boxed{P_{cr} = \frac{\pi^2 EI}{L^2}} \tag{4.24}$$

This is the critical buckling load given without proof in Chapter 1. Note the dependency on $L^2$, so the buckling load drops with the square of the length.

This strong dependency on length shows why crossbracing is so important for preventing buckling. If a brace is added at the beam's midpoint, as shown in Fig. 4.19, to eliminate deflection there, the buckling shape is forced to adopt a wavelength of $L$ rather than $2L$. This is equivalent to making the beam half as long, which increases the critical buckling load by a factor of 4.

Similar reasoning can be used to assess the result of having different support conditions. If, for instance, the beam is clamped at one end but unsupported at the other, its buckling shape will be a quarter sine wave. This is equivalent to making the

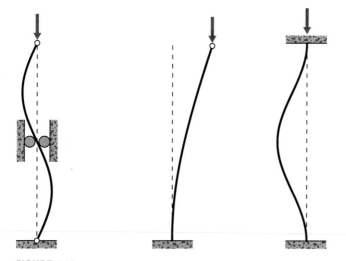

**FIGURE 4.19** Effect of lateral support and end conditions on beam buckling.

beam twice as long as the case with both ends pinned, so the buckling load will go *down* by a factor of 4. Clamping *both* ends forces a full-wave shape, with the same buckling load as the pinned beam with a midpoint support.

## 4.4 SHEAR STRESSES

Transverse loads bend beams by inducing axial tensile and compressive normal strains in the beam's x-direction, as discussed in Section 4.2. In addition, they cause shear effects that tend to slide vertical planes tangentially to one another as depicted in Fig. 4.20, much like sliding playing cards past one another. The stresses $\tau_{xy}$ associated with this shearing effect add up to the vertical shear force that we have been calling $V$, and we now seek to understand how these stresses are distributed over the beam's cross section. The shear stress on vertical planes must be accompanied by an equal stress on horizontal planes, because $\tau_{xy} = \tau_{yx}$, and these horizontal shearing stresses must become zero at the upper and lower surfaces of the beam unless a traction is applied there to balance them. Hence, they must reach a maximum somewhere within the beam.

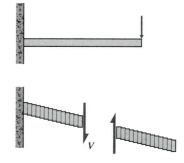
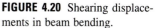

**FIGURE 4.20** Shearing displacements in beam bending.

The variation of this horizontal shear stress with vertical position $y$ can be determined by examining a free body of width $dx$ cut from the beam a distance $y$ above neutral axis, as shown in Fig. 4.21. The moment on the left vertical face is $M(x)$, and on the right face it has increased to $M + dM$. Since the horizontal normal stresses are directly proportional to the moment ($\sigma_x = My/I$), any increment in moment $dM$ over the distance $dx$ produces an imbalance in the horizontal force arising from the normal stresses. This imbalance must be compensated by a shear stress $\tau_{xy}$ on the horizontal plane at $y$. The horizontal force balance is written as

$$\tau_{xy}b\,dx = \int_{A'} \frac{dM\,\xi}{I}\,dA'$$

where $b$ is the width of the beam at $y$, $\xi$ is a dummy height variable ranging from $y$ to the outer surface of the beam, and $A'$ is the cross-sectional area between the plane at

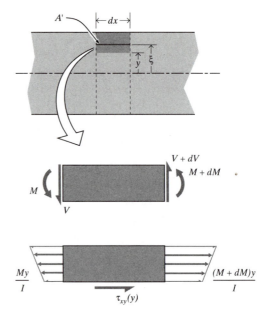

**FIGURE 4.21** Shear and bending moment in a differential length of beam.

$y$ and the outer surface. Using $dM = V\,dx$ from Eq. 4.8, this relation becomes

$$\tau_{xy} = \frac{V}{Ib}\int_{A'}\xi\,dA' = \frac{VQ}{Ib}$$

(4.25)

where here $Q(y) = \int_{A'}\xi\,dA' = \bar{\xi}A'$ is the first moment of the area above $y$ about the neutral axis.

The parameter $Q(y)$ is notorious for confusing persons new to beam theory. To determine it for a given height $y$ relative to the neutral axis, begin by sketching the beam cross section, and draw a horizontal line at the position $y$ at which $Q$ is sought (Fig. 4.22 shows a rectangular beam of constant width $b$ and height $h$ for illustration). Note the area $A'$ between this line and the outer surface (indicated by cross-hatching in Fig. 4.22). Now compute the distance $\bar{\xi}$ from the neutral axis to the centroid of $A'$. The parameter $Q(y)$ is the product of $A'$ and $\bar{\xi}$; this is the first moment of the area $A'$ with respect to the centroidal axis. For the rectangular beam, it is

$$Q = A'\bar{\xi} = \left[b\left(\frac{h}{2}-y\right)\right]\left[y+\frac{1}{2}\left(\frac{h}{2}-y\right)\right] = \frac{b}{2}\left(\frac{h^2}{4}-y^2\right)$$

Note that $Q(y)$, and therefore $\tau_{xy}(y)$ as well, is parabolic, being maximum at the neutral axis ($y = 0$) and zero at the outer surface ($y = h/2$). Using $I = bh^3/12$ for the rectangular beam, the maximum shear stress, as given by Eq. 4.25, is

$$\tau_{xy,\text{max}} = \tau_{xy}\big|_{y=0} = \frac{3V}{2bh}$$

(Keep in mind that these two expressions for $Q$ and $\tau_{xy,\text{max}}$ are for rectangular cross section only; sections of other shapes will have different results.) These shear stresses are most important in beams that are short relative to their height, because the bending

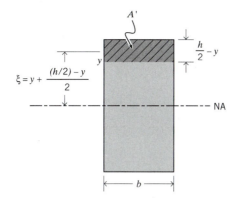

**FIGURE 4.22** Section of a rectangular beam.

moment usually increases with length and the shear force does not (see Problem 18 in this chapter). One standard test for interlaminar shear strength[2] is to place a short beam in bending and observe the load at which cracks develop along the midplane.

## EXAMPLE 4.6

Since the normal stress is maximum where the horizontal shear stress is zero (at the outer fibers), and the shear stress is maximum where the normal stress is zero (at the neutral axis), it is often possible to consider them one at a time. However, the juncture of the web and the flange in I and T beams is often a location of special interest, because here both stresses can take on substantial values.

Consider the T beam seen previously in Example 4.5, and examine the location at point $A$ shown in Fig. 4.23, in the web immediately below the flange. Here the width $b$ in Eq. 4.25 is the dimension labeled $c$; because the beam is thin here, the shear stress $\tau_{xy}$ will tend to be large, but it will drop dramatically in the flange as the width jumps to the larger value $a$. The normal stress at point $A$ is computed from $\sigma_x = My/I$, using $y = d - \bar{y}$. This value will be almost as large as the outer-fiber stress if the flange thickness $b$ is small compared with the web height $d$. The Mohr's circle for the stress state at point $A$ would then have appreciable contributions from both $\sigma_x$ and $\tau_{xy}$, and the principal stress might be larger than that at either the outer fibers or the neutral axis.

This problem provides a good review of the governing relations for normal and shear stresses in beams and is also a natural application for symbolic-manipulation computer methods. Using MAPLE™ V software (available from Brooks/Cole Publishing Co., 51 Forest Lodge Road, Pacific Grove, CA 93950; Maple V © 1981–1992 University of Waterloo), we might begin by computing the location of the centroidal axis:

```
> ybar := ((d/2)*c*d) + ((d+(b/2))*a*b)/(c*d + a*b);
```

Here the > symbol is the MAPLE prompt, and the ; is needed by MAPLE to end the command. The maximum shear force and bending moment (present at the wall)

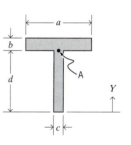

**FIGURE 4.23** Section of T beam.

[2]"Apparent Horizontal Shear Strength of Reinforced Plastics by Short Beam Method," ASTM D2344, American Society for Testing and Materials.

are defined in terms of the distributed load and the beam length as

```
> V := w*L;
> M := -(w*L)*(L/2);
```

For plotting purposes it will be convenient to have a height variable Y measured from the bottom of the section. The relations for normal stress, shear stress, and the first principal stress are functions of Y; these are defined using the MAPLE procedure command:

```
> sigx := proc (Y) -M*(Y-ybar)/Iz end;
> tauxy := proc (Y) V*Q(Y)/(Iz*B(Y)) end;
> sigp1 := proc (Y) (sigx(Y)/2) + sqrt((sigx(Y)/2)^2
> + (tauxy(Y))^2) end;
```

The moment of inertia Iz is computed as

```
> I1 := (a*b^3)/12 + a*b* (d+(b/2)-ybar)^2;
> I2 := (c*d^3)/12 + c*d* ((d/2)-ybar)^2;
> Iz := I1+I2;
```

The beam width B is defined to take the appropriate value depending on whether the variable Y is in the web or the flange:

```
> B:= proc (Y) if Y<d then B:=c else B:=a fi end;
```

The command fi (if spelled backwards) is used to end an if-then loop. The function Q(Y) is defined for the web and the flange separately:

```
> Q:= proc (Y) if Y<d then
>   int( (yy-ybar)*c,yy=Y..d) + int( (yy-ybar)*a,yy=d..(d+b) )
> else
>   int( (yy-ybar)*a,yy=Y.. (d+b) )
> fi end;
```

Here int is the MAPLE command for integration, and yy is used as the dummy height variable. The numerical values of the various parameters are defined as

```
> a:=3;
> b:=1/4;
> c:=1/4;
> d:=3-b;
> L:=8;
> w:=100;
```

Finally, the stresses can be graphed using the MAPLE plot command:

```
> plot({sigx,tauxy,sigp1},Y=0..3,sigx=-500..2500);
```

The resulting plot is shown in Fig. 4.24.

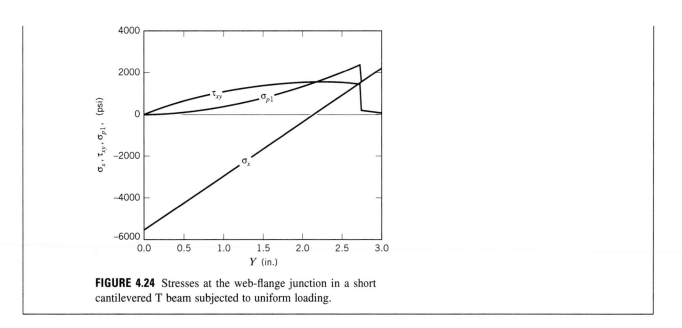

**FIGURE 4.24** Stresses at the web-flange junction in a short cantilevered T beam subjected to uniform loading.

## EXAMPLE 4.7

In the previous example, we were interested in the variation of stress as a function of height in a beam of irregular cross section. Another common design or analysis problem is that of the variation of stress, not only as a function of height but also of distance along the span dimension of the beam. The shear and bending moments $V(x)$ and $M(x)$ vary along this dimension, and so naturally do the stresses $\sigma_x(x, y)$ and $\tau_{xy}(x, y)$, which depend on the moments according to Eqs. 4.20 and 4.25.

Consider a short beam of rectangular cross section subjected to four-point loading as seen in Fig. 4.25. The loading, shear, and bending moment functions are

$$q(x) = P\langle x \rangle_{-1} - P\langle x - a \rangle_{-1} - P\langle x - 2a \rangle_{-1} + P\langle x - 3a \rangle_{-1}$$

$$V(x) = -\int q(x)\, dx = -P\langle x \rangle^0 + P\langle x - a \rangle^0 + P\langle x - 2a \rangle^0 - P\langle x - 3a \rangle^0$$

$$M(x) = -\int V(x)\, dx\, P\langle x \rangle^1 - P\langle x - a \rangle^1 - P\langle x - 2a \rangle^1 + P\langle x - 3a \rangle^1$$

The shear and normal stresses can be determined as functions of $x$ and $y$ directly from these functions, as well as such parameters as the principal stress. Because $\sigma_y$ is zero everywhere, the principal stress is

$$\sigma_{p1} = \frac{\sigma_x}{2} + \sqrt{\left(\frac{\sigma_x}{2}\right)^2 + \tau_{xy}^2}$$

One way to visualize the $x$–$y$ variation of $\sigma_{p1}$ is by means of a three-dimensional surface plot, which can be prepared easily by MAPLE. For the numerical values $P = 100, a = h = 10, b = 3$ we could use the expressions (MAPLE responses removed for brevity):

```
> # use Heaviside for singularity functions
> readlib(Heaviside);
```

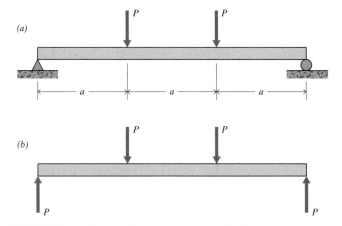

(a)

(b)

**FIGURE 4.25** (*a*) Beam in four-point bending. (*b*) Free-body diagram.

```
> sfn := proc(x,a,n) (x-a)^n * Heaviside(x-a) end;
> # define shear and bending moment functions
> V:=(x)-> -P*sfn(x,0,0)+P*sfn(x,a,0)+P*sfn(x,2*a,0)
> -P*sfn(x,3*a,0);
> M:=(x)-> P*sfn(x,0,1)-P*sfn(x,a,1)-P*sfn(x,2*a,1)
> +P*sfn(x,3*a,1);
> # define shear stress function
> tau:=V(x)*Q/(Iz*b);
> Q:=(b/2)*( (h^2/4) -y^2);
> Iz:=b*h^3/12;
> # define normal stress function
> sig:=M(x)*y/Iz;
> # define principal stress
> sigp:= (sig/2) + sqrt( (sig/2)^2 + tau^2 );
> # define numerical parameters
> P:=100;a:=10;h:=10;b:=3;
> # make plot
> plot3d(sigp,x=0..3*a,y=-h/2 .. h/2);
```

The resulting plot is shown in Fig. 4.26. The dominance of the parabolic shear stress is evident near the beam ends, because here the shear force is at its maximum value

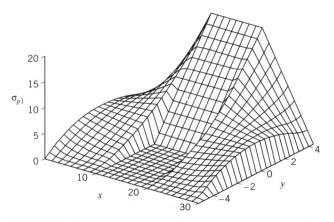

**FIGURE 4.26** Variation of principal stress $\sigma_{p1}$ in four-point bending.

but the bending moment is small (plot the shear and bending moment diagrams to confirm this). In the central part of the beam, where $a < x < 2a$, the shear force vanishes and the principal stress is governed only by the normal stress $\sigma_x$, which varies linearly from the beam's neutral axis. The first principal stress is zero in the compressive lower part of this section, because here the normal stress $\sigma_x$ is negative and the right edge of the Mohr's circle must pass through the zero value of the other normal stress $\sigma_y$. Working through the plot of Fig. 4.26 is a good review of the beam stress formulas.

## 4.5 DISPLACEMENTS

We want to be able to predict the deflection of beams in bending, because many applications have limitations on the amount of deflection that can be tolerated. Another common need for deflection analysis arises from materials testing, in which the transverse deflection induced by a bending load is measured. If we know the relation expected between the load and the deflection, we can "back out" the material properties (specifically the modulus) from the measurement. We will show, for instance, that the deflection at the midpoint of a beam subjected to three-point bending (in which the beam is loaded at its center and simply supported at its edges) is

$$\delta_P = \frac{PL^3}{48EI}$$

where the length $L$ and the moment of inertia $I$ are geometrical parameters. If the ratio of $\delta_P$ to $P$ is measured experimentally, the modulus $E$ can be determined. A stiffness measured this way is called the *flexural modulus*.

There are a number of approaches to the beam deflection problem, and some texts spend a good amount of print on this subject. The following treatment outlines only a few of the more straightforward methods, more with a goal of understanding the general concepts than with developing a lot of facility for doing them manually. In practice, design engineers will usually consult handbook tabulations of deflection formulas as needed, so even before the computer age many of these methods were a bit academic.

### 4.5.1 Multiple Integration

We have earlier seen how two integrations of the loading function $q(x)$ produce first the shear function $V(x)$ and then the moment function $M(x)$:

$$V = -\int q(x)\,dx + c_1 \tag{4.26}$$

$$M = -\int V(x)\,dx + c_2 \tag{4.27}$$

where the constants of integration are evaluated from suitable boundary conditions on $V$ and $M$. If singularity functions are used, the constants are identically zero. From Eq. 4.19, the curvature $v_{,xx}(x)$ is just the moment $M$ divided by the section modulus $EI$. Another two integrations then give

$$v_{,x}(x) = \frac{1}{EI}\int M(x)\,dx + c_3 \tag{4.28}$$

$$v(x) = \int v_{,x}(x)\,dx + c_4 \tag{4.29}$$

where $c_3$ and $c_4$ are determined from boundary conditions on slope or deflection.

## EXAMPLE 4.8

As an illustration of this process, consider the case of three-point bending shown in Fig. 4.27. This geometry is often used in materials testing because it avoids the need to clamp the specimen to the testing apparatus. If the load $P$ is applied at the midpoint, the reaction forces at $A$ and $B$ are equal to half the applied load. The loading function is then

$$q(x) = \frac{P}{2}\langle x\rangle_{-1} - P\left\langle x - \frac{L}{2}\right\rangle_{-1}$$

Integrating according to the above scheme just presented,

$$V(x) = -\frac{P}{2}\langle x\rangle^0 + P\left\langle x - \frac{L}{2}\right\rangle^0$$

$$M(x) = \frac{P}{2}\langle x\rangle^1 - P\left\langle x - \frac{L}{2}\right\rangle^1 \tag{4.30}$$

$$EIv_{,x}(x) = \frac{P}{4}\langle x\rangle^2 - \frac{P}{2}\left\langle x - \frac{L}{2}\right\rangle^2 + c_3$$

From symmetry, the beam has zero slope at the midpoint. Hence $v_{,x} = 0$ for $x = L/2$, so $c_3$ can be found to be $-PL^2/16$. Integrating again,

$$EIv(x) = \frac{P}{12}\langle x\rangle^3 - \frac{P}{6}\left\langle x - \frac{L}{2}\right\rangle^2 - \frac{PL^2 x}{16} + c_4$$

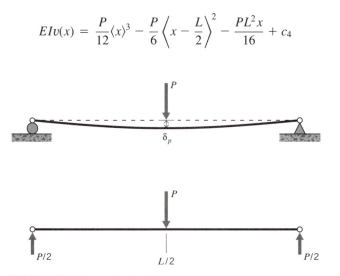

**FIGURE 4.27** Three-point bending.

The deflection is zero at the left end, so $c_4 = 0$. Rearranging, the beam deflection is given by

$$v = \frac{P}{48EI}\left[4x^3 - 3L^2 x - 8\left\langle x - \frac{L}{2} \right\rangle^3\right] \qquad (4.31)$$

The maximum deflection occurs at $x = L/2$, which we can evaluate just before the singularity term activates. Then

$$\boxed{\delta_{max} = \frac{PL^3}{48EI}} \qquad (4.32)$$

This expression is much used in flexural testing and is the example used to begin this section.

Before the loading function $q(x)$ can be written, the reaction forces at the beam supports must be determined. If the beam is statically determinate, as in the foregoing example, this can be done by invoking the equations of static equilibrium. Static determinacy means that only two reaction forces or moments can be present, because we have only a force balance in the direction transverse to the beam axis and one moment equation available. A simply supported beam (one resting on only two supports) or a cantilevered beam are examples of such determinate beams; in the former case there is one reaction force at each support, and in the latter case there is one transverse force and one moment at the clamped end.

Of course, there is no stringent engineering reason to limit the number of beam supports to those sufficient for static equilibrium. Adding "extra" supports will limit deformations and stresses, when will often be worthwhile in spite of the extra construction expense. However, the analysis is now a bit more complicated, because not all of the unknown reactions can be found from the equations of static equilibrium. In these statically indeterminate cases it will be necessary to invoke geometrical constraints to develop enough equations to solve the problem.

This is done by writing the slope and deflection equations, carrying the unknown reaction forces and moments as undetermined parameters. The slopes and deflections are then set to their known values at the supports, and the resulting equations are solved for the unknowns. If, for instance, a beam is resting on three supports, there will be three unknown reaction forces, and we will need a total of five equations: three for the unknown forces and two more for the constants of integration that arise when the slope and deflection equations are written. Two of these equations are given by static equilibrium, and three more are obtained by setting the deflections at the supports to zero. The following example illustrates the procedure, which is straightforward, although tedious if done manually.

## EXAMPLE 4.9

Consider a triply supported beam of length $L = 15$, as shown in Fig. 4.28, carrying a constant uniform load of $w = -10$. There are not sufficient equilibrium equations to determine the reaction forces $R_a$, $R_b$, and $R_c$, so these are left as unknowns while

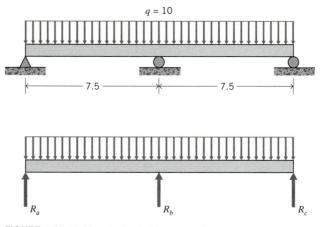

**FIGURE 4.28** Uniformly loaded beam resting on three supports.

multiple integration is used to develop a deflection equation:

$$q(x) = R_a \langle x \rangle_{-1} + R_b \langle x - 7.5 \rangle_{-1} + R_c \langle x - 15 \rangle_{-1} - 10 \langle x \rangle^0$$

$$V(x) = -\int q(x)\,dx = -R_a \langle x \rangle^0 - R_b \langle x - 7.5 \rangle^0 - R_c \langle x - 15 \rangle^0 + 10 \langle x \rangle^1$$

$$M(x) = -\int V(x)\,dx = R_a \langle x \rangle^1 + R_b \langle x - 7.5 \rangle^1 + R_c \langle x - 15 \rangle^1 - \frac{10}{2} \langle x \rangle^2$$

$$EIy'(x) = \int M(x)\,dx = \frac{R_a}{2} \langle x \rangle^2 + \frac{R_b}{2} \langle x - 7.5 \rangle^2 + \frac{R_c}{2} \langle x - 15 \rangle^2 - \frac{10}{6} \langle x \rangle^3 + c_1$$

$$EIy(x) = \int EIy'(x)\,dx = \frac{R_a}{6} \langle x \rangle^3 + \frac{R_b}{6} \langle x - 7.5 \rangle^3 + \frac{R_c}{6} \langle x - 15 \rangle^3 - \frac{10}{24} \langle x \rangle^4$$
$$+ c_1 x + c_2$$

These equations have five unknowns: $R_a$, $R_b$, $R_c$, $c_1$, and $c_2$. These must be obtained from the two equilibrium equations,

$$\sum F_y = 0 = R_a + R_b + R_c - qL$$

$$\sum M_a = 0 = qL\frac{L}{2} - R_b\frac{L}{2} - R_c L$$

and the three known zero displacements at the supports,

$$y(0) = y(L/2) = y(L) = 0$$

Although the process is straightforward, there is a lot of algebra to wade through. As a result, statically indeterminate beams are infamous as being among the very most tedious of problems in introductory mechanics of materials.

Fortunately for students and engineers today, this is the sort of problem symbolic manipulation software is really made for. Follow how easily this example is handled by the MAPLE V package (some of the MAPLE responses have been removed for brevity):

```
> # read the library containing the Heaviside function
> readlib(Heaviside);
> # use the Heaviside function to define singularity
> # functions;
> #  sfn(x,a,n) is same as <x-a>^n
> sfn := proc(x,a,n) (x-a)^n * Heaviside(x-a) end;
> # define the deflection function:
> y := (x)->(Ra/6)*sfn(x,0,3)+(Rb/6)*sfn(x,7.5,3)
> +(Rc/6)*sfn(x,15,3)-(10/24)*sfn(x,0,4)+c1*x+c2;
> # Now define the five constraint equations;
> # first vertical equilibrium:
> eq1 := 0=Ra+Rb+Rc-(10*15);
> # rotational equilibrium:
> eq2 := 0=(10*15*7.5)-Rb*7.5-Rc*15;
> # Now the three zero displacements at the supports:
> eq3 := y(0)=0;
> eq4 := y(7.5)=0;
> eq5 := y(15)=0;
> # set precision; 4 digits is enough:
> Digits:=4;
> # solve the 5 equations for the 5 unknowns:
> solve({eq1,eq2,eq3,eq4,eq5},{Ra,Rb,Rc,c1,c2});
         {c2 = 0, c1 = -87.82, Rb = 93.78, Ra = 28.11, Rc = 28.11}
> # assign the known values for plotting purposes:
> c1:=-87.82;c2:=0;Ra:=28.11;Rb:=93.78;Rc:=28.11;
> # the equation of the deflection curve is:
> y(x);
        3                                    3
  4.686 x  Heaviside(x) + 15.63 (x - 7.5)  Heaviside(x - 7.5)
                      3                        4
   + 4.686 (x - 15)  Heaviside(x - 15) - 5/12 x  Heaviside(x) - 87.82 x
> # plot the deflection curve:
> plot(y(x),x=0..15);
> # The maximum deflection occurs at the quarter points:
> y(15/4);
                              -164.7
```

The plot of the deflection curve is shown in Fig. 4.29.

## 4.5.2 Energy Method

The strain energy in bending as given by Eq. 4.21 can be used to find deflections, and this may be more convenient than successive integration if the deflection at only a single point is desired. Castigliano's theorem gives the deflection congruent to a load $Q$ as

$$\delta_Q = \frac{\partial U}{\partial Q} = \frac{\partial}{\partial Q} \int_L \frac{M^2 \, dx}{2EI}$$

It is usually more convenient to do the differentiation before the integration, because this lowers the order of the expression in the integrand:

$$\delta_Q = \int_L \frac{M}{EI} \frac{\partial M}{\partial Q} \, dx$$

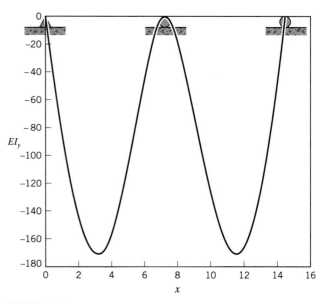

**FIGURE 4.29** Deflection curve for uniformly loaded, triply supported beam (note difference in horizontal and vertical scales).

The shear contribution to bending can be obtained similarly. Knowing the shear stress $\tau = VQ/Ib$ (omitting the $xy$ subscript on $\tau$ for now), the strain energy due to shear $U_s$ can be written

$$U_s = \int_V \frac{\tau^2}{2G}\, dV = \int_L \frac{V^2}{2GI}\left[\int_A \frac{Q^2}{L^2}\, dA\right] dx$$

The integral over the cross-sectional area $A$ is a purely geometrical factor, and we can write

$$U_s = \int_L \frac{V^2 f_s}{2GA}\, dA \tag{4.33}$$

where the $f_s$ is a dimensionless *form factor for shear*, defined as

$$f_s = \frac{A}{I^2}\int_A \frac{Q^2}{b^2}\, dA \tag{4.34}$$

Evaluating $f_s$ for rectangular sections for illustration (see Fig. 4.30), we have in that case

$$A = bh, \qquad I = \frac{bh^3}{12}$$

$$Q = \left[y + \frac{(h/2) - y}{2}\right]\left[b\left(\frac{h}{2} - y\right)\right]$$

$$f_s = \frac{(bh)}{(bh^3/12)^2}\int_{-h/2}^{h/2} \frac{1}{b^2} Q\, dy = \frac{6}{5}$$

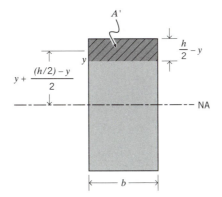

**FIGURE 4.30** Rectangular beam section.

Hence $f_s$ is the same for all rectangular sections, regardless of their particular dimensions. Similarly, it can be shown (see Problem 23 in this chapter) that for solid circular sections $f_s = \frac{10}{9}$, and for hollow circular sections $f_s = 2$.

## EXAMPLE 4.10

If, for instance, we are seeking the deflection under the load $P$ in the three-point bending example given earlier, we can differentiate the moment given in Eq. 4.30 to obtain

$$\frac{\partial M}{\partial P} = \frac{1}{2}\langle x \rangle^1 - \left\langle x - \frac{L}{2} \right\rangle^1$$

Then

$$\delta_P = \frac{1}{EI}\int_L \left(\frac{P}{2}\langle x \rangle^1 - P\left\langle x - \frac{L}{2} \right\rangle^1 \right)\left(\frac{1}{2}\langle x \rangle^1 - \left\langle x - \frac{L}{2} \right\rangle^1 \right) dx$$

Expanding this expression and adjusting the limits of integration to account for singularity functions that have not been activated, we have

$$\delta_P = \frac{P}{EI}\left\{ \int_0^L \frac{x^2}{4}\,dx + \int_{L/2}^L \left[ -x\left(x - \frac{L}{2}\right) + \left(x - \frac{L}{2}\right)^2 \right] dx \right\}$$

$$= -\frac{PL^3}{48EI}$$

as before.

The contribution of shear to the deflection can be found by using $V = P/2$ in the equation for strain energy. For the case of a rectangular beam, with $f_s = \frac{6}{5}$, we have

$$U_s = \frac{(P/2)^2(6/5)}{2GA}L$$

$$\delta_{P,s} = \frac{\partial U_s}{\partial P} = \frac{6PL}{20GA}$$

The shear contribution can be compared with the bending contribution by replacing $A$ with $12I/h^2$ (since $A = bh$ and $I = bh^3/12$). Then the ratio of the shear to bending contributions is

$$\frac{PLh^2/40GI}{PL^3/24EI} = \frac{3h^2E}{5L^2G}$$

Hence the importance of the shear term scales as $(h/L)^2$, that is, quadratically as the span-to-depth ratio.

---

The energy method is often convenient for systems having complicated geometries and combined loading. For slender shafts transmitting axial, torsional, bending, and shearing loads, the strain energy is

$$U = \int_L \left( \frac{P^2}{2EA} + \frac{T^2}{2GJ} + \frac{M^2}{2EI} + \frac{V^2 f_s}{2GA} \right) dx \qquad (4.35)$$

## EXAMPLE 4.11

Consider a cantilevered circular beam as shown in Fig. 4.31 that tapers from radius $r_1$ to $r_2$ over the length $L$. We wish to determine the deflection caused by a force $F$ applied to the free end of the beam, at an angle $\theta$ from the horizontal. Turning to MAPLE to avoid the algebraic tedium, the dimensional parameters needed in Eq. 4.35 are defined as

```
> r := proc (x) r1 + (r2-r1)*(x/L) end;
> A := proc (r) Pi*(r(x))^2 end;
> Iz := proc (r) Pi*(r(x))^4 /4 end;
> Jp := proc (r) Pi*(r(x))^4 /2 end;
```

where $r(x)$ is the radius, $A(r)$ is the section area, $Iz$ is the rectangular moment of inertia, and $Jp$ is the polar moment of inertia. The axial, bending, and shear loads are given in terms of $F$ as

```
> P := F* cos(theta);
> V := F* sin(theta);
> M := proc (x) -F* sin(theta) * x end;
```

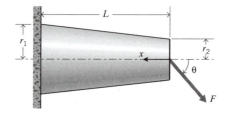

**FIGURE 4.31** Tapered circular beam.

The strain energies corresponding to tension, bending, and shear are

```
> U1 := P^2/(2*E*A(r));
> U2 := (M(x))^2/(2*E*Iz(r));
> U3 := V^2*(10/9)/(2*G*A(r));
> U := int( U1+U2+U3, x=0..L);
```

Finally, the deflection congruent to the load $F$ is obtained by differentiating the total strain energy:

```
> dF := diff(U,F);
```

The result of these manipulations yields

$$\delta_F = \frac{LF\left[12L^2G - 12GL^2\cos^2\theta + 9Gr_2^2\cos^2\theta + 10r_2^2E - 10r_2^2E\cos^2\theta\right]}{9r_1r_2^3E\pi G}$$

This displacement is in the direction of the applied force $F$; the horizontal and vertical deflections of the end of the beam are then

$$\delta_x = \delta_F\cos\theta$$
$$\delta_y = \delta_F\sin\theta$$

## 4.5.3 Superposition

In practice, many beams will be loaded in a complicated manner consisting of several concentrated or distributed loads acting at various locations along the beam. Although these multiple-load cases can be solved from scratch using the methods already described, it is often easier to solve the problem by superposing solutions of simpler problems whose solutions are tabulated. Fig. 4.32 gives an abbreviated collection of deflection formulas[3] that will suffice for many problems. The superposition approach is valid because the governing equations are linear; therefore, the response to a combination of loads is the sum of the responses that would be generated by each separate load acting alone.

## EXAMPLE 4.12

We wish to find the equation of the deflection curve for a simply supported beam loaded in symmetric four-point bending as shown in Fig. 4.33. From Fig. 4.32, the deflection of a beam with a single load at a distance $a$ from the left end is

$$\delta(x) = \frac{Pb}{6LEI}\left[\frac{L}{b}\langle x - a\rangle^3 - x^3 + \left(L^2 - b^2\right)x\right]$$

---

[3] A more exhaustive listing is available in W.C. Young, *Roark's Formulas for Stress and Strain,* McGraw-Hill, New York, 1989.

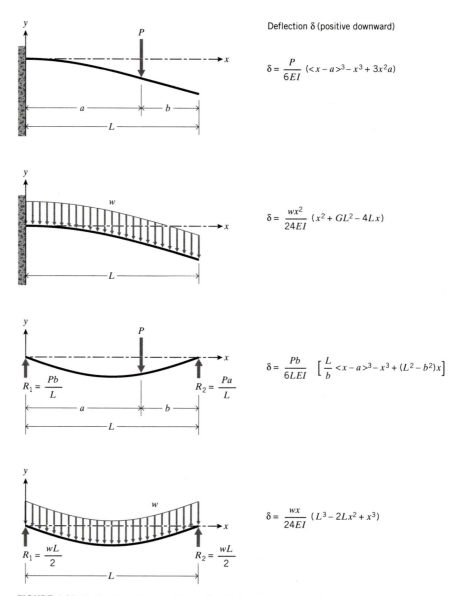

Deflection δ (positive downward)

$$\delta = \frac{P}{6EI}\ (<x-a>^3 - x^3 + 3x^2 a)$$

$$\delta = \frac{wx^2}{24EI}\ (x^2 + GL^2 - 4Lx)$$

$$\delta = \frac{Pb}{6LEI}\ \left[ \frac{L}{b} <x-a>^3 - x^3 + (L^2 - b^2)x \right]$$

$$\delta = \frac{wx}{24EI}\ (L^3 - 2Lx^2 + x^3)$$

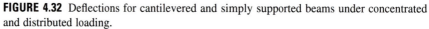

**FIGURE 4.32** Deflections for cantilevered and simply supported beams under concentrated and distributed loading.

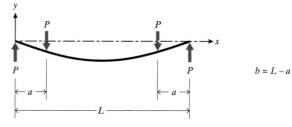

$$b = L - a$$

**FIGURE 4.33** Four-point bending.

Our present problem is just two such loads acting simultaneously, so we have

$$\delta(x) = \frac{Pb}{6LEI}\left[\frac{L}{L-a}\langle x - a\rangle^3 - x^3 + \left(L^2 - (L-a)^2\right)x\right]$$

$$+ \frac{Pb}{6LEI}\left[\frac{L}{a}\langle x - (L-a)\rangle^3 - x^3 + \left(L^2 - a^2\right)x\right]$$

In some cases the designer may not need the entire deflection curve, and super-position of tabulated results for maximum deflection and slope is equally valid.

## 4.6 PLATE BENDING

A *plate* is the two-dimensional analog of a beam in that it can be bent around two axes rather than one. A floor is a natural example, although in practice most floors are plates resting on a number of beams; the beams really carry the more substantial loads, with the flooring having just enough strength and stiffness to permit us to stand between the beams.

The bending of plates is a well-established specialty in applied mechanics, with several texts devoted to it.[4] We will outline only a few introductory aspects of plate bending here, partly to show how beam theory can be generalized to two-dimensional bending. The general approach is the same direct method we have used previously, but now we consider the possibility of three moments $M_x$, $M_y$, and $M_{xy}$, as shown in Fig. 4.34. It will be convenient to normalize these moments by the width of the plate, so they have units of N-m/m, or simply N. Coordinates $x$ and $y$ are the directions in the plane of the plate, and $z$ is customarily taken as positive downward. The deflection in the $z$ direction is termed $w$, also taken as positive downward. The plate is assumed to have a constant thickness $h$.

In seeking relations between the applied moments and the resulting stresses and deformations, the now-familiar direct approach proceeds as follows:

1. *Geometrical statement:* Analogously with the Euler assumption for beams, the *Kirshchoff assumption* takes initially straight vertical lines to remain straight

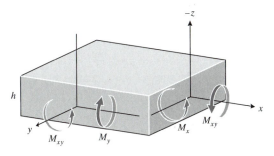

**FIGURE 4.34** Plate bending.

[4]For example, S. Timoshenko and S. Woinowsky-Krieger, *Theory of Plates and Shells,* McGraw-Hill, New York, 1959.

but rotate around the midplane ($z = 0$). The horizontal displacements $u$ and $v$ in the $x$ and $y$ directions can be taken to a reasonable approximation from the rotation angle and distance from midplane:

$$u = -zw_{,x} \tag{4.36}$$

$$v = -zw_{,y} \tag{4.37}$$

These are the two-dimensional analogs to Eq. 4.14 for beam bending.

**2.** *Kinematic equation:* The strains are just the gradients of the displacements; using pseudovector notation this can be written

$$\boldsymbol{\epsilon} = \mathbf{Lu} = \left\{ \begin{array}{c} -zw_{,xx} \\ -zw_{,yy} \\ -z2w_{,xy} \end{array} \right\} = z\boldsymbol{\kappa} \tag{4.38}$$

where $\boldsymbol{\kappa}$ is the pseudovector of second derivatives of the displacement, called the *curvature*:

$$\boldsymbol{\kappa} = \left\{ \begin{array}{c} \kappa_x \\ \kappa_x \\ \kappa_x \end{array} \right\} = \left\{ \begin{array}{c} \partial^2 w/\partial x^2 \\ \partial^2 w/\partial y^2 \\ \partial^2 w/\partial x\,\partial y \end{array} \right\}$$

The component $\kappa_{xy}$ is a *twisting* curvature, stating how the $x$-direction slope changes with $y$ (or equivalently, how the $y$-direction slope changes with $x$). Since $\boldsymbol{\epsilon}$ is a second-rank tensor (written here in pseudovector form) and $z$ is a simple scalar, the curvature $\boldsymbol{\kappa}$ must be a second-rank tensor as well. It therefore transforms similarly to stresses and strains. Mohr's circle constructions can be used with $\boldsymbol{\kappa}$ as well, for instance to find principal axes. Such axes would be $x'$-$y'$ axes in the plane of the plate on which the twisting curvature $\kappa_{xy}$ vanishes and the normal curvatures become a maximum and a minimum. A "saddle" surface has equal and opposite principal curvatures, and the maximum twist will occur at $\pm 45°$ to the principal curvature directions.

**3.** *Constitutive equation:* The stresses are obtained from the strains as

$$\boldsymbol{\sigma} = \mathbf{D}\boldsymbol{\epsilon} = z\mathbf{D}\boldsymbol{\kappa} \tag{4.39}$$

where $\mathbf{D}$ is the stiffness matrix for plane strain (Eqs. 3.49 and 3.50):

$$\mathbf{D} = \frac{E}{1-\nu^2} \begin{bmatrix} 1 & \nu & 0 \\ \nu & 1 & 0 \\ 0 & 0 & \frac{1-\nu}{2} \end{bmatrix}$$

**4.** *Equilibrium equation:* The stresses must balance the applied moments just as for beams, so we have

$$\mathbf{M} = \left\{ \begin{array}{c} M_x \\ M_y \\ M_{xy} \end{array} \right\} = \int_{-h/2}^{h/2} z\boldsymbol{\sigma}\,dz = \int_{-h/2}^{h/2} z^2 \mathbf{D}\boldsymbol{\kappa}\,dz$$

$$\mathbf{M} = \frac{h^3}{12} \mathbf{D}\boldsymbol{\kappa} \tag{4.40}$$

In expanded form this is

$$\begin{Bmatrix} M_x \\ M_y \\ M_{xy} \end{Bmatrix} = \frac{Eh^3}{12(1-\nu^2)} \begin{bmatrix} 1 & \nu & 0 \\ \nu & 1 & 0 \\ 0 & 0 & \frac{1-\nu}{2} \end{bmatrix} \begin{Bmatrix} \kappa_x \\ \kappa_y \\ \kappa_{xy} \end{Bmatrix}$$

The quantity $Eh^3/12(1-\nu^2)$ is the *flexural rigidity* of the plate. Note the strong dependency on plate thickness. Inverting, the plate curvature is given in terms of the applied moments as

$$\boxed{\kappa = \frac{12}{h^3} \mathbf{D}^{-1} \mathbf{M}} \tag{4.41}$$

or, in expanded form,

$$\begin{Bmatrix} \kappa_x \\ \kappa_y \\ \kappa_{xy} \end{Bmatrix} = \frac{12}{Eh^3} \begin{Bmatrix} M_x - \nu M_y \\ M_y - \nu M_x \\ 2(1+\nu)M_z \end{Bmatrix}$$

This equation is the two-dimensional analog to Eq. 4.19 for beam curvature. Substituting this back into Eq. 4.39, the stresses are given in terms of the applied moments as

$$\boxed{\sigma = \frac{12z}{h^3} \mathbf{M}} \tag{4.42}$$

The maximum stresses occur at the outer surface ($z = h/2$):

$$\sigma_{max} = \frac{6}{h^2} \mathbf{M} \tag{4.43}$$

## EXAMPLE 4.13

Consider a wide rectangular plate as shown in Fig. 4.35, supported so as to enable free rotation at its clamped edges, and subjected to a bending moment $M_x$. With $M_y = M_{xy} = 0$, the curvatures are

$$\kappa_x = \frac{12}{Eh^3} M_x$$

This is the same curvature given by Eq. 4.19 for beams, for unit width $b = 1$. The other curvatures are

$$\kappa_y = -\nu \frac{12}{Eh^3} M_x$$

$$\kappa_{xy} = 0$$

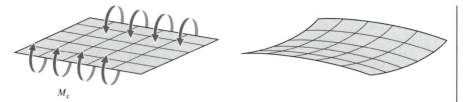

**FIGURE 4.35** Cylindrical bending of a plate.

Hence the application of a couple along one edge of the plate produces an anticlastic curvature, in which the plate develops a reverse curvature along the axis perpendicular to the applied moment.

If on the other hand the plate is clamped to prevent curvature along the $y$-axis, the bending is then constrained to be cylindrical. The edges will then have to supply a moment $M_y$ to prevent the anticlastic curvature; this can be calculated as

$$\kappa_y = 0 = \frac{12}{Eh^3}(M_y - \nu M_x) \rightarrow M_y = \nu M_x$$

The $x$-curvature is then

$$\kappa_x = \frac{12}{Eh^3}(M_x - \nu M_y) = \frac{12(1 - \nu^2)}{Eh^3}M_x$$

The theory just outlined must be expanded considerably before most typical engineering plate problems can be treated. In general, a lateral load $p$ is imposed on each unit area in addition to bending moments applied at the boundaries, and the internal bending moment $\mathbf{M}(x, y)$ needed in Eq. 4.42 is not known. When the lateral load $p$ is included, a fourth-order differential equation for the transverse deflection $w$ can be developed; this can be written

$$\frac{\partial^4 w}{\partial x^4} + 2\frac{\partial^4 w}{\partial x^2 \, \partial y^2} + \frac{\partial^4 w}{\partial y^4} = \frac{p}{D} \tag{4.44}$$

where $D = Eh^3/[12(1 - \nu^2)]$ is the plate flexural rigidity introduced earlier.

Solutions to plate problems consist of finding solutions to Eq. 4.44 that also satisfy the appropriate boundary conditions on deflection and load. Once $w(x, y)$ is known, the curvatures $\kappa$ can be obtained by differentiation, and from these the strains $\epsilon$ and therefore the stresses $\sigma$ can be found. These solutions are often given in terms of series approximations and are developed in texts dedicated to this important subject. Also, compendia of stress solutions such as that of Roark[5] provide solutions for a number of plate problems. But these solutions move beyond the scope of this text, and with the exception of the extension of the simple theory to composite laminates in the next section, we will not treat plate theory further here.

[5] W.C. Young, *Roark's Formulas for Stress and Strain*, McGraw-Hill, New York, 1989.

## 4.7 LAMINATED COMPOSITE PLATES

One of the most common forms of fiber-reinforced composite materials is the cross-plied laminate, in which the fabricator *lays up* a sequence of unidirectionally reinforced *plies*, each being a thin (approximately 0.2 mm) sheet of collimated fibers impregnated with an uncured epoxy or other thermosetting polymer matrix material. The orientation of each ply is arbitrary, and the layup sequence is tailored to achieve the properties desired of the laminate. In this section we outline how such laminates are designed and analyzed.

We follow the scheme used above for the bending of homogeneous plates, but with two differences: We now allow for the presence of in-plane tractions in addition to bending moments, and we must include the varying stiffness of each ply in the analysis. The first of these is accomplished simply be adding the midplane strains $\boldsymbol{\epsilon}^0$ to Eq. 4.38:

$$\boldsymbol{\epsilon} = \boldsymbol{\epsilon}^0 + z\boldsymbol{\kappa} \tag{4.45}$$

The stresses relative to the $x$-$y$ axes are now determined from the strains, and this must take the second complication into consideration. Each ply will in general have a different stiffness, depending on its own properties and also its orientation with respect to the $x$-$y$ axes. This is accounted for by computing the transformed stiffness matrix $\overline{\mathbf{D}}$ as described in Chapter 3 (Eq. 3.56). Recall that the ply stiffnesses given by Eq. 3.55 are those along the fiber and transverse directions of that particular ply. The properties of each ply must be transformed to common $x$-$y$ axes, chosen arbitrarily for the entire laminate. The stresses at any vertical position are then

$$\boldsymbol{\sigma} = \overline{\mathbf{D}}\boldsymbol{\epsilon} = \overline{\mathbf{D}}\boldsymbol{\epsilon}^0 + z\overline{\mathbf{D}}\boldsymbol{\kappa} \tag{4.46}$$

where here $\overline{\mathbf{D}}$ is the transformed stiffness of the ply at the position at which the stresses are being computed.

Each of these ply stresses must add to balance the traction per unit width $\mathbf{N}$:

$$\mathbf{N} = \int_{-h/2}^{+h/2} \boldsymbol{\sigma} \, dz \tag{4.47}$$

where $\mathbf{N}$ is the column vector

$$\mathbf{N} = \left\{ \begin{array}{c} N_x \\ N_y \\ N_{xy} \end{array} \right\} \tag{4.48}$$

The integral in Eq. 4.47 can be performed by summing over the individual plies:

$$\mathbf{N} = \sum_{i=1}^{n} \int_{z_i}^{z_{i+1}} \boldsymbol{\sigma}_i \, dz \tag{4.49}$$

where $\boldsymbol{\sigma}_i$ is the stress in the $i$th ply and $z_i$ is the distance from the laminate midplane to the bottom of the $i$th ply. Using Eq. 4.46 to write the stresses in terms of the midplane strains and curvatures

$$N = \sum_{i=1}^{n} \left( \int_{z_i}^{z_{i+1}} \overline{D}\epsilon^0 \, dz + \int_{z_i}^{z_{i+1}} \overline{D}\kappa z \, dz \right) \tag{4.50}$$

The curvature $\kappa$ and midplane strain $\epsilon^0$ are constant throughout $z$, and the transformed stiffness $\overline{D}$ does not change within a given ply. Removing these quantities from within the integrals, we have

$$N = \sum_{i=1}^{n} \left( \overline{D}\epsilon^0 \int_{z_i}^{z_{i+1}} dz + \overline{D}\kappa \int_{z_i}^{z_{i+1}} z \, dz \right) \tag{4.51}$$

After evaluation of the integrals, this expression can be written in the compact form

$$N = \mathcal{A}\epsilon^0 + \mathcal{B}\kappa \tag{4.52}$$

where $\mathcal{A}$ is an *extensional* stiffness matrix, defined as

$$\mathcal{A} = \sum_{i=1}^{n} \overline{D}(z_{i+1} - z_i) \tag{4.53}$$

and $\mathcal{B}$ is a *coupling* stiffness matrix, defined as

$$\mathcal{B} = \frac{1}{2} \sum_{i=1}^{n} \overline{D}(z_{i+1}^2 - z_i^2) \tag{4.54}$$

The rationale for the names *extensional* and *coupling* is suggested by Eq. 4.52. The $\mathcal{A}$ matrix gives the influence of an extensional midplane strain $\epsilon^0$ on the in-plane traction $N$, and the $\mathcal{B}$ matrix gives the contribution of a curvature $\kappa$ to the traction. It may not be obvious why bending the plate will require an in-plane traction or, conversely, why pulling the plate in its plane will cause it to bend. However, visualize the plate containing plies all of the same stiffness, except for some very-low-modulus plies somewhere above its midplane. When the plate is pulled, the more compliant plies above the midplane will tend to stretch more than the stiffer plies below the midplane. The top half of the laminate stretches more than the bottom half, so the plate takes on a concave-downward curvature.

Similarly, the moment resultants per unit width must be balanced by the moments contributed by the internal stresses:

$$M = \int_{-h/2}^{+h/2} \sigma z \, dz = \mathcal{B}\epsilon^0 + \mathcal{D}\kappa \tag{4.55}$$

where $\mathcal{D}$ is a *bending* stiffness matrix, defined as

$$\mathcal{D} = \frac{1}{3} \sum_{i=1}^{n} \overline{D}(z_{i+1}^3 - z_i^3) \tag{4.56}$$

The complete set of relations between applied forces and moments, and the result-ing midplane strains and curvatures, can be summarized as a single matrix equation:

$$\left\{ \begin{array}{c} \mathbf{N} \\ \mathbf{M} \end{array} \right\} = \left[ \begin{array}{cc} \mathcal{A} & \mathcal{B} \\ \mathcal{B} & \mathcal{D} \end{array} \right] \left\{ \begin{array}{c} \boldsymbol{\epsilon}^0 \\ \boldsymbol{\kappa} \end{array} \right\} \qquad (4.57)$$

The presence of nonzero elements in the coupling matrix $\mathcal{B}$ indicates that the ap-plication of an in-plane traction will lead to a curvature or warping of the plate or that an applied bending moment will also generate an extensional strain. These ef-fects are usually undesirable. However, they can be avoided by making the lami-nate *symmetric* about the midplane, as examination of Eq. 4.54 can reveal. (In some cases, this extension-curvature coupling can be used as an interesting design feature. For instance, it is possible to design a composite propeller blade whose angle of at-tack changes automatically with its rotational speed: Increased speed increases the in-plane centripetal loading, which induces a twist into the blade.)

The foregoing relations provide a straightforward (although tedious, unless a computer is used) means of determining stresses and displacements in laminated com-posites subjected to in-plane traction or bending loads:

1. For each ply in the stacking sequence, obtain by measurement or microme-chanical estimation the four independent anisotropic parameters appearing in Eq. 3.55: ($E_1$, $E_2$, $\nu_{12}$, and $G_{12}$).

2. Using Eq. 3.56, transform the compliance matrix from the principal material directions to some convenient reference axes that will be used for each ply in the laminate.

3. Invert the transformed compliance matrix to obtain the transformed (relative to $x$-$y$ axes) stiffness matrix $\overline{\mathbf{D}}$.

4. Add each ply's contribution to the $\mathcal{A}$, $\mathcal{B}$, and $\mathcal{D}$ matrices as prescribed by Eqs. 4.53, 4.54, and 4.56.

5. Input the prescribed tractions $\mathbf{N}$ and bending moments $\mathbf{M}$, and form the system equations given by Eq. 4.57.

6. Solve the resulting system for the unknown values of in-plane strain $\boldsymbol{\epsilon}^0$ and curvature $\boldsymbol{\kappa}$.

7. Use Eq. 4.46 to determine the ply stresses for each ply in the laminate in terms of $\boldsymbol{\epsilon}^0$, $\boldsymbol{\kappa}$, and $z$. These will be the stresses relative to the $x$-$y$ axes.

8. Use Eq. 3.27 to transform the $x$-$y$ stresses back to the principal material axes (parallel and transverse to the fibers).

9. If desired, the individual ply stresses can be used in a suitable failure criterion to assess the likelihood of that ply failing. The *Tsai-Hill* criterion is popularly used for this purpose:

$$\left( \frac{\sigma_1}{\hat{\sigma}_1} \right)^2 - \frac{\sigma_1 \sigma_2}{\hat{\sigma}_1^2} + \left( \frac{\sigma_2}{\hat{\sigma}_2} \right)^2 + \left( \frac{\tau_{12}}{\hat{\tau}_{12}} \right)^2 = 1 \qquad (4.58)$$

Here $\hat{\sigma}_1$ and $\hat{\sigma}_2$ are the ply tensile strengths parallel to and along the fiber direction, and $\hat{\tau}_{12}$ is the intralaminar ply strength. This criterion predicts failure whenever the left-hand side of Eq. 4.58 equals or exceeds unity.

EXAMPLE 4.14

We wish to use fiber-reinforced composites to design a recreational downhill ski, and as a preliminary step we assume the weight of a 150-lb (668 N) skier is distributed uniformly over the snow by a ski 180 cm in length and 7.5 cm in width as shown in Fig. 4.36. This distribution is highly approximate, but suitable for getting the design started. The bending moment at the midpoint is then half the total skier weight times one-quarter of the ski length. Dividing by the ski width to obtain moment per unit width, we have

$$M_x = \frac{(668/2)(1.8/4)}{0.075} = 2090 \text{ N}$$

We will carry out an initial analysis considering only this bending moment, although in reality the shear resultant due to the transverse loading can be expected to produce stresses on the same order as the bending. As a preliminary design choice, we try layers of glass/epoxy on the top and bottom surfaces, arranged in a 0/45/0/−45/0 symmetric sequence, separated by a pine core (Fig. 4.37). These parameters are entered into the `plate` analysis code (described in Appendix G and included on the diskette accompanying the text) in the following exercise:

```
> plate
assign properties for lamina type 1...

  enter modulus in fiber direction...
    (enter -1 to stop): 55e9
  enter modulus in transverse direction: 16e9
  enter principal Poisson ratio: .26
  enter shear modulus: 7.6e9
  enter ply thickness: .15e-3

assign properties for lamina type 2...

  enter modulus in fiber direction...
    (enter -1 to stop): 11.5e9
  enter modulus in transverse direction: .7e9
  enter principal Poisson ratio: .3
  enter shear modulus: .7e9
  enter ply thickness: 1.25e-2
```

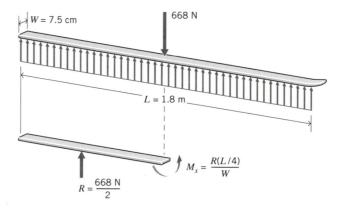

**FIGURE 4.36** Baseline ski design.

**FIGURE 4.37** Many modern skis use composite construction, with a lightweight core and stiff fiber-reinforced face sheets. (Marshall/The Stock Market.)

```
assign properties for lamina type 3...

  enter modulus in fiber direction...
    (enter -1 to stop): -1
  define layup sequence, starting at bottom...
    (use negative material set number to stop)

  enter material set number for ply number 1: 1
   enter ply angle: 0

  enter material set number for ply number 2: 1
   enter ply angle: 45

  enter material set number for ply number 3: 1
   enter ply angle: 0

  enter material set number for ply number 4: 1
   enter ply angle: -45

  enter material set number for ply number 5: 1
   enter ply angle: 0

  enter material set number for ply number 6: 2
   enter ply angle: 0

  enter material set number for ply number 7: 1
   enter ply angle: 0

  enter material set number for ply number 8: 1
   enter ply angle: -45

  enter material set number for ply number 9: 1
   enter ply angle: 0
```

```
           enter material set number for ply number 10: 1
             enter ply angle: 45

           enter material set number for ply number 11: 1
             enter ply angle: 0

           enter material set number for ply number 12: -1

           laminate stiffness matrix:

   0.2117e+09  0.1404e+08  0.0000e+00   -0.4922e+00 -0.1025e-01 -0.9766e-03
   0.1404e+08  0.4018e+08  0.0000e+00   -0.9277e-02 -0.1094e+00  0.0000e+00
   0.0000e+00  0.0000e+00  0.2518e+08   -0.9766e-03  0.0000e+00 -0.3467e-01

  -0.4922e+00 -0.1025e-01 -0.9766e-03    0.4835e+04  0.5350e+03  0.1186e+02
  -0.9277e-02 -0.1094e+00  0.0000e+00    0.5350e+03  0.1493e+04  0.1186e+02
  -0.9766e-03  0.0000e+00 -0.3467e-01    0.1186e+02  0.1186e+02  0.8357e+03

           laminate compliance matrix:

   0.4835e-08 -0.1689e-08  0.2318e-22    0.5196e-12 -0.2766e-12  0.2202e-14
  -0.1689e-08  0.2548e-07 -0.4792e-22   -0.3417e-12  0.1977e-11 -0.2518e-13
   0.2324e-22 -0.4780e-22  0.3971e-07    0.5652e-14 -0.1511e-13  0.1648e-11

   0.5196e-12 -0.3365e-12  0.5652e-14    0.2154e-03 -0.7714e-04 -0.1962e-05
  -0.2798e-12  0.1976e-11 -0.1511e-13   -0.7714e-04  0.6973e-03 -0.8801e-05
   0.2246e-14 -0.2524e-13  0.1648e-11   -0.1962e-05 -0.8801e-05  0.1197e-02

           input tractions and moments...

             Nx: 0
             Ny: 0
            Nxy: 0
             Mx: 2100
             My: 0
            Mxy: 0

           midplane strains:

            eps-xx = 0.1091e-08
            eps-yy = -0.7176e-09
            eps-xy = 0.1187e-10

           rotations:

            kappa-xx = 0.4523e+00
            kappa-yy= -0.1620e+00
            kappa-xy = -0.4120e-02

           stresses:
            ply      sigma-1      sigma-2      sigma-12

              1 -0.1710e+09   0.5017e+07   0.2168e+06
              2 -0.2039e+08  -0.5862e+08  -0.3163e+08
              3 -0.1636e+09   0.4800e+07   0.2074e+06
              4 -0.1917e+08  -0.5741e+08   0.3023e+08
              5 -0.1562e+09   0.4582e+07   0.1980e+06
              6  0.1609e+02  -0.2876e+00   0.6294e-02
```

```
 7   0.1562e+09  -0.4582e+07  -0.1980e+06
 8   0.1917e+08   0.5741e+08  -0.3023e+08
 9   0.1636e+09  -0.4800e+07  -0.2074e+06
10   0.2039e+08   0.5862e+08   0.3163e+08
11   0.1710e+09  -0.5017e+07  -0.2168e+06
```

The resulting ply stresses (given here in Pa) are well below the failure levels of the materials, and the curvature doesn't seem excessive. This might therefore be a reasonable baseline starting point in the design. Of course, there is much left to be done, including the construction and resulting stresses at other points along the ski length; treatment of sideward and combined bending moments during various skiing maneuvers; and consideration of weight, manufacturability, and cost.[6] And perhaps every bit as important as the technical aspects of design, it's vital that a company starting up in the ski business be aware of the ins and outs of this very competitive market. One very competent high-tech composites company found they just couldn't make money in skis because they "didn't know the territory," but a smaller company of dyed-in-the wool skiers who bought the technology did all right.

## EXAMPLE 4.15

A *composite beam* is one composed of two or more materials having different moduli of elasticity. These beams can be analyzed using an approach similar to that of Section 4.2, but adjusting Eq. 4.16 to let $E$ be a function of $y$. This produces explicit but somewhat tedious formulae for stress in terms of the various moduli and geometries of the component materials.[7]

An alternate approach is to use the `plate` code, which works for fully isotropic as well as transversely isotropic materials. If the same value is input for both "fiber" and "transverse" directions, the code assumes isotropy and computes the shear modulus $G$ from $E$ and $\nu$ rather than prompting the user for it.

Consider a laminate designed for use as dent-resistant automotive body panels, having an inner core, 0.8 mm thick, of polypropylene ($E = 2.5$ GPa, $\nu = 0.4$), clad by two 0.2-mm-thick sheets of aluminum ($E = 70$ GPa, $\nu = 0.3$). In this application, we wish to calculate the overall flexural modulus of the panel, such as would be determined by performing a beam flexure test on a simple noncomposite beam. Equation 4.19 gives the effective modulus $E$ in terms of the curvature as $\kappa_{xx} = M/EI$, so $E$ is determined if the curvature $\kappa_{xx}$ is known for a given bending moment $M$. We obtain this by "building" the laminate in `plate`, and imposing only a unit moment $M_x = 1$ N-m/m:

```
1> plate
 assign properties for lamina type 1...

 enter modulus in fiber direction...
  (enter -1 to stop): 70e9
 enter modulus in transverse direction: 70e9
 enter principal Poisson ratio: .3
```

---

[6]H.C. Boehm. "Influence of Composite Materials on Alpine Ski Design," *SAMPE J.*, Sep./Oct. 1979, pp. 14–20.

[7]Cf. A.C. Ugural and S. K. Fenster, *Advanced Strength and Applied Elasticity,* 2d ed., Elsevier, 1987, p. 156.

```
                enter ply thickness: .2e-3
                assign properties for lamina type 2...

                enter modulus in fiber direction...
                 (enter -1 to stop): 2.5e9
                enter modulus in transverse direction: 2.5e9
                enter principal Poisson ratio: .4
                enter ply thickness: .8e-3
                assign properties for lamina type 3...

                enter modulus in fiber direction...
                 (enter -1 to stop): -1
                define layup sequence, starting at bottom...
                 (use negative material set number to stop)

                enter material set number for ply number 1: 1
                enter ply angle: 0

                enter material set number for ply number 2: 2
                enter ply angle: 0

                enter material set number for ply number 3: 1
                enter ply angle: 0

                enter material set number for ply number 4: -1

                laminate stiffness matrix:

  0.3315e+08  0.1018e+08  0.0000e+00   0.0000e+00   0.0000e+00   0.0000e+00
  0.1018e+08  0.3315e+08  0.0000e+00   0.0000e+00   0.0000e+00   0.0000e+00
  0.0000e+00  0.0000e+00  0.1148e+08   0.0000e+00   0.0000e+00   0.0000e+00

  0.0000e+00  0.0000e+00  0.0000e+00   0.7922e+01   0.2389e+01   0.0000e+00
  0.0000e+00  0.0000e+00  0.0000e+00   0.2389e+01   0.7922e+01   0.0000e+00
  0.0000e+00  0.0000e+00  0.0000e+00   0.0000e+00   0.0000e+00   0.2766e+01

                laminate compliance matrix:

  0.3331e-07 -0.1023e-07  0.0000e+00   0.0000e+00   0.0000e+00   0.0000e+00
 -0.1023e-07  0.3331e-07  0.0000e+00   0.0000e+00   0.0000e+00   0.0000e+00
  0.0000e+00  0.0000e+00  0.8708e-07   0.0000e+00   0.0000e+00   0.0000e+00

  0.0000e+00  0.0000e+00  0.0000e+00   0.1389e+00  -0.4188e-01   0.0000e+00
  0.0000e+00  0.0000e+00  0.0000e+00  -0.4188e-01   0.1389e+00   0.0000e+00
  0.0000e+00  0.0000e+00  0.0000e+00   0.0000e+00   0.0000e+00   0.3615e+00

                input tractions and moments...

                    Nx: 0
                    Ny: 0
                    Nxy: 0
                    Mx: 1  <---- Applied bending moment
                    My: 0
                    Mxy: 0

                midplane strains:

                    eps-xx = 0.0000e+00
                    eps-yy = 0.0000e+00
                    eps-xy = 0.0000e+00
```

172

rotations:

```
   kappa-xx = 0.1389e+00 <--- resulting curvature
   kappa-yy= -0.4188e-01 <--- anticlastic curvature
   kappa-xy = 0.0000e+00
```

stresses:

```
  ply      sigma-1      sigma-2     sigma-12

    1 -0.4858e+07   0.8561e+04   0.0000e+00
    2  0.0000e+00   0.0000e+00   0.0000e+00
    3  0.4858e+07  -0.8561e+04   0.0000e+00
```

The $M_x$ here is the moment per unit width, so we calculate a moment of inertia for a unit width of beam ($b = 1$) as $I = bh^3/12 = (1)(1.2 \times 10^{-3})^3/12$, where $h = 0.2 + 0.8 + 0.2 = 1.2$ mm is the beam's height. Solving for the effective modulus $E$, we obtain

$$E = \frac{M}{\kappa_{xx} I} = \frac{1}{(0.1389)\frac{(1.2 \times 10^{-3})^3}{12}} = 50 \times 10^9 \text{ Pa}$$

Note that a negative curvature $\kappa_{yy}$ develops in the $y$ direction; this is the anticlastic curvature due to transverse Poisson contraction in the tensile portion of the beam and expansion in the compressive portion.

Note also that the flexural modulus is higher than a simple rule of mixtures analysis for tensile loading, such as led to Eq. 1.30, predicts:

$$E = V_1 E_1 + V_2 E_2 = \left(\frac{0.2 + 0.2}{0.2 + 0.2 + 0.8}\right) 70 + \left(\frac{0.8}{0.2 + 0.2 + 0.8}\right) 2.5 = 25.9 \text{ GPa}$$

Here the beam is substantially stiffer in bending than in tension, because in bending the outer surfaces have more influence than the inner regions, and the designers have naturally placed the stiffer material on the outside.

## 4.8 PROBLEMS

**1.** (a)–(c) Locate the magnitude and position of the force equivalent to the loading distributions shown in Fig. P.4.1.

**2.** (a)–(c) Determine the reaction forces at the supports of the cases in Problem 1.

**3.** (a)–(h) Sketch the shear and bending moment diagrams for the load cases shown in Fig. P.4.3.

**4.** (a)–(h) Write singularity-function expressions for the shear and bending moment distributions for the cases in Problem 3.

**5.** (a)–(h) Use MAPLE V (or other) software to plot the shear and bending moment distributions for the cases in Problem 3, using the values (as needed) $L = 25$ in, $a = 15$ in, $w = 10$ lb/in, $P = 150$ lb.

**6.** The transverse deflection of a beam under an axial load $P$, as shown in Fig. P.4.6, is taken to be $\delta(y) = \delta_0 \sin(y\pi/L)$. Determine the bending moment $M(y)$ along the beam.

**7.** Determine the bending moment $M(\theta)$ along the circular curved beam shown in Fig. P.4.7.

**8.** Derive the composite area theorem for determining the centroid of a compound area:

$$\bar{y} = \frac{\sum_i A_i \bar{y}_i}{\sum_i A_i}$$

**9.** (a)–(d) Locate the centroids of the areas shown in Fig. P.4.9.

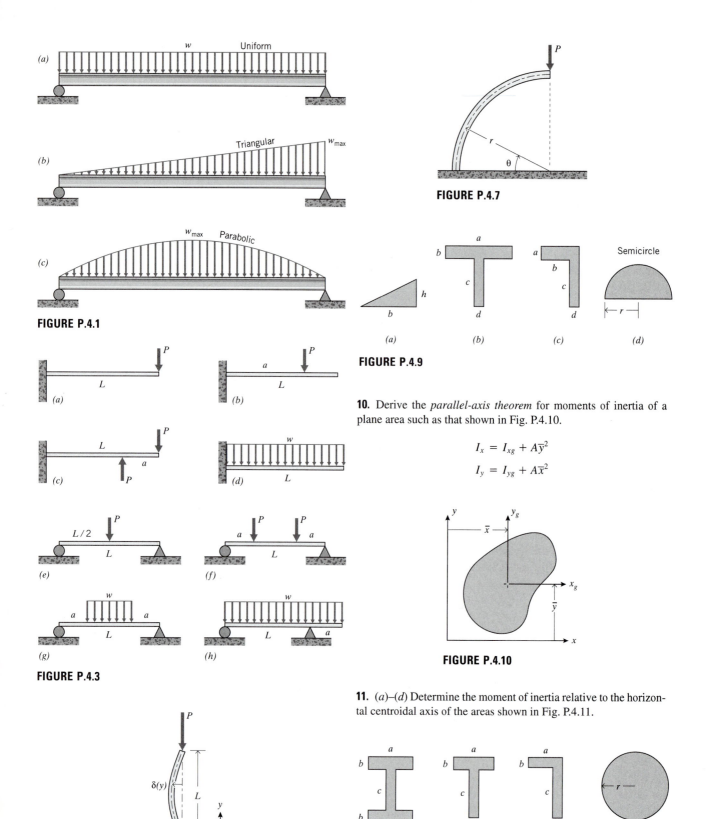

FIGURE P.4.1

FIGURE P.4.3

FIGURE P.4.6

FIGURE P.4.7

FIGURE P.4.9

**10.** Derive the *parallel-axis theorem* for moments of inertia of a plane area such as that shown in Fig. P.4.10.

$$I_x = I_{xg} + A\bar{y}^2$$

$$I_y = I_{yg} + A\bar{x}^2$$

FIGURE P.4.10

**11.** (*a*)–(*d*) Determine the moment of inertia relative to the horizontal centroidal axis of the areas shown in Fig. P.4.11.

FIGURE P.4.11

**12.** Show that the moment of inertia transforms with respect to axis rotations exactly as does the stress:

$$I_{x'} = I_x \cos^2 \theta + I_y \sin^2 \theta - 2I_{xy} \sin \theta \cos \theta$$

where $I_x$ and $I_y$ are the moments of inertia relative to the $x$ and $y$ axes, respectively, and $I_{xy}$ is the *product of inertia,* defined as

$$I_{xy} = \int_A xy \, dA$$

**13.** (*a*)–(*h*) Determine the maximum normal stress $\sigma_x$ in the beams of Problem 3, using the values (as needed) $L = 25$ in, $a = 15$ in, $w = 10$ lb/in, $P = 150$ lb. Assume a rectangular cross section of width $b = 1$ in and height $h = 2$ in.

**14.** Justify the following statement in ASTM test D790, "Standard Test Methods for Flexural Properties of Unreinforced and Reinforced Plastics and Electrical Insulating Materials":

When a beam of homogeneous, elastic material is tested in flexure as a simple beam supported at two points and loaded at the midpoint, the maximum stress in the outer fibers occurs at midspan. This stress may be calculated for any point on the load-deflection curve by the following equation:

$$S = 3PL/2bd^2$$

where $S$ = stress in the outer fibers at midspan, MPa; $P$ = load at a given point on the load-deflection curve; $L$ = support span, mm; $b$ = width of beam tested, mm; and $d$ = depth of beam tested, mm.

**15.** Justify the following statement in ASTM test D790, "Standard Test Methods for Flexural Properties of Unreinforced and Reinforced Plastics and Electrical Insulating Materials":

The tangent modulus of elasticity, often called the "modulus of elasticity," is the ratio, within the elastic limit of stress to corresponding strain and shall be expressed in megapascals. It is calculated by drawing a tangent to the steepest initial straight-line portion of the load-deflection curve and using [the expression]

$$E_b = L^3 m/4bd^3$$

where $E_b$ = modulus of elasticity in bending, MPa; $L$ = support span, mm; $d$ = depth of beam tested, mm; and $m$ = slope of the tangent to the initial straight-line portion of the load-deflection curve, N/mm of deflection.

**16.** A rectangular beam is to be milled from circular stock as shown in Fig. P.4.16. What should be the ratio of height to width ($b/h$) to minimize the stresses when the beam is put into bending?

**17.** (*a*)–(*h*) Determine the maximum shear $\tau_{xy}$ in the beams of Problem 3, using the values (as needed) $L = 25$ in, $a = 15$ in, $w = 10$ lb/in, $P = 150$ lb. Assume a rectangular cross section of width $b = 1$ in and height $h = 2$ in.

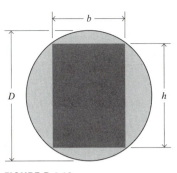

**FIGURE P.4.16**

**18.** Show that the ratio of maximum shearing stress to maximum normal stress in a beam subjected to three-point bending as shown in Fig. P.4.18 is

$$\frac{\tau}{\sigma} = \frac{h}{2L}$$

Hence, the importance of shear stress increases as the beam becomes shorter in comparison with its height.

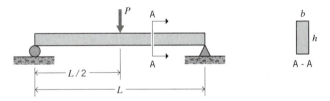

**FIGURE P.4.18**

**19.** Read ASTM test D4475, "Standard Test Method for Apparent Horizontal Shear Strength of Pultruded Reinforced Plastic Rods by the Short-Beam Method," and justify the expression given there for the apparent shear strength:

$$S = 0.849P/d^2$$

where $S$ = apparent shear strength, N/m² (or psi); $P$ = breaking load, N (or lbf); and $d$ = diameter of specimen, m (or in.).

**20.** For the T beam shown in Fig. P.4.20, with dimensions $L = 3$, $a = 0.05, b = 0.005, c = 0.005, d = 0.7$ (all in m) and a loading distribution of $w = 5000$ N/m, determine the principal and maximum shearing stresses at point $A$.

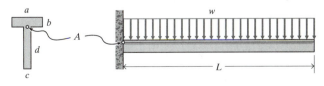

**FIGURE P.4.20**

**21.** (a)–(h) Write expressions for the slope and deflection curves of the beams in Problem 3.

**22.** (a)–(h) Use MAPLE V (or other) software to plot the slope and deflection curves for the beams in Problem 3, using the values (as needed) $L = 25$ in, $a = 15$ in, $w = 10$ lb/in, $P = 150$ lb.

**23.** Show that the shape factor for shear for a circular cross section is

$$f_s = \frac{A}{I^2} \int_A \frac{Q^2}{b^2} dA = \frac{10}{9}$$

**24.** (a)–(b) Determine the deflection curves for the beams shown in Fig. P.4.24. Plot these curves for the values as needed: $L = 25$ in, $a = 15$ in, $w = 10$ lb/in, $P = 150$ lb.

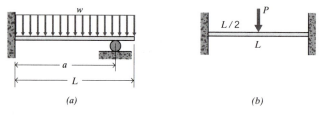

**FIGURE P.4.24**

**25.** Determine the maximum normal stress in a cantilevered beam of circular cross section, shown in Fig. P.4.25, whose radius varies linearly from $4r_0$ to $r_0$ in a distance $L$, and that is loaded with a force $P$ at the free end.

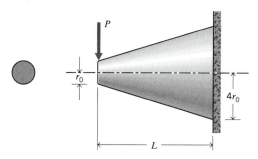

**FIGURE P.4.25**

**26.**

(a) Determine the deflection of the coil spring shown in Fig. P.4.26a under the influence of an axial force $F$, including the contribution of bending, direct shear, and torsional shear effects. Using $r = 1$ mm and $R = 10$ mm, compute the relative magnitudes of the three contributions.

(b) Repeat the solution in part (a), but take the axial load to be placed at the outer radius of the coil as shown in Fig. 4.26b.

**27.** A carbon steel column has a length $L = 1$ m and a circular cross section of diameter $d = 20$ mm. Determine the critical buckling load $P_c$ for the cases shown in Fig. P.4.27: (a) both ends pinned, (b) one end pinned, (c) both ends pinned but supported laterally at the midpoint.

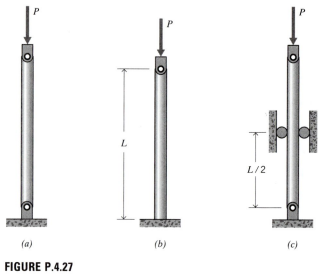

**FIGURE P.4.27**

**28.** A carbon steel column has a length $L = 1$ m and a circular cross section. Determine the diameter $d$ at which the column has an equal probability of buckling or yielding in compression.

**29.**

(a) For columns of circular cross section with length $L$ and diameter $d$, develop an expression for the critical buckling load per unit of column weight $P_c/W$.

(b) Rank the materials in Appendix A (given in the file **props.csv**) in terms of $P_c/W$.

**30.**

(a) In Problem 29 $d$ was specified and the critical buckling load $P_c$ was a dependent variable. Now consider the case where $P_c$ is specified, and $d$ is to be determined for a given length $L$ and modulus $E$. Develop an expression for the buckling efficiency $P_c/W$ for these conditions.

(b) Rank the materials in Appendix A (given in the file **props.csv**) in terms of $P_c/W$ using this expression.

**31.** (a)–(c) Use the method of superposition to write expressions for the deflection curve $\delta(x)$ for the cases shown in Fig. P.4.31.

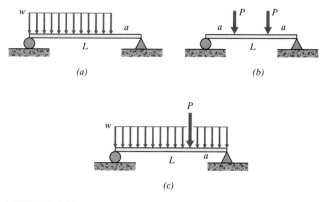

**FIGURE P.4.31**

**32.** The deflection surface of a plate is given by $w = 2x^2 - y^2$. Plot or sketch this function. What are the principal curvatures of the plate? What is the maximum value of the twisting curvature $k_{x'y'}$, and along what directions does it occur?

**33.** Show that the normal bending stress $\sigma_x$ and the curvatures $\kappa_{xx}$, $\kappa_{yy}$ developed by a plate subjected to a moment $M$ per unit width are the same as in a beam of unit width ($b = 1$) subjected to a moment $M$.

**34.** A $60°/0°/-60°$ layup gives an example of what are called *quasi-isotropic* laminates, which have equal stiffness in the $x$ and $y$ directions regardless of the laminate orientation. Verify that this is so for two laminate orientations: one having the middle ply fibers oriented along the $x$ axis and the other with the middle ply fibers oriented at $30°$ from the $x$ axis. Examine the $\mathcal{B}$ submatrices (Eq. 4.57) for these cases, and explain how the laminates do *not* act as isotropic materials.

**35.** Is a $0°/90°$ two-ply laminate isotropic? Does it have the same effective modulus in both the $x$ and $y$ directions? Is this true for laminate orientations other than with the fibers along the $x$ and $y$ directions? Is there a coupling between bending and extension?

**36.** Compute the strain ratio $\epsilon_x/\epsilon_y$ induced in a single ply of composite material, aligned with its fiber direction inclined $54.7°$ from the $x$ axis, and subjected to biaxial loading of the sort expected in a closed-end pressure vessel; that is, $N_y/N_x = 2$. Use values of $E_2$ and $G_{12}$ much less than $E_1$, so the assumptions of Example 2.8 apply (fibers carry all the load, with no influence of matrix).

**37.** Plot the effective Young's modulus, measured along the $x$-axis, of a single unidirectional ply of graphite-epoxy composite as a function of the angle between the ply fiber direction and the $x$-axis.

# 5 GENERAL STRESS ANALYSIS

The results presented in earlier chapters for trusses, beams, and other simple shapes provide much of the information needed in the design of load-bearing structures. However, structural and materials engineers routinely need to estimate stresses and deflections in geometrically more irregular articles. This is the function of *stress analysis,* by which we mean the collection of theoretical and experimental techniques that goes beyond the direct-analysis approach used up to now. This is a career field in its own right, and this chapter will limit itself to outlining only a few of its principal features.

## 5.1 CLOSED-FORM SOLUTIONS

### 5.1.1 Governing Equations

We have earlier shown how the spatial gradients of the six Cauchy stresses are related by three *equilibrium* equations that can be written in pseudovector form as

$$\mathbf{L}^{\mathrm{T}}\boldsymbol{\sigma} = \mathbf{0} \tag{5.1}$$

These are augmented by six *constitutive* equations

$$\boldsymbol{\sigma} = \mathbf{D}\boldsymbol{\epsilon} \tag{5.2}$$

and six *kinematic* or strain-displacement equations

$$\boldsymbol{\epsilon} = \mathbf{L}\mathbf{u} \tag{5.3}$$

These fifteen equations must be satisfied by the fifteen unknowns (three displacements **u**, six strains $\boldsymbol{\epsilon}$, and six stresses $\boldsymbol{\sigma}$). These functions must also satisfy boundary conditions on displacement

$$\mathbf{u} = \hat{\mathbf{u}} \qquad \text{on } \Gamma_u \tag{5.4}$$

where $\Gamma_u$ is the portion of the boundary on which the displacements $\mathbf{u} = \hat{\mathbf{u}}$ are prescribed. The remainder of the boundary must then have prescribed tractions $\mathbf{T} = \hat{\mathbf{T}}$, on which the stresses must satisfy Cauchy's relation:

$$\boldsymbol{\sigma}\hat{\mathbf{n}} = \hat{\mathbf{T}} \qquad \text{on } \Gamma_T \qquad (5.5)$$

In the familiar cantilevered beam shown in Fig. 5.1, the region of the beam at the wall constitutes $\Gamma_u$, having specified (zero) displacement and slope. All other points on the beam boundary make up $\Gamma_T$, with a load of $P$ at the loading point $A$ and a specified load of zero elsewhere.

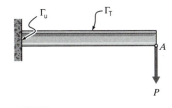

**FIGURE 5.1** Cantilevered beam.

With structures such as a beam that have simple geometries, solutions can be obtained by the direct method we have used up to now: An expression for the displacements is written, from which the strains and stresses can be obtained, and the stresses are then balanced against the externally applied loads. In situations not having this geometrical simplicity, the analyst must carry out a mathematical solution, seeking functions of stress, strain, and displacement that satisfy both the governing equations and the boundary conditions.

Currently, practical problems are likely to be solved by computational approximation, but it is almost always preferable to obtain a closed-form solution if at all possible. The mathematical result will show the functional importance of the various parameters, such as loading conditions or material properties, in a way a numerical solution cannot, and it is therefore more useful for guiding design decisions. For this reason the designer should always begin an analysis of load-bearing structures by searching for closed-form solutions of the given, or similar, problem. Several compendia of such solutions are available.[1]

However, there is always a danger in performing that sort of "handbook engineering" blindly. This section is intended partly to illustrate the mathematical concepts that underlie many of these published solutions. It is probably true that most of the problems that can be solved mathematically have already been completed; these are the classical problems of applied mechanics, and they often require a rather high level of mathematical sophistication. This text cannot describe the mathematical approach in anything approaching completeness, but some particularly useful methods will be outlined briefly to give a sense of the process. This will also give us the opportunity to develop certain expressions we will need in our later discussion of fracture mechanics. The classic text by Timoshenko and Goodier[2] is an excellent source for further reading in this area.

## 5.1.2 The Viscoelastic Correspondence Principle

In elastic materials the boundary tractions and displacements may depend on time as well as position without affecting the solution: Time is carried only as a parameter, because no time derivatives appear in the governing equations. With viscoelastic materials the constitutive equation (Eq. 5.2) is replaced by a time-differential equation, which complicates the subsequent solution. In many cases, however, the field equations possess certain mathematical properties that permit a solution to be obtained relatively easily.[3] The viscoelastic correspondence principle to be outlined here works

[1] W.C. Young, *Roark's Formulas for Stress and Strain,* McGraw-Hill, New York, 1989, is a useful example.

[2] S. Timoshenko and J. N. Goodier, *Theory of Elasticity,* McGraw-Hill, New York, 1951.

[3] E.H. Lee, "Viscoelasticity," *Handbook of Engineering Mechanics,* W. Flugge, ed., McGraw-Hill, New York, 1962, Chapter 53.

by adapting a previously available elastic solution to make it applicable to viscoelastic materials as well, so that a new solution from scratch is unnecessary.

We turn to Laplace transforms as a natural attempt to remove the time dependency from viscoelastic problems. The transformed field equations are

$$\mathbf{L}^T \overline{\boldsymbol{\sigma}} = 0$$
$$\overline{\boldsymbol{\sigma}} = \mathscr{D}\overline{\boldsymbol{\epsilon}}$$
$$\overline{\boldsymbol{\epsilon}} = \mathbf{L}\overline{\mathbf{u}} \tag{5.6}$$
$$\overline{\mathbf{u}} = \overline{\hat{\mathbf{u}}} \qquad \text{on } \Gamma_u$$
$$\overline{\boldsymbol{\sigma}}\hat{\mathbf{n}} = \overline{\hat{\mathbf{T}}} \qquad \text{on } \Gamma_T$$

where the overline represents the transformed variable and $\mathscr{D}$ is the stiffness matrix of viscoelastic materials functions.

Since the parameters not depending explicitly on time are not altered spatially by the transformation, these equations can be interpreted as representing a stress analysis problem for an elastic body of the same shape as the viscoelastic body, with one proviso: Although the physical shape of the body is unchanged upon passing to the Laplace plane, the transformed boundary constraints $\hat{\mathbf{f}}$, $\hat{\mathbf{T}}$, and $\overline{\hat{\mathbf{u}}}$ may exhibit an altered spatial distribution. For instance, if $\hat{T} = \cos(xt)$, then $\overline{\hat{T}} = s/(s^2 + x^2)$; this is obviously of a different spatial form than the original untransformed function.

However, functions that can be written as separable space and time factors will not change spatially on transformation:

$$\hat{T}(x, t) = f(x)g(t) \Rightarrow \overline{\hat{T}} = f(x)\overline{g}(s)$$

This means that the stress analysis problems whose boundary constraints are independent of time or, at worst, are separable functions of space and time will *look* the same in both the actual and Laplace planes. In the Laplace plane, the problem is then geometrically identical with an *associated* elastic problem.

Having reduced the viscoelastic problem to an associated elastic one by taking transforms, we may use the vast library of elastic solutions: We look up the solution to the associated elastic problem and then perform a Laplace inversion to return to the time plane. The process of viscoelastic stress analysis employing transform methods is usually called the *correspondence principle*, which can be stated as the following recipe:

1. Determine the nature of the associated elastic problem. If the spatial distribution of the boundary and body force conditions is unchanged on transformation—a common occurrence—then the associated elastic problem appears exactly like the original viscoelastic one.

2. Determine the solution to this associated elastic problem. This can often be done by reference to standard handbooks or texts on the theory of elasticity.

3. Recast the elastic constants appearing in the elastic solution in terms of suitable viscoelastic operators. It is often easiest to replace $E$ and $\nu$ with $G$ and $K$ and then replace the $G$ and $K$ by their viscoelastic analogs:

$$\left.\begin{array}{c} E \\ \nu \end{array}\right\} \longrightarrow \left\{\begin{array}{c} G \longrightarrow \mathscr{G} \\ K \longrightarrow \mathscr{K} \end{array}\right.$$

4. Replace the applied boundary and body force constraints by their transformed counterparts:

$$\hat{T} \Rightarrow \overline{T}$$
$$\hat{u} \Rightarrow \overline{u}$$

5. Invert the expression found in step 4 to obtain the solution to the viscoelastic problem in the time plane.

## EXAMPLE 5.1

In Chapter 2 we obtained the elastic solution for the radial expansion of a closed-end cylindrical pressure vessel of radius $r$ and thickness $b$ as

$$\delta_r = \frac{pr^2}{bE}\left(1 - \frac{\nu}{2}\right)$$

Following that recipe, the associated solution in the Laplace plane is found to be

$$\overline{\delta}_r = \frac{\overline{p}r^2}{b\mathcal{E}}\left(1 - \frac{\mathcal{N}}{2}\right)$$

In terms of hydrostatic and shear response functions, we have from Chapter 3

$$\mathcal{E}(s) = \frac{9\mathcal{G}(s)\mathcal{H}(s)}{3\mathcal{H}(s) + \mathcal{G}(s)}$$

$$\mathcal{N}(s) = \frac{3\mathcal{H}(s) - 2\mathcal{G}(s)}{6\mathcal{H}(s) + 2\mathcal{G}(s)}$$

In Example 3.16 we considered a PVC material at 75°C that to a good approximation was elastic in hydrostatic response and viscoelastic in shear. Using the Standard Linear Solid model, we had

$$\mathcal{H} = K_e, \qquad \mathcal{G} = G_r + \frac{(G_g - G_r)s}{s + \dfrac{1}{\tau}}$$

where $K_e = 1.33$ GPa, $G_g = 800$ MPa, $G_r = 1.67$ MPa, and $\tau = 100$ s.

For constant internal pressure $p(t) = p_0$, $\overline{p} = p_0/s$. All these expressions must be combined, and the result inverted. The algebra involved is daunting, and in an earlier time the problem might have been given up as intractable, but, as we have seen several times in this text already, symbolic computer methods have made such calculations simple and convenient. The MAPLE commands for this problem might be

```
> # define shear operator
> G:=Gr+((Gg-Gr)*s)/(s+(1/tau));
> # define Poisson operator
> N:=(3*K-2*G)/(6*K+2*G);
> # define modulus operator
```

```
> Eop:=(9*G*K)/(3*K+G);
> # define pressure operator
> pbar:=p0/s;
> # get d1, radial displacement (in Laplace plane)
> d1:=(pbar*r^2)*(1-(N/2))/(b*Eop);
> # read MAPLE library for Laplace transforms
> readlib(laplace);
> # invert transform to get d2, radial displacement
> # in real plane
> d2:=invlaplace(d1,s,t);
```

After some manual rearrangement, the radial displacement $\delta_r(t)$ can be written in the form

$$\delta_r(t) = \frac{r^2 p_0}{b}\left[\left(\frac{1}{4G_r} + \frac{1}{6K}\right) - \left(\frac{1}{4G_r} - \frac{1}{4G_g}\right)e^{-t/\tau_c}\right]$$

where the creep retardation time is $\tau_c = \tau(G_g/G_r)$. Continuing the MAPLE session, we have

```
> # define numerical parameters
> Gg:=800*10^6; Gr:=1.67*10^6; tau:=100; K:=1.33*10^9;
> r:=.05; b:=.005; p0:=2*10^5;
> # resulting expression for radial displacement
> d2;
              - .01494 exp( - .00002088 t) + .01498
```

A log-log plot of this function is shown in Fig. 5.2. Note that for this problem the effect of the small change in Poisson's ratio $\nu$ during the transition is negligible in comparison with the very large change in the modulus $E$, so a nearly identical result would have been obtained simply by letting $\nu$ = constant = 0.5. On the other hand, it isn't appreciably more difficult to include the time dependence of $\nu$ if symbolic manipulation software is available.

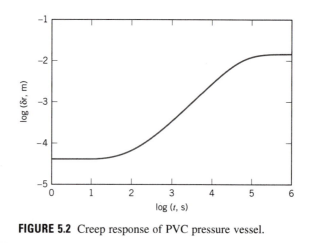

**FIGURE 5.2** Creep response of PVC pressure vessel.

## 5.1.3 The Airy Stress Function

Proceeding to cases in which we wish to obtain elastic solutions from scratch, we recall the kinematic or strain-displacement equations in two dimensions:

$$\epsilon_x = \frac{\partial u}{\partial x}$$

$$\epsilon_y = \frac{\partial v}{\partial y} \tag{5.7}$$

$$\gamma_{xy} = \frac{\partial v}{\partial x} + \frac{\partial u}{\partial y}$$

Since three strains ($\epsilon_x$, $\epsilon_y$, $\gamma_{xy}$) are written in terms of only two displacements ($u$, $v$), they cannot be specified arbitrarily; a relation must exist between the three strains. If $\epsilon_x$ is differentiated twice by $dx$, $\epsilon_y$ twice by $dy$, and $\gamma_{xy}$ by $dx$ and then $dy$, we have directly

$$\frac{\partial^2 \epsilon_x}{\partial y^2} + \frac{\partial^2 \epsilon_y}{\partial x^2} = \frac{\partial^2 \gamma_{xy}}{\partial x \, \partial y} \tag{5.8}$$

In order for the displacements to be so differentiable, they must be continuous functions, which means physically that the body must deform in a *compatible* manner: that is, without developing cracks or overlaps. For this reason Eq. 5.8 is called the *compatibility* equation for strains; the continuity of the body is guaranteed if the strains satisfy it.

The compatibility equation can be written in terms of the stresses rather than the strains by recalling the constitutive equations for elastic plane stress:

$$\epsilon_x = \frac{1}{E}(\sigma_x - \nu\sigma_y)$$

$$\epsilon_y = \frac{1}{E}(\sigma_y - \nu\sigma_x) \tag{5.9}$$

$$\gamma_{xy} = \frac{1}{G}\tau_{xy} = \frac{2(1 + \nu)}{E}\tau_{xy}$$

Substituting these in Eq. 5.8 gives

$$\frac{\partial^2}{\partial y^2}(\sigma_x - \nu\sigma_y) + \frac{\partial^2}{\partial x^2}(\sigma_y - \nu\sigma_x) = 2(1 + \nu)\frac{\partial^2 \tau_{xy}}{\partial x \, \partial y} \tag{5.10}$$

Stresses satisfying this relation guarantee compatibility of strain.

The stresses must also satisfy the equilibrium equations, which in two dimensions can be written

$$\frac{\partial \sigma_x}{\partial x} + \frac{\partial \tau_{xy}}{\partial y} = 0$$

$$\frac{\partial \tau_{xy}}{\partial x} + \frac{\partial \sigma_y}{\partial y} = 0 \tag{5.11}$$

As a means of simplifying the search for functions whose derivatives obey these rules, Airy[4] defined a *stress function* $\phi$, from which the stresses could be obtained by differentiation:

---

[4]Sir George Biddell Airy (1801–1892) was an English astronomer and mathematician; he was the British Astronomer Royal (1835–1881) and President of the Royal Society (1871–1873).

$$\sigma_x = \frac{\partial^2 \phi}{\partial y^2}$$

$$\sigma_y = \frac{\partial^2 \phi}{\partial x^2} \tag{5.12}$$

$$\tau_{xy} = -\frac{\partial^2 \phi}{\partial x \, \partial y}$$

Direct substitution will show that stresses obtained by this procedure will automatically satisfy the equilibrium equations. This maneuver is essentially limited to two-dimensional problems, but with that proviso it provides a great simplification in searching for valid functions for the stresses.

Now substituting these into Eq. 5.10, we have

$$\frac{\partial^4 \phi}{\partial x^4} + 2\frac{\partial^4 \phi}{\partial x^2 \, \partial y^2} + \frac{\partial^4 \phi}{\partial y^4} \equiv \nabla^2(\nabla^2 \phi) \equiv \nabla^4 \phi = 0 \tag{5.13}$$

Any function $\phi(x, y)$ that satisfies this relation will satisfy the governing relations for equilibrium, geometric compatibility, and linear elasticity. Of course, many functions could be written that satisfy the compatibility equation; for instance setting $\phi = 0$ would always work. However, to make the solution be applicable to a particular stress analysis, the boundary conditions on stress and displacement must be satisfied as well. This is usually a much more difficult undertaking, and no general solution that works for all cases exists. It can be shown, however, that a solution satisfying both the compatibility equation and the boundary conditions is unique; in other words, that it is the *only* correct solution.

### 5.1.4  Stresses around a Circular Hole

To illustrate the use of the Airy function approach, we will outline the important work of Kirsch,[5] who obtained a solution for the influence of a hole in the material on the stresses. This is vitally important in analyzing such problems as rivet holes used in joining, and the effect of a manufacturing void in initiating failure. Consider a thin sheet as illustrated in Fig. 5.3, infinite in lateral dimensions but containing a circular hole of radius $a$ and subjected to a uniaxial stress $\sigma$. Using circular $r$, $\theta$ coordinates centered on the hole, the compatibility equation for $\phi$ is

$$\nabla^4 \phi = \left(\frac{\partial^2}{\partial r^2} + \frac{1}{r}\frac{\partial}{\partial r} + \frac{1}{r^2}\frac{\partial^2}{\partial \theta^2}\right)\left(\frac{\partial^2 \phi}{\partial r^2} + \frac{1}{r}\frac{\partial \phi}{\partial r} + \frac{1}{r^2}\frac{\partial^2 \phi}{\partial \theta^2}\right) = 0 \tag{5.14}$$

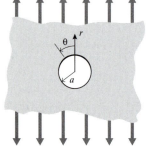

**FIGURE 5.3** Circular hole in a uniaxially stressed plate.

In these circular coordinates the stresses are obtained from $\phi$ as

$$\sigma_r = \frac{1}{r}\frac{\partial \phi}{\partial r} + \frac{1}{r^2}\frac{\partial^2 \phi}{\partial \theta^2}$$

$$\sigma_\theta = \frac{\partial^2 \phi}{\partial r^2} \tag{5.15}$$

$$\tau_{r\theta} = -\frac{\partial}{\partial r}\left(\frac{1}{r}\frac{\partial \phi}{\partial \theta}\right)$$

[5]G. Kirsch, *VDI*, vol. 42, 1898.

We now seek a function $\phi(r, \theta)$ that satisfies Eq. 5.14 and also the boundary conditions of the problem. On the periphery of the hole the radial and shearing stresses must vanish, because no external tractions exist there:

$$\sigma_r = \tau_{r\theta} = 0, \qquad r = a \tag{5.16}$$

Far from the hole the stresses must become the far-field value $\sigma$; the Mohr procedure gives the radial and tangential stress components in circular coordinates as

$$\left.\begin{array}{l} \sigma_r = \dfrac{\sigma}{2}(1 + \cos 2\theta) \\[2mm] \sigma_\theta = \dfrac{\sigma}{2}(1 - \cos 2\theta) \\[2mm] \tau_{r\theta} = \dfrac{\sigma}{2} \sin 2\theta \end{array}\right\} \qquad r \to \infty \tag{5.17}$$

Since the normal stresses vary circumferentially as $\cos 2\theta$ (temporarily removing the first $\sigma/2$ term) and the shear stresses vary as $\sin 2\theta$, an acceptable stress function could be of the form

$$\phi = f(r) \cos 2\theta \tag{5.18}$$

When this is substituted into Eq. 5.14, an ordinary differential equation in $f(r)$ is obtained:

$$\left(\frac{d^2}{dr^2} + \frac{1}{r}\frac{d}{dr} - \frac{4}{r^2}\right)\left(\frac{d^2 f}{dr^2} + \frac{1}{r}\frac{df}{dr} - \frac{4f}{r^2}\right) = 0$$

This has the general solution

$$f(r) = Ar^2 + Br^4 + C\frac{1}{r^2} + D \tag{5.19}$$

The stress function obtained from Eqs. 5.18 and 5.19 is now used to write expressions for the stresses according to Eq. 5.15, and the constants are determined using the boundary conditions in Eqs. 5.16 and 5.17; this gives

$$A = -\frac{\sigma}{4}, \qquad B = 0, \qquad C = -\frac{a^4\sigma}{4}, \qquad D = \frac{a^2\sigma}{2}$$

Substituting these values into the expressions for stress and restoring the $\sigma/2$ factor that was temporarily removed, the final expressions for the stresses are

$$\sigma_r = \frac{\sigma}{2}\left(1 - \frac{a^2}{r^2}\right) + \frac{\sigma}{2}\left(1 + \frac{3a^4}{r^4} - \frac{4a^2}{r^2}\right)\cos 2\theta$$

$$\sigma_\theta = \frac{\sigma}{2}\left(1 + \frac{a^2}{r^2}\right) - \frac{\sigma}{2}\left(1 + \frac{3a^4}{r^4}\right)\cos 2\theta \tag{5.20}$$

$$\tau_{r\theta} = -\frac{\sigma}{2}\left(1 - \frac{3a^4}{r^4} + \frac{2a^2}{r^2}\right)\sin 2\theta$$

As seen in the plot of Fig. 5.4, the stress reaches a maximum value of $(\sigma_\theta)_{max} = 3\sigma$ at the periphery of the hole ($r = a$), at a diametral position transverse to the loading direction ($\theta = \pi/2$). The *stress concentration factor*, or SCF, for this problem is therefore 3. The $x$-direction stress falls to zero at the position $\theta = \pi/2$, $r = a$, as it must to satisfy the stress-free boundary condition at the periphery of the hole.

Note that in the case of a circular hole the SCF does not depend on the size of the hole: Any hole, no matter how small, increases the local stresses near the hole by a factor of three. This is a very serious consideration in the design of structures that must be drilled and riveted in assembly. This is the case in construction of most jetliner fuselages, the skin of which must withstand substantial stresses as the differential cabin pressure is cycled by approximately 10 psig during each flight. The high-stress region near the rivet holes has a dangerous propensity to incubate fatigue cracks, and several catastrophic aircraft failures have been traced to exactly this cause.

Note also that the stress concentration effect is confined to the region quite close to the hole, with the stresses falling to their far-field values within three or so hole diameters. This is a manifestation of *St. Venant's principle,*[6] which is a common-sense statement that the influence of a perturbation in the stress field is largely confined to the region of the disturbance. This principle is extremely useful in engineering approximations, but of course the stress concentration near the disturbance itself must be kept in mind.

When at the beginning of this section we took the size of the plate to be "infinite in lateral extent," we really meant that the stress conditions at the plate edges were far enough away from the hole that they did not influence the stress state near the hole. With the Kirsch solution now in hand, we can be more realistic about this: The plate must be at least three or so times larger than the hole, or the Kirsch solution will be unreliable.

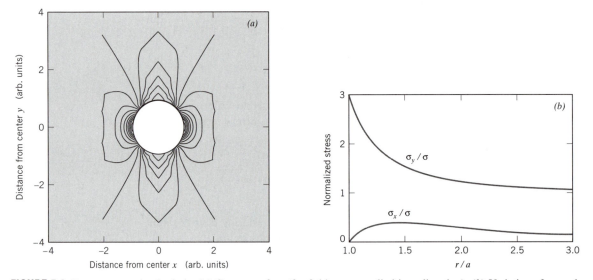

**FIGURE 5.4** Stresses near circular hole. (*a*) Contours of $\sigma_y$ (far-field stress applied in $y$-direction). (*b*) Variation of $\sigma_y$ and $\sigma_x$ along $\theta = \pi/2$ line.

---

[6]The French scientist Barré de Saint-Venant (1797–1886) is one of the great pioneers in mechanics of materials.

## 5.1.5 Complex Functions

In many problems of practical interest, it is convenient to use stress functions as complex functions of two variables. We will see that these have the ability to satisfy the governing equations automatically, leaving only adjustments needed to match the boundary conditions. For this reason, complex-variable methods play an important role in theoretical stress analysis, and even in this introductory text we wish to illustrate the power of the method. To outline a few necessary relations, consider $z$ to be a complex number in Cartesian coordinates $x$ and $y$ or polar coordinates $r$ and $\theta$ as

$$z = x + iy = re^{i\theta} \tag{5.21}$$

where $i = \sqrt{-1}$. An *analytic function* $f(z)$ is one whose derivatives depend on $z$ only, and takes the form

$$f(z) = \alpha + i\beta \tag{5.22}$$

where $\alpha$ and $\beta$ are real functions of $x$ and $y$. It is easily shown that $\alpha$ and $\beta$ satisfy the *Cauchy-Riemann* equations:

$$\frac{\partial \alpha}{\partial x} = \frac{\partial \beta}{\partial y} \qquad \frac{\partial \alpha}{\partial y} = -\frac{\partial \beta}{\partial x} \tag{5.23}$$

If the first of these is differentiated with respect to $x$ and the second with respect to $y$, and the results added, we obtain

$$\frac{\partial^2 \alpha}{\partial x^2} + \frac{\partial^2 \alpha}{\partial y^2} \equiv \nabla^2 \alpha = 0 \tag{5.24}$$

This is *Laplace's equation,* and any function that satisfies this equation is termed a *harmonic* function. Equivalently, $\alpha$ could have been eliminated in favor of $\beta$ to give $\nabla^2 \beta = 0$, so both the real and imaginary parts of any complex function provide solutions to Laplace's equation. Now consider a function of the form $x\psi$, where $\psi$ is harmonic; it can be shown by direct differentiation that

$$\nabla^4(x\psi) = 0 \tag{5.25}$$

That is, any function of the form $x\psi$, where $\psi$ is harmonic, satisfies Eq. 5.13, and may thus be used as a stress function. Similarly, it can be shown that $y\psi$ and $(x^2 + y^2)\psi = r^2\psi$ are also suitable, as is $\psi$ itself. In general, a suitable stress function can be obtained from any two analytic functions $\psi$ and $\chi$ according to

$$\phi = \mathrm{Re}\left[(x - iy)\psi(z) + \chi(z)\right] \tag{5.26}$$

where Re indicates the real part of the complex expression. The stresses corresponding to this function $\phi$ are obtained as

$$\sigma_x + \sigma_y = 4\,\mathrm{Re}\,\psi'(z)$$
$$\sigma_y - \sigma_x + 2\,i\tau_{xy} = 2\left[\bar{z}\psi''(z) + \chi''(z)\right] \tag{5.27}$$

where the primes indicate differentiation with respect to $z$ and the overbar indicates the *conjugate function,* obtained by replacing $i$ with $-i$; hence, $\bar{z} = x - iy$.

## 5.1.6 Stresses around an Elliptical Hole

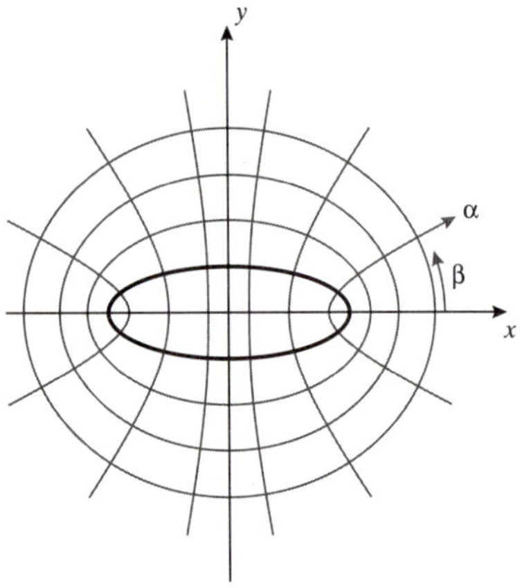

In a development very important to the theory of fracture, Inglis[7] used complex potential functions to extend Kirsch's work to treat the stress field around a plate containing an elliptical rather than a circular hole. This permits cracklike geometries to be treated by making the minor axis of the ellipse small. It is convenient to work in elliptical $\alpha, \beta$ coordinates, as shown in Fig. 5.5, defined as

$$x = c \cosh \alpha \cos \beta, \qquad y = c \sinh \alpha \sin \beta \tag{5.28}$$

where $c$ is a constant. If $\beta$ is eliminated, this is seen in turn to be equivalent to

$$\frac{x^2}{\cosh^2 \alpha} + \frac{y^2}{\sinh^2 \alpha} = c^2 \tag{5.29}$$

On the boundary of the ellipse $\alpha = \alpha_0$, so we can write

$$c \cosh \alpha_0 = a, \qquad c \sinh \alpha_0 = b \tag{5.30}$$

where $a$ and $b$ are constants. On the boundary, then

$$\frac{x^2}{a^2} + \frac{y^2}{b^2} = 1 \tag{5.31}$$

which is recognized as the Cartesian equation of an ellipse, with $a$ and $b$ being the major and minor radii. The elliptical coordinates can be written in terms of complex variables as

$$z = c \cosh \zeta, \qquad \zeta = \alpha + i\beta \tag{5.32}$$

As the boundary of the ellipse is traversed, $\alpha$ remains constant at $\alpha_0$ while $\beta$ varies from 0 to $2\pi$. Hence the stresses must be periodic in $\beta$ with period $2\pi$, becoming equal to the far-field uniaxial stress $\sigma_y = \sigma, \sigma_x = \tau_{xy} = 0$ far from the ellipse; Eq. 5.27 then gives

$$\left. \begin{array}{r} 4 \operatorname{Re} \psi'(z) = \sigma \\ 2[\bar{z}\psi''(z) + \chi''(z)] = \sigma \end{array} \right\} \quad \zeta \to \infty \tag{5.33}$$

These boundary conditions can be satisfied by potential functions in the forms

$$4\psi(z) = Ac \cosh \zeta + Bc \sinh \zeta$$

$$4\chi(z) = Cc^2 \zeta + Dc^2 \cosh 2\zeta + Ec^2 \sinh 2\zeta$$

where $A, B, C, D, E$ are constants to be determined from the boundary conditions. When this is done, the complex potentials are given as

$$4\psi(z) = \sigma c[(1 + e^{2\alpha_0}) \sinh \zeta - e^{2\alpha_0} \cosh \zeta]$$

$$4\chi(z) = -\sigma c^2 \left[ (\cosh 2\alpha_0 - \cosh \pi)\zeta + \frac{1}{2} e^{2\alpha_0} - \cosh 2\left(\zeta - \alpha_0 - i\frac{\pi}{2}\right) \right]$$

The stresses $\sigma_x, \sigma_y,$ and $\tau_{xy}$ can be obtained by using these in Eq. 5.27. However, the amount of labor in carrying out these substitutions isn't to be sneezed at. Before

**FIGURE 5.5** Elliptical coordinates.

[7]C.E. Inglis, "Stresses in a Plate Due to the Presence of Cracks and Sharp Corners," *Transactions of the Institution of Naval Architects,* Vol. 55, London, 1913, pp. 219–230.

in probing the nature of the stress field near crack tips.

Figure 5.6 shows stress contours computed by Cook and Gordon[8] from the Inglis equations. A strong stress concentration of the stress $\sigma_y$ is noted at the periphery of the hole, as would be expected. The horizontal stress $\sigma_x$ goes to zero at this same position, as it must to satisfy the boundary conditions there. Note, however, that $\sigma_x$ exhibits a mild stress concentration (one-fifth of that for $\sigma_y$, it turns out) a little distance away from the hole. If the material has planes of weakness along the $y$ direction, for instance as between the fibrils in wood or many other biological structures, the stress $\sigma_x$ could cause a split to open up in the $y$ direction just ahead of the main crack. This would act to blunt and arrest the crack and thus impart a measure of toughness to the material. This effect is sometimes called the *Cook-Gordon toughening mechanism.*

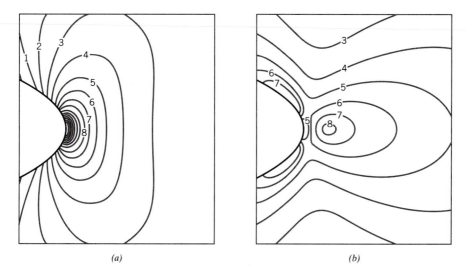

(a)                                                                    (b)

**FIGURE 5.6** Stress field in the vicinity of an elliptical hole, with uniaxial stress applied in $y$-direction. (*a*) Contours of $\sigma_y$, (*b*) contours of $\sigma_x$. (From: THE SCIENCE OF STRUCTURES AND MATERIALS by J. E. Gordon. Copyright © 1988 by Scientific American Books, Inc. Used with permission of W. H. Freeman and Company.)

The mathematics of the Inglis solution are simpler at the surface of the elliptical hole, because here the normal component $\sigma_\alpha$ must vanish. The tangential stress component can then be computed directly:

$$(\sigma_\beta)_{\alpha = \alpha_0} = \sigma e^{2\alpha_0} \left[ \frac{\sinh 2\alpha_0 (1 + e^{-2\alpha_0})}{\cosh 2\alpha_0 - \cos 2\beta} - 1 \right]$$

The greatest stress occurs at the end of the major axis ($\cos 2\beta = 1$):

$$(\sigma_\beta)_{\beta = 0, \pi} = \sigma_y = \sigma \left( 1 + 2\frac{a}{b} \right) \tag{5.34}$$

This can also be written in terms of the radius of curvature $\rho$ at the tip of the major axis as

$$\boxed{\sigma_y = \sigma \left( 1 + 2\sqrt{\frac{a}{\rho}} \right)} \tag{5.35}$$

[8]J.E. Gordon, *The Science of Structures and Materials,* Scientific American Library, New York, 1988.

This result is immediately useful. It is clear that large cracks are worse than small ones (the local stress increases with crack size $a$), and it is also obvious that sharp voids (decreasing $\rho$) are worse than rounded ones. Note also that the stress $\sigma_y$ increases without limit as the crack becomes sharper ($\rho \to 0$), so the concept of a stress concentration factor becomes difficult to use for very sharp cracks. When the major and minor axes of the ellipse are the same ($b = a$), the result becomes identical to that for the circular hole outlined earlier.

### 5.1.7 Stresses near a Sharp Crack

The Inglis solution is difficult to apply, especially as the crack becomes sharp. A more tractable and now more widely used approach was developed by Westergaard,[9] which treats a sharp crack of length $2a$ in a thin but infinitely wide sheet (see Fig. 5.7). The stresses that act perpendicularly to the crack-free surfaces (the crack "flanks") must be zero, while at distances far from the crack they must approach the far-field imposed stresses. Consider a harmonic function $\phi(z)$, with first and second derivatives $\phi'(z)$ and $\phi''(z)$ and first and second integrals $\overline{\phi}(z)$ and $\overline{\overline{\phi}}(z)$. Westergaard constructed a stress function as

$$\Phi = \operatorname{Re} \overline{\overline{\phi}}(z) + y \operatorname{Im} \overline{\phi}(z) \tag{5.36}$$

It can be shown directly that the stresses derived from this function satisfy the equilibrium, compatibility, and constitutive relations. The function $\phi(z)$ needed here is a harmonic function such that the stresses approach the far-field value of $\sigma$ at infinity, but are zero at the crack flanks except at the crack tip where the stress becomes unbounded:

$$\sigma_y = \begin{cases} \sigma, & x \to \pm\infty \\ \infty, & x = \pm a, y = 0 \end{cases}$$

These conditions are satisfied by complex functions of the form

$$\phi(z) = \frac{\sigma}{\sqrt{1 - a^2/z^2}} \tag{5.37}$$

This gives the needed singularity for $z = \pm a$, and the other boundary conditions can be verified directly as well. The stresses are now found by suitable differentiations of the stress function; for instance

$$\sigma_y = \frac{\partial^2 \Phi}{\partial x^2} = \operatorname{Re} \phi(z) + y \operatorname{Im} \phi'(z)$$

In terms of the distance $r$ from the crack tip, this becomes

$$\sigma_y = \sigma \sqrt{\frac{a}{2r}} \cdot \cos \frac{\theta}{2} \left(1 + \sin \frac{\theta}{2} \sin \frac{3\theta}{2}\right) + \cdots \tag{5.38}$$

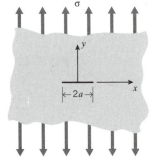

**FIGURE 5.7** Sharp crack in an infinite sheet.

[9]Westergaard, H.M., "Bearing Pressures and Cracks," *Transactions,* Am. Soc. Mech. Engrs., *Journal of Applied Mechanics,* Vol. 5, p. 49, 1939.

where these are the initial terms of a series approximation. Near the crack tip, when $r \ll a$, we can write

$$(\sigma_y)_{y=0} = \sigma \sqrt{\frac{a}{2r}} \equiv \frac{K}{\sqrt{2\pi r}} \qquad (5.39)$$

where $K = \sigma \sqrt{\pi a}$ is the *stress intensity factor*, with units of N-m$^{-3/2}$ or psi $\sqrt{\text{in}}$. (The factor $\pi$ seems redundant here, because it appears to the same power in both the numerator and denominator, but it is usually included as written here for agreement with the older literature.) We will see in Chapter 7 that the stress intensity factor is a commonly used measure of the driving force for crack propagation, and thus underlies much of modern fracture mechanics. The dependency of the stress on distance from the crack is singular, with a $1/\sqrt{r}$ dependency. The $K$ factor scales the intensity of the overall stress distribution, with the stress always becoming unbounded as the crack tip is approached.

## 5.2 EXPERIMENTAL SOLUTIONS

### 5.2.1 Introduction

As was seen in the previous section, stress analysis even of simple-appearing geometries can lead to complicated mathematical maneuvering. Actual articles—engine crankshafts, medical prostheses, tennis rackets, and so forth—have boundary shapes that cannot easily be described mathematically, and even if they could be described, it would be extremely difficult to fit solutions of the governing equations to them. One approach to this impasse is the experimental one, in which we seek to construct a physical laboratory model that somehow reveals the stresses in a measurable way.

It is the nature of forces and stresses that they cannot be measured directly. The *effect* of a force is what is measurable. When we weigh an object on a spring scale, we are actually measuring the stretching of the spring and then calculating the force from Hooke's law. Experimental stress analysis, then, is actually experimental *strain* analysis. The difficulty is that strains in the linear elastic regime are almost always small, on the order of 1 percent or less, and the art in this field is that of detecting and interpreting small displacements. We look for phenomena that exhibit large and measurable changes due to small and difficult-to-measure displacements. There are a number of such techniques, and three of these will be outlined briefly in this section. A good deal of methodology has been developed around these and other experimental methods, and both further reading[10] and laboratory practice would be required for an engineer who wishes to become competent in this area.

### 5.2.2 Strain Gages

The term *strain gage* usually refers to a thin wire or foil, folded back and forth on itself and bonded to the specimen surface as seen in Fig. 5.8, that is able to generate an electrical measure of strain in the specimen. As the wire is stretched along with the specimen, the wire's electrical resistance $R$ changes, both because its length $L$ is

**FIGURE 5.8** Wire resistance strain gage. (Courtesy BHL Electronics, Inc.)

[10]A.S. Kobayashi, Ed., *Manual on Experimental Stress Analysis,* 4th ed., Society for Experimental Stress Analysis, 1983.

increased and because its cross-sectional area $A$ is reduced. For many resistors these variables are related by the simple expression discovered in 1856 by Lord Kelvin:

$$R = \frac{\rho L}{A}$$

where here $\rho$ is the material's *resistivity*. To express the effect of a strain $\epsilon = dL/L$ in the wire's long direction on the electrical resistance, assume a circular wire with $A = \pi r^2$ and take logarithms:

$$\ln R = \ln \rho + \ln L - (\ln \pi + 2 \ln r)$$

The total differential of this expression gives

$$\frac{dR}{R} = \frac{d\rho}{\rho} + \frac{dL}{L} - 2\frac{dr}{r}$$

Since

$$\epsilon_r = \frac{dr}{r} = -\nu\frac{dL}{L}$$

then

$$\frac{dR}{R} = \frac{d\rho}{\rho} + (1 + 2\nu)\frac{dL}{L}$$

Bridgeman[11] in 1929 studied the effect of volume change on electrical resistivity and found these to vary proportionally:

$$\frac{d\rho}{\rho} = \alpha\frac{dV}{V}$$

where $\alpha$ is the constant of proportionality between resistivity change and volume change. Writing the volume change in terms of changes in length and area, this becomes

$$\frac{d\rho}{\rho} = \alpha\left(\frac{dL}{L} + \frac{dA}{A}\right) = \alpha(1 - 2\nu)\frac{dL}{L}$$

Hence

$$\frac{dR/R}{\epsilon} = (1 + 2\nu) + \alpha(1 - 2\nu) \tag{5.40}$$

This quantity is called the *gage factor*, GF. Constantan, a 45/55 nickel/copper alloy, has $\alpha = 1.13$ and $\nu = 0.3$, giving GF $\approx 2.0$. This material also has a low temperature coefficient of resistivity, which reduces the temperature sensitivity of the strain gage.

---

[11]P.W. (Percy Williams) Bridgeman (1882–1961) is known for his studies in the physics of high-pressure effects.

A change in resistance of only 2 percent, which would be generated by a gage with GF = 2 at 1 percent strain, would not be noticeable on a simple ohmmeter. For this reason, strain gages are almost always connected to a Wheatstone bridge circuit, as seen in Fig. 5.9. The circuit can be adjusted by means of the variable resistance $R_2$ to produce a zero output voltage $V_{out}$ before strain is applied to the gage. Typically the gage resistance is approximately 350 $\Omega$ and the excitation voltage is near 10 V. When the gage resistance is changed by strain, the bridge is unbalanced, and a voltage appears on the output according to the relation

$$\frac{V_{out}}{V_{in}} = \frac{\Delta R}{2R_0}$$

where $R_0$ is the nominal resistance of the four bridge elements. The output voltage is easily measured, because it is a deviation from zero rather than being a relatively small change superimposed on a much larger quantity; it can thus be amplified to suit the needs of the data acquisition system.

Temperature compensation can be achieved by making a bridge element on the opposite side of the bridge from the active gage, say $R_3$, an inactive gage that is placed near the active gage but not bonded to the specimen. Resistance changes in the active gage due to temperature will then be offset by an equal resistance change in the other arm of the bridge.

It is often difficult to mount a tensile specimen in the testing machine without inadvertently applying bending in addition to tensile loads. If a single gage were applied to the convex-outward side of the specimen, its reading would be erroneously high. Similarly, a gage placed on the concave-inward or compressive-tending side would read low. These bending errors can be eliminated by using an active gage on each side of the specimen, as shown in Fig. 5.10, and wiring them on the same side of the Wheatstone bridge, for example, as $R_1$ and $R_4$. The tensile component of bending on one side of the specimen is accompanied by an equal but compressive component on the other side, and these will generate equal but opposite resistance changes in $R_1$ and $R_4$. The effects of bending will therefore cancel, and the gage combination will measure only the tensile strain (with doubled sensitivity because both $R_1$ and $R_4$ are active).

The strain in the gage direction can be found directly from the gage factor (Eq. 5.40). When the direction of principal stress is unknown, strain gage *rosettes* are useful; these employ multiple gages on the same film backing, oriented in different directions. The rectangular three-gage rosette shown in Fig. 5.11 uses two gages oriented perpendicularly and a third gage oriented at 45° to the first two.

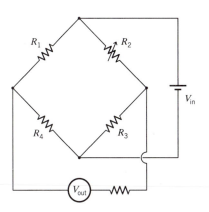

**FIGURE 5.9** Wheatstone bridge circuit for strain gages.

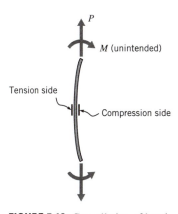

**FIGURE 5.10** Cancellation of bending effects.

---

## EXAMPLE 5.2

A three-gage rosette gives readings $\epsilon_0 = 150~\mu$, $\epsilon_{45} = 200~\mu$, and $\epsilon_{90} = -100~\mu$ (here the $\mu$ symbol indicates micrometers per meter). If we align the $x$ and $y$ axis along the 0° and 90° gage directions, then $\epsilon_x$ and $\epsilon_y$ are measured directly, since these are $\epsilon_{0°}$ and $\epsilon_{90°}$ respectively. To determine the shear strain $\gamma_{xy}$, we use Eq. 3.32 to write the normal strain at 45°:

$$\epsilon_{45} = 200~\mu = \epsilon_x \cos^2 45° + \epsilon_y \sin^2 45° + \gamma_{xy} \sin 45° \cos 45°$$

Substituting the known values for $\epsilon_x$ and $\epsilon_y$ and solving, we obtain

$$\gamma_{xy} = 350~\mu$$

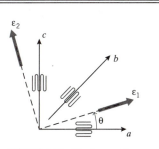

**FIGURE 5.11** Strain rosette.

The principal strains can now be found as

$$\epsilon_{1,2} = \frac{\epsilon_x + \epsilon_y}{2} \pm \sqrt{\left(\frac{\epsilon_x - \epsilon_y}{2}\right)^2 + \left(\frac{\tau_{xy}}{2}\right)^2} = 240 \, \mu, \, 190 \, \mu$$

The angle from the $x$-axis to the principal plane is

$$\tan 2\theta_p = \frac{\gamma_{xy}/2}{(\epsilon_x - \epsilon_y)/2} \rightarrow \theta_p = 27.2°$$

The stresses can be found from the strains using Eq. 3.49; for instance for steel with $E = 205$ GPa and $\nu = 0.3$ the principal stress is

$$\sigma_1 = \frac{E}{1 - \nu^2}(\epsilon_1 - \nu\epsilon_2) = 66.9 \text{ GPa}$$

For the specific case of a 0°–45°–90° rosette, the orientation of the principal strain axis can be given directly by[12]

$$\tan 2\theta = \frac{2\epsilon_b - \epsilon_a - \epsilon_c}{\epsilon_a - \epsilon_c} \tag{5.41}$$

and the principal strains are

$$\epsilon_{1,2} = \frac{\epsilon_a + \epsilon_c}{2} \pm \sqrt{\frac{(\epsilon_a - \epsilon_b)^2 + (\epsilon_b - \epsilon_c)^2}{2}} \tag{5.42}$$

Graphical solutions based on Mohr's circles are also useful for reducing gage output data.

Strain gages are used very extensively, and critical structures such as aircraft may be instrumented with hundreds of gages during testing. Each gage must be bonded carefully to the structure and connected by its two leads to the signal-conditioning unit, which includes the excitation voltage source and the Wheatstone bridge. This can obviously be a major instrumentation chore, with computer-aided data acquisition and reduction a practical necessity.

### 5.2.3 Photoelasticity

Wire or foil resistance strain gages are probably the principal device used in experimental stress analysis today, but they have the disadvantage of monitoring strain only at a single point. The photoelasticity and moiré methods, to be outlined in the following sections, are more complicated in concept and application but have the ability to provide *full-field* displays of the strain distribution. The intuitive insight from these displays can be so valuable that it may be unnecessary to convert them to numerical values, although the conversion can be done if desired.

---

[12]M. Hetenyi, ed., *Handbook of Experimental Stress Analysis,* Wiley, New York, 1950.

Photoelasticity employs a property of many transparent polymers and inorganic glasses, called *birefringence*. To explain this phenomenon, recall the definition of *refractive index, n*, which is the ratio of the speed of light $v$ in the medium to that in vacuum, $c$:

$$n = \frac{v}{c} \qquad (5.43)$$

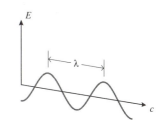

As the light beam travels in space (see Fig. 5.12), its electric field vector $E$ oscillates up and down at an angular frequency $\omega$ in a fixed plane, termed the *plane of polarization* of the beam. (The wavelength of the light is $\lambda = 2\pi c/\omega$.) A birefringent material is one in which the refractive index depends on the orientation of the plane of polarization; the magnitude of the birefringence is the difference in indices:

**FIGURE 5.12** Light propagation.

$$\Delta n = n_\perp - n_\parallel$$

where $n_\perp$ and $n_\parallel$ are the refractive indices on the two planes. Those two planes that produce the maximum $\Delta n$ are the *principal optical planes*. As shown in Fig. 5.13, a birefringent material can be viewed simplistically as a Venetian blind that resolves an arbitrarily oriented electric field vector into two components, one on each of the two principal optical planes, after which each component will transit the material at a different speed according to Eq. 5.43. After the two components eventually exit the material, they will again be traveling at the same speed but will be shifted in phase from one another by an amount related to the difference in transit times.

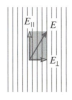

A *photoelastic* material is one in which the birefringence depends on the applied stress, and many such materials can be described to a good approximation by the *stress-optical law*

**FIGURE 5.13**
Venetian-blind model of birefringence.

$$\Delta n = C(\lambda)(\sigma_1 - \sigma_2) \qquad (5.44)$$

where $C$ is the *stress-optical coefficient* and the quantity in the second parentheses is the difference between the two principal stresses in the plane normal to the light propagation direction; this is just twice the maximum shear stress in that plane. The stress-optical coefficient is generally a function of the wavelength $\lambda$.

The stress distribution in an irregularly shaped body can be viewed by replicating the actual structure (probably scaled up or down in size for convenience) in a birefringent material such as epoxy. If the structure is statically determinate, the stresses in the model will be the same as that in the actual structure, in spite of the differences in modulus. To make the birefringence effect visible, the model is placed between crossed polarizers in an apparatus known as a *polariscope*. (Polarizers such as Polaroid, a polymer sheet containing oriented inorganic crystals, are essentially just birefringent materials that pass only light polarized in the polarizer's principal optical plane.)

The radiation source can produce either conventional white (polychromatic) or filtered (monochromatic) light. The electric field vector of light striking the first polarizer with an arbitrary orientation can be resolved into two components as shown in Fig. 5.14: one in the polarization direction and the other perpendicular to it. The polarizer will block the transverse component, allowing the parallel component to pass through to the specimen. This polarized component can be written

$$u_P = A \cos \omega t$$

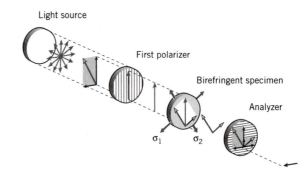

**FIGURE 5.14** The circular polariscope.

where $u_P$ is the field intensity at time $t$. The birefringent specimen will resolve this component into two further components, one along each of the principal stress directions; these can be written as

$$u_1 = A \cos \alpha \cos \omega t$$

$$u_2 = A \sin \alpha \cos \omega t$$

where $\alpha$ is the (unknown) angle the principal stress planes make with the polarization direction. Both of these new components pass through the specimen, but at different speeds, as given by Eq. 5.44. After traveling through the specimen a distance $h$ with velocities $v_1$ and $v_2$, they emerge as

$$u_1' = A \cos \alpha \cos \omega [t - (h/v_1)]$$

$$u_2' = A \sin \alpha \cos \omega [t - (h/v_2)]$$

These two components then fall on the second polarizer, oriented at 90° to the first and known as the *analyzer.* Each is again resolved into further components parallel and perpendicular to the analyzer axis, and the perpendicular components are blocked while the parallel components pass through. The transmitted component can be written as

$$u_A = -u_1' \sin \alpha + u_2' \cos \alpha$$

$$= -A \sin \alpha \cos \alpha \left[ \cos \omega \left( t - \frac{h}{v_1} \right) - \cos \omega \left( t - \frac{h}{v_2} \right) \right]$$

$$= A \sin 2\alpha \sin \omega \left( \frac{h}{2v_1} - \frac{h}{2v_2} \right) \sin \omega \left( t - \frac{h}{2v_1} - \frac{h}{2v_2} \right)$$

This is of the form $u_A = A' \sin(\omega t - \delta)$, where $A'$ is an amplitude and $\delta$ is a phase angle. Note that the amplitude is zero, so that no light will be transmitted either if $\alpha = 0$ or if

$$\frac{2\pi c}{\lambda} \left( \frac{h}{2v_1} - \frac{h}{2v_2} \right) = 0, \pi, 2\pi, \ldots \tag{5.45}$$

The case for which $\alpha = 0$ occurs when the principal stress planes are aligned with the polarizer-analyzer axes. All positions on the model at which this is true thus produce an extinction of the transmitted light. These are seen as dark bands called *isoclinics,* since they map out lines of constant inclination of the principal stress axes. These contours can be photographed at a sequence of polarization orientations if desired, giving an even more complete picture of stress directions.

Positions of zero stress produce extinction as well, because then the retardation is zero and the two light components exiting the analyzer cancel one another. The neutral axis of a beam in bending, for instance, shows as a black line in the observed field. As the stress at a given location is increased from zero, the increasing phase shift between the two components causes the cancellation to be incomplete, and light is observed. Eventually, as the stress is increased still further, the retardation will reach $\delta = \pi$, and extinction occurs again. This produces another dark fringe in the observed field. In general, alternating light areas and dark fringes are seen, corresponding to increasing orders of extinction.

Close fringe spacing indicates a steep stress gradient, similar to elevation contour lines on a topographic map. Figure 5.15 shows the patterns around circular and sharp-crack stress risers. It may suffice simply to observe the locations of high fringe density to note the presence of stress concentrations, which could then be eliminated by suitable design modifications (such as rounding corners or relocating abrupt geometrical discontinuities from high-stress regions). If white rather than monochromatic light is used, brightly colored lines rather than dark fringes are observed, with each color being the complement of that color that has been brought into extinction according to Eq. 5.43. These bands of constant color are termed *isochromatics*.

Converting the fringe patterns to numerical stress values is usually straightforward but tedious, because the fringes are related to the stress difference $\sigma_1 - \sigma_2$ rather than a single stress. At a free boundary, however, the stress components normal to the boundary must be zero, which means that the stress tangential to the boundary is a principal stress and is therefore given directly by the fringe order there. The reduction of photoelastic patterns to numerical values usually involves beginning at these free surfaces and then working gradually into the interior of the body using a graphical procedure.

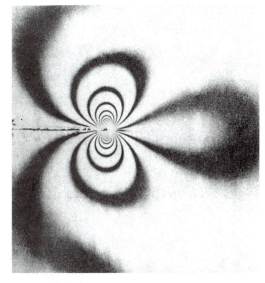

(*a*)                                           (*b*)

**FIGURE 5.15** Photoelastic patterns for stress around (*a*) a circular hole and (*b*) a sharp crack. (Miklos Heteni; *Handbook of Experimental Stress Analysis;* John Wiley & Sons, 1950.)

## 5.2.4 Moiré

The term *moiré* derives not from someone's name but from the name of a silk fabric that shows patterns of light and dark bands. Bands of this sort are also developed by the superposition of two almost-identical gratings, such as might be seen when looking through two window screens slightly rotated from one another. Figure 5.16 demonstrates that fringes are developed if the two grids have different spacing as well as different orientations. The fringes change dramatically for even small motions or strains in the gratings, and this visual amplification of motion can be used in detecting and quantifying strain in the specimen.

As a simple illustration of moiré strain analysis, assume that a grating of vertical lines of spacing $p$ (the *specimen* grating) is bonded to the specimen and that this is observed by looking through another *reference* grating of the same period but not bonded to the specimen. Now let the specimen undergo a strain, so that the specimen grating is stretched to a period of $p'$. A dark fringe will appear when the lines from the two gratings superimpose, and this will occur where $N(p' - p) = p$, since after $N$ lines on the specimen grid the incremental gap $(p' - p)$ will have accumulated to one reference pitch distance $p$. The distance $S$ between the fringes is then

$$S = Np' = \frac{pp'}{p' - p} \tag{5.46}$$

The normal strain $\epsilon_x$ in the horizontal direction is now given directly from the fringe spacing as

$$\epsilon_x = \frac{p' - p}{p} = \frac{p}{S} \tag{5.47}$$

Fringes will also develop if the specimen grid undergoes a rotation relative to the reference grid; if the rotation is small, then

$$\frac{p}{S} = \tan \theta \approx \theta$$

$$S = \frac{p}{\theta}$$

This angle is also the shear strain $\gamma_{xy}$, so

$$\gamma_{xy} = \theta = \frac{p}{\theta} \tag{5.48}$$

More generally, consider the interference fringes that develop between a vertical reference grid and an arbitrarily displaced specimen grid (originally vertical). The *zeroth-order* $(N = 0)$ fringe is that corresponding to positions having zero horizontal

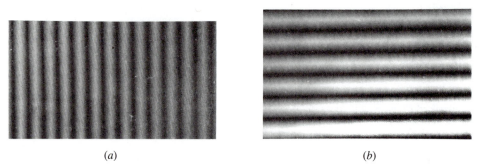

(*a*)                                   (*b*)

**FIGURE 5.16** Moiré fringes developed by difference in line pitch (*a*) and line orientation (*b*). (Prof. Fu-Pen Chiang, SUNY—Stony Brook.)

displacement; the first-order ($N = 1$) fringe corresponds to horizontal displacements of exactly one pitch distance; and so forth. The horizontal displacement is given directly by the fringe order as $u = Np$, from which the strain is given by

$$\epsilon_x = \frac{\partial u}{\partial x} = p\frac{\partial N}{\partial x} \tag{5.49}$$

so the strain is given as the slope of the fringe.

Similarly, a moiré pattern developed between two originally horizontal grids, characterized by fringes $N' = 0, 1, 2, \ldots$ gives the vertical strains:

$$\epsilon_y = \frac{\partial v}{\partial y} = \frac{\partial(N'p)}{\partial y} = p\frac{\partial N'}{\partial y} \tag{5.50}$$

The shearing strains are found from the slopes of both the $u$-field and $v$-field fringes:

$$\gamma_{xy} = p\left(\frac{\partial N}{\partial y} + \frac{\partial N'}{\partial x}\right) \tag{5.51}$$

Figure 5.17 shows the fringes corresponding to vertical displacements around a circular hole in a plate subjected to loading in the $y$-direction. The vertical strain $\epsilon_y$ is proportional to the $y$-distance between these fringes, each of which is a contour of constant vertical displacement. This strain is largest along the $x$-axis at the periphery of the hole and smallest along the $y$-axis at the periphery of the hole.

**FIGURE 5.17** Moiré patterns of the vertical displacements of a bar with a hole under pure tension. (Prof. Fu-Pen Chiang, SUNY—Stony Brook.)

## 5.3 FINITE ELEMENT ANALYSIS

### 5.3.1 Introduction

Finite element analysis (FEA) has become commonplace in recent years and is now the basis of a multibillion-dollar-per-year industry. Numerical solutions to even very complicated stress problems can now be obtained routinely using FEA. However, as was mentioned earlier, the disadvantages to computer solutions must be kept in mind when using this and similar methods. They do not necessarily reveal how the stresses are influenced by important problem variables such as materials properties and geometrical features, and errors in input data can produce wildly incorrect results that may be overlooked by the analyst. Perhaps the most important function of theoretical modeling is that of sharpening the designer's intuition; users of finite element codes should plan their strategy toward this end, supplementing the computer simulation with as much closed-form and experimental analysis as possible.

Finite element codes are less complicated than many of the word-processing and spreadsheet packages found on modern microcomputers. Nevertheless, they are complex enough that most users do not find it effective to program their own code. A number of prewritten commercial codes are available, representing a broad price range and compatible with machines from microcomputers to supercomputers.[13] However, users with specialized needs should not necessarily shy away from code development and may find the code sources available in such texts as that by Zienkiewicz[14] to be

[13]C.A. Brebbia, ed., *Finite Element Systems, A Handbook,* Springer-Verlag, Berlin, 1982.

[14]O.C. Zienkiewicz and R.L. Taylor, *The Finite Element Method,* McGraw-Hill Co., London, 1989.

a useful starting point. The `felt` program mentioned in Chapter 2, freely available through the Internet, also includes the source code. Most finite element software is written in Fortran, but some newer codes such as `felt` are in C.

In practice, a finite element analysis usually consists of three principal steps:

1. *Preprocessing:* The user constructs a *model* of the part to be analyzed in which the geometry is divided into a number of discrete subregions, or *elements,* connected at discrete points, called *nodes.* Certain of these nodes will have fixed displacements, and others will have prescribed loads. These models can be extremely time-consuming to prepare, and commercial codes vie with one another to have the most user-friendly graphical "preprocessor" to assist in this rather tedious chore. Some of these preprocessors can overlay a mesh on a preexisting CAD file, so that finite element analysis can be done conveniently as part of the computerized drafting and design process.

2. *Analysis:* The dataset prepared by the preprocessor is used as input to the finite element code itself, which constructs and solves a system of linear or nonlinear algebraic equations

$$\mathbf{K}_{ij}\mathbf{u}_j = \mathbf{f}_i$$

   where $\mathbf{u}$ and $\mathbf{f}$ are the displacements and externally applied forces at the nodal points. This is the same approach used in the matrix analysis of truss problems presented in Chapter 2, the only difference being the formation of the stiffness matrix $\mathbf{K}$. Finite element codes are therefore supersets of truss analysis codes, with trusses being just one element subroutine in the element library.

3. *Postprocessing:* In the earlier days of finite element analysis the user would pore through reams of numbers generated by the code, listing displacements and stresses at discrete positions within the model. It is easy to miss important trends and hot spots this way, and modern codes use graphical displays to assist in visualizing the results. A typical postprocessor display overlays colored contours, representing stress levels, on the model, showing a full-field picture similar to that of photoelastic or moiré experimental results.

The operation of a specific code is usually detailed in the documentation accompanying the software, and vendors of the more expensive codes will often offer workshops or training sessions as well to help users learn the intricacies of code operation. One problem users may have even after this training is that the code tends to be a "black box" whose inner workings are not understood. In the next section, we will outline the principles underlying most current finite element stress analysis codes, limiting the discussion to linear elastic stress analysis for now. Understanding this theory helps dissipate the black-box syndrome and also serves to summarize the analytical foundations of solid mechanics.

### 5.3.2 Theoretical Background

In the usual "displacement formulation" of the finite element method, the governing equations are combined so as to have only displacements appearing as unknowns; this can be done by using Eq. 5.2 to replace the stresses in Eq. 5.1 by the strains and then using Eq. 5.3 to replace the strains by the displacements. This gives

$$\mathbf{L}^{\mathrm{T}}\boldsymbol{\sigma} = \mathbf{L}^{\mathrm{T}}\mathbf{D}\boldsymbol{\epsilon} = \mathbf{L}^{\mathrm{T}}\mathbf{D}\mathbf{L}\mathbf{u} = \mathbf{0} \tag{5.52}$$

As was noted in the previous section, it is often impossible to solve these equations in closed form for irregular boundary conditions of the sort commonly encountered in engineering practice. However, the equations are amenable to discretization and solution by numerical techniques such as finite differences or finite elements.

The finite element method is one of several approximate numerical techniques available for the solution of engineering boundary value problems. Problems in the mechanics of materials often lead to equations of this type, and finite element methods have a number of advantages in handling them. The method is particularly well suited to problems with irregular geometries and boundary conditions, and it can be implemented in general computer codes that can be used for many different problems.

To obtain a numerical solution for the stress analysis problem, let us postulate a function $\tilde{\mathbf{u}}(x, y)$ as an approximation to $\mathbf{u}$:

$$\tilde{\mathbf{u}}(x, y) \approx \mathbf{u}(x, y) \tag{5.53}$$

Many different forms might be adopted for the approximation $\tilde{\mathbf{u}}$. The finite element method discretizes the solution domain into an assemblage of subregions, or elements, each of which has its own approximating functions. Specifically, the approximation for the displacement $\tilde{\mathbf{u}}(x, y)$ within an element is written as a combination of the (as yet unknown) displacements at the nodes belonging to that element:

$$\tilde{\mathbf{u}}(x, y) = N_j(x, y)\mathbf{u}_j \tag{5.54}$$

Here the index $j$ ranges over the element's nodes, the $\mathbf{u}_j$ are the nodal displacements, and the $N_j$ are *interpolation functions*, which are usually simple polynomials (generally linear, quadratic, or, occasionally, cubic polynomials) that are chosen to become unity at node $j$ and zero at the other element nodes. The interpolation functions can be evaluated at any position within the element by means of standard subroutines, so the approximate displacement at any position within the element can be obtained in terms of the nodal displacements directly from Eq. 5.54.

The interpolation concept can be illustrated by asking how we might guess the value of a function $u(x)$ at an arbitrary point $x$ located between two nodes at $x = 0$ and $x = 1$, assuming we know somehow the nodal values $u(0)$ and $u(1)$. We might assume that as a reasonable approximation $u(x)$ simply varies linearly between these two values, as shown in Fig. 5.18, and write

$$u(x) \approx \tilde{u}(x) = u_0(1 - x) + u_1(x)$$

or

$$\tilde{u}(x) = u_0 N_0(x) + u_1 N_1(x), \quad \begin{cases} N_0(x) = (1 - x) \\ N_1(x) = x \end{cases}$$

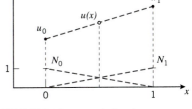

**FIGURE 5.18** Interpolation in one dimension.

Here the $N_0$ and $N_1$ are the linear interpolation functions for this one-dimensional approximation. Finite element codes have subroutines that extend this interpolation concept to two and three dimensions.

Approximations for the strain and stress follow directly from the displacements:

$$\tilde{\boldsymbol{\epsilon}} = \mathbf{L}\tilde{\mathbf{u}} = \mathbf{L}N_j\mathbf{u}_j \equiv \mathbf{B}_j\mathbf{u}_j \tag{5.55}$$

$$\tilde{\boldsymbol{\sigma}} = \mathbf{D}\tilde{\boldsymbol{\epsilon}} = \mathbf{D}\mathbf{B}_j\mathbf{u}_j \tag{5.56}$$

where $\mathbf{B}_j(x, y) = \mathbf{L}N_j(x, y)$ is an array of derivatives of the interpolation functions:

$$\mathbf{B}_j = \begin{bmatrix} N_{j,x} & 0 \\ 0 & N_{j,y} \\ N_{j,y} & N_{j,x} \end{bmatrix} \tag{5.57}$$

A "virtual work" argument can now be invoked to relate the nodal displacement $\mathbf{u}_j$ appearing at node $j$ to the forces applied externally at node $i$: If a small, or "virtual," displacement $\delta\mathbf{u}_i$ is superimposed on node $i$, the increase in strain energy $\delta U$ within an element connected to that node is given by

$$\delta U = \int_V \delta\boldsymbol{\epsilon}^\mathrm{T} \boldsymbol{\sigma} \, dV \tag{5.58}$$

where $V$ is the volume of the element. Using the approximate strain obtained from the interpolated displacements, $\delta\tilde{\boldsymbol{\epsilon}} = \mathbf{B}_i\delta\mathbf{u}_i$ is the approximate virtual increase in strain induced by the virtual nodal displacement. Using Eq. 5.56 and the matrix identity $(\mathbf{AB})^\mathrm{T} = \mathbf{B}^\mathrm{T}\mathbf{A}^\mathrm{T}$, we have

$$\delta U = \delta\mathbf{u}_i^\mathrm{T} \int_V \mathbf{B}_i^\mathrm{T}\mathbf{DB}_j \, dV \, \mathbf{u}_j \tag{5.59}$$

(The nodal displacements $\delta\mathbf{u}_i^\mathrm{T}$ and $\mathbf{u}_j$ are not functions of $x$ and $y$, and so can be brought from inside the integral.) The increase in strain energy $\delta U$ must equal the work done by the nodal forces; this is

$$\delta U = \delta W = \delta\mathbf{u}_i^\mathrm{T}\mathbf{f}_i \tag{5.60}$$

Equating Eqs. 5.59 and 5.60 and canceling the common factor $\delta\mathbf{u}_i^\mathrm{T}$, we have

$$\left[ \int_V \mathbf{B}_i^\mathrm{T}\mathbf{DB}_j \, dV \right] \mathbf{u}_j = \mathbf{f}_i \tag{5.61}$$

This is of the desired form $\mathbf{k}_{ij}\mathbf{u}_j = \mathbf{f}_i$, where $\mathbf{k}_{ij} = \int_V \mathbf{B}_i^\mathrm{T}\mathbf{DB}_j \, dV$ is the element stiffness.

Finally, the integral in Eq. 5.61 must be replaced by a numerical equivalent acceptable to the computer. Gauss-Legendre numerical integration is commonly used in finite element codes for this purpose, because that technique provides a high ratio of accuracy to computing effort. Stated briefly, the integration consists of evaluating the integrand at optimally selected integration points within the element and forming a weighted summation of the integrand values at these points. In the case of integration over two-dimensional element areas this can be written

$$\int_A f(x, y) \, dA \approx \sum_l f(x_l, y_l)w_l \tag{5.62}$$

The location of the sampling points $x_l$, $y_l$ and the associated weights $w_l$ are provided by standard subroutines. In most modern codes these routines map the element into a convenient shape, determine the integration points and weights in the transformed coordinate frame, and then map the results back to the original frame.

The functions $N_j$ used earlier for interpolation can be used for the mapping as well, achieving a significant economy in coding. This yields what are known as *numerically integrated isoparametric elements,* and these are a mainstay of the finite element industry.

Equation 5.61, with the integral replaced by numerical integrations of the form in Eq. 5.62, is the finite element counterpart of Eq. 5.52, the differential governing equation. The computer will carry out the analysis by looping over each element and, within each element, looping over the individual integration points. At each integration point the components of the element stiffness matrix $\mathbf{k}_{ij}$ are computed according to Eq. 5.61 and added into the appropriate positions of the $\mathbf{K}_{ij}$ global stiffness matrix, as was done in the assembly step of the matrix truss method described in Chapter 2. It can be appreciated that a good deal of computation is involved just in forming the terms of the stiffness matrix and that the finite element method could never have been developed without convenient and inexpensive access to a computer.

### 5.3.3 Stresses around a Circular Hole

We have considered the problem of a uniaxially loaded plate containing a circular hole several times previously, including the theoretical Kirsch solution and experimental determinations using both photoelastic and moiré methods. This problem is of practical importance—for instance, we have noted the dangerous stress concentration that appears near rivet holes—and it is also quite demanding in both theoretical and numerical analyses. Since the stresses rise sharply near the hole, a finite element grid must be refined there in order to produce acceptable results.

Figure 5.19 shows a mesh of elements developed by the `felt-velvet` graphical FEA package that can be used to approximate the displacements and stresses around

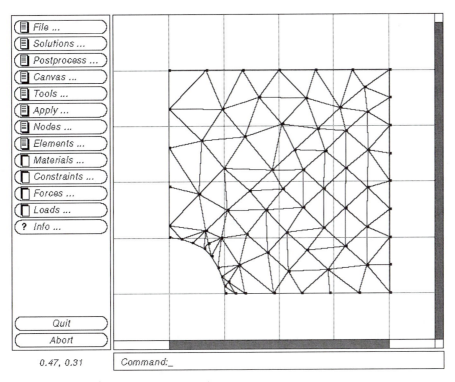

**FIGURE 5.19** Mesh for circular-hole problem.

a uniaxially loaded plate containing a circular hole. Since both theoretical and experimental results for this stress field are available, as already mentioned, the circular hole problem is a good one for becoming familiar with code operation.

The user should take advantage of symmetry to reduce problem size whenever possible. In this case only one quadrant of the problem need be meshed. The center of the hole is kept fixed, so the symmetry requires that nodes along the left edge be allowed to move vertically but not horizontally. Similarly, nodes along the lower edge are constrained vertically but left free to move horizontally. Loads are applied to the nodes along the upper edge, with each load being the resultant of the far-field stress acting along half of the element boundaries between the given node and its neighbors. (The far-field stress is taken as unity.) Portions of the `felt` input dataset for this problem are

```
problem description
nodes=76 elements=116

nodes
1   x=1 y=-0 z=0 constraint=slide_x
2   x=1.19644 y=-0 z=0
3   x=0.984562 y=0.167939 z=0 constraint=free
4   x=0.940634 y=0.335841 z=0
5   x=1.07888 y=0.235833 z=0
    .
    .
    .
72  x=3.99602 y=3.01892 z=0
73  x=3.99602 y=3.51942 z=0
74  x=3.33267 y=4 z=0
75  x=3.57706 y=3.65664 z=0
76  x=4 y=4 z=0

CSTPlaneStress elements
1   nodes=[13,12,23] material=steel
2   nodes=[67,58,55]
    .
    .
    .
6   nodes=[50,41,40]
7   nodes=[68,67,69] load=load_case_1
8   nodes=[68,58,67]
9   nodes=[57,58,68] load=load_case_1
10  nodes=[57,51,58]
11  nodes=[52,51,57] load=load_case_1
12  nodes=[37,39,52] load=load_case_1
13  nodes=[39,51,52]
    .
    .
    .
116 nodes=[2,3,1]

material properties
steel E=2.05e+11 nu=0.33 t=1

distributed loads
load_case_1 color=red direction=GlobalY values=(1,1) (3,1)
```

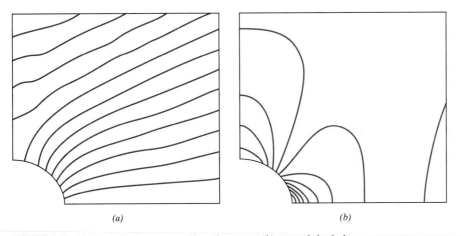

(a)                                              (b)

**FIGURE 5.20** Vertical displacements (*a*) and stresses (*b*) around the hole.

```
constraints
free Tx=u Ty=u Tz=u Rx=u Ry=u Rz=u
slide_x color=red Tx=u Ty=c Tz=c Rx=u Ry=u Rz=u
slide_y color=red Tx=c Ty=u Tz=c Rx=u Ry=u Rz=u

end
```

The *y*-displacements and vertical stresses $\sigma_y$ are contoured in Figs. 5.20*a* and *b* respectively; these should be compared with Figs. 5.17 and 5.15*a*. (These FEA results were obtained with the Zienkiewicz-Taylor `feap` code rather than `felt`.) The stress at the integration point closest to the *x*-axis at the hole is computed to be $\sigma_{y,\max} = 3.26$, 9 percent larger than the theoretical value of 3.00. In drawing the contours of Fig. 5.17*b*, the postprocessor extrapolated the stresses to the nodes by fitting a least-squares plane through the stresses at all four integration points within the element. This produces an even higher value for the stress concentration factor, 3.593. The user must be aware that graphical postprocessors smooth out results that are themselves only approximations, so numerical inaccuracy is a real possibility. Refining the mesh, especially near the region of highest stress gradient at the hole meridian, would reduce this error.

## 5.4 PROBLEMS

**1.** Expand the governing equations (5.1 through 5.3) in two Cartesian dimensions. Identify the unknown functions. How many equations and unknowns are there?

**2.** Justify the boundary conditions given in Eqs. 5.15 for stress in circular coordinates ($\sigma_r$, $\sigma_\theta$, $\tau_{xy}$) appropriate to a uniaxially loaded plate containing a circular hole as shown in Fig. P.5.2.

**3.** Develop an expression for the hoop stress $\sigma_\theta(t)$ and radial expansion $\delta_r(t)$ in a thin-walled pressure vessel subjected to a constantly increasing internal pressure $P(t) = R_P t$. The elastic solutions are

$$\sigma_\theta = \frac{Pr}{b} \qquad \delta_r = \frac{Pr^2}{bE}$$

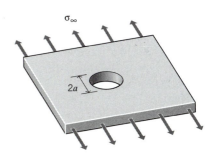

**FIGURE P.5.2**

where $P$ is the pressure, $r$ is the vessel radius, $b$ is the wall thickness, and $E$ is Young's modulus. The material is a Voigt solid (a Hookean spring of stiffness $k$ is parallel with a Newtonian dashpot of viscosity $\eta$). Express your answer in terms of the relaxation time $\tau = \eta/k$ and the spring stiffness $k$.

**4.** Develop an equation for the time-dependent midpoint deflection $\delta_P(t)$ of a viscoelastic beam in symmetric three-point bending, subject to a constant load $P$, as shown in Fig. P.5.4. The beam is constructed of a Standard Linear Solid material whose creep compliance can be written $C_{\text{crp}} = C_g + (C_r - C_g)(1 - e^{-t/\tau})$ (Eq. 1.29). The elastic solution is $\delta_P = PL^3/48EI$, where $L$ and $I$ are the length and the moment of inertia.

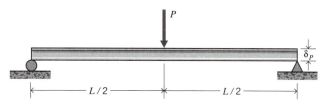

**FIGURE P.5.4**

**5.** Repeat Problem 4 but with a constantly increasing load $P(t) = R_P t$, where $R_P$ is the rate of load increase.

**6.** The deflection curve of a simply supported beam subjected to a uniform load of $w_0$, as shown in Fig. P.5.6, is given in handbooks as $\delta(x) = w_0 x/24EI)(L^3 - 2Lx^2 + x^3)$, where $L$ is the beam's length and $I$ is the cross-sectional moment of inertia about the centroid. Obtain an expression for the time-dependent deflection $\delta(x, t)$ of a beam constructed from a material that can be modeled as the Standard Linear Solid as in Problem 4.

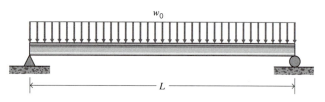

**FIGURE P.5.6**

**7.** A pressure vessel with a diameter of $d = 18''$ and a length of $L = 6'$ is to be constructed of a viscoelastic material that can be modeled as a Standard Linear Solid with a glassy compliance $C_g = 10^{-6}$ in$^2$/lb, a rubbery compliance $C_r$ of twice that, and a retardation time $\tau_{\text{crp}} = 6$ months. The vessel is to be capable of withstanding an internal pressure of $p = 1000$ psi.

(a) What should be the wall thickness $b$ in order to keep the hoop stresses under 2500 psi?

(b) Using the value of $b$ from part (a), what will be the creep deformation of the vessel radius after one year?

**8.** A linear viscoelastic material is placed in a rigid die and loaded in uniaxial compression ($\sigma_z = 0$), as shown in Fig. P.5.8. The material can be modeled as being elastic in dilatation (bulk modulus $K = 3 \times 10^{10}$ Pa), while the deviatoric components of the constitutive response are those of a Standard Linear Solid with $k_e = 3.81 \times 10^9$ Pa, $k_1 = 1.22 \times 10^{10}$ Pa, $\eta = 1.22 \times 10^{10}$ Pa-s. Determine the transverse stress $\sigma_x(t)$ and justify the form of the result.

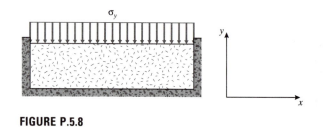

**FIGURE P.5.8**

**9.** The radial stress $\sigma_r$ developed by a load $P$ acting on the surface of a semi-infinite solid, directed normally to the surface toward the interior of the solid, as shown in Fig. P.5.9, is

$$\sigma_r = \frac{P}{2\pi} \left\{ (1 - 2\nu)\left[ \frac{1}{r^2} - \frac{z}{r^2} \sqrt{r^2 + z^2} \right] - 3r^2 z \sqrt{r^2 + z^2} \right\}$$

Develop the solution for the case of a semi-infinite viscoelastic body, assuming the material behaves as a Voigt solid in shear but is elastic under hydrostatic compression.

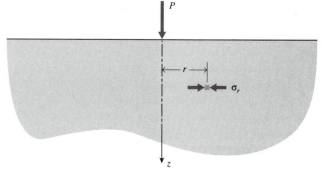

**FIGURE P.5.9**

**10.** The thick-walled pressure vessel shown in Fig. P.5.10 has inner radius $r_i$ and outer radius $r_o$ and is subjected to an internal pressure $p_i$ and an external pressure $p_o$. Assume a trial solution for the radial displacement of the form $u(r) = Ar + B/r$; this relation can be shown to satisfy the governing equations for equilibrium, strain-displacement, and stress-strain governing equations.

(a) Evaluate the constants $A$ and $B$ using the boundary conditions

$$\sigma_r = -p_i \text{ at } r = r_i, \qquad \sigma_r = -p_o \text{ at } r = r_o$$

(b) Then show that

$$\sigma_r(r) = -\frac{p_i[(r_o/r)^2 - 1] + p_o[(r_o/r_i)^2 - (r_o/r)^2]}{(r_o/r_i)^2 - 1}$$

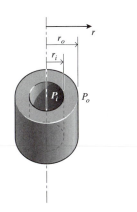

**FIGURE P.5.10**

**11.** Show that the Airy function $\phi(x, y)$ defined by Eqs. 5.12 satisfies the equilibrium equations.

**12.** Show that stress functions in the form of quadratic ($\phi = a_2x^2 + b_2xy + c_2y^2$) or cubic ($\phi = a_3x^3 + b_3x^2y + c_3xy^2 + d_3y^3$) polynomials automatically satisfy the governing relation $\nabla^4\phi = 0$.

**13.** Write the stresses $\sigma_x, \sigma_y, \tau_{xy}$ corresponding to the quadratic and cubic stress functions of Problem 12.

**14.** Choose the constants in the quadratic stress functions of Problems 12 and 13 so as to represent the conditions shown by the free-body diagrams in Fig. P.5.14: (a) simple tension, (b) biaxial tension, and (c) pure shear of a rectangular plate.

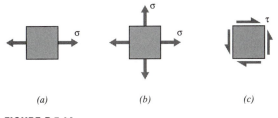

(a)          (b)          (c)

**FIGURE P.5.14**

**15.** Choose the constants in the cubic stress function of Problem 12 so as to represent pure bending induced by couples applied to vertical sides of a rectangular plate as shown in Fig. P.5.15.

**FIGURE P.5.15**

**16.** The cantilevered beam shown in Fig. P.5.16, with rectangular cross section, is loaded at the free end ($x = 0$) with a force $P$. At the free end, the boundary conditions on stress can be written $\sigma_x = \sigma_y = 0$, and

$$\int_{-h/2}^{h/2} \tau_{xy}\, dy = P$$

The horizontal edges are not loaded, so we also have $\sigma_x = \tau_{xy} = 0$ at $y = \pm h/2$.

(a) Show that these conditions are satisfied by a stress function of the form

$$\phi = b_2xy + d_4xy^3$$

(b) Evaluate the constants to show that the stresses can be written as

$$\sigma_x = \frac{Pxy}{I} \qquad \sigma_y = 0 \qquad \tau_{xy} = \frac{P}{2I}\left[\left(\frac{h}{2}\right)^2 - y^2\right]$$

in agreement with the elementary theory results given in Chapter 4.

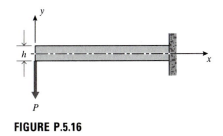

**FIGURE P.5.16**

**17.** A $0°$–$45°$–$90°$ three-arm strain gage rosette bonded to a steel specimen gives readings $\epsilon_0 = 175\ \mu$, $\epsilon_{45} = 150\ \mu$, and $\epsilon_{90} = -120\ \mu$. Determine the principal stresses and the orientation of the principal planes at the gage location.

**18.** Repeat Problem 17, but with gage readings $\epsilon_0 = 150\ \mu$, $\epsilon_{45} = 200\ \mu$, and $\epsilon_{90} = 125\ \mu$.

**19.** Obtain a plane-stress finite element solution for a cantilevered beam with a single load at the free end as shown in Fig. P.5.19. Use arbitrarily chosen (but reasonable) dimensions and material properties. Plot the stresses $\sigma_x$ and $\tau_{xy}$ as functions of $y$ at an arbitrary station along the span; also plot the stresses given by the elementary theory of Chapter 4 and assess the magnitude of the numerical error.

**FIGURE P.5.19**

**20.** Repeat Problem 20, but with the symmetrically-loaded beam in three-point bending shown in Fig. P.5.20.

**21.** Use axisymmetric elements to obtain a finite element solution for the radial stress in a thick-walled pressure vessel (using arbitrary geometry and material parameters). Compare the results with the theoretical solution of Problem 10.

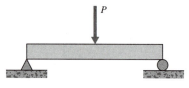

**FIGURE P.5.20**

# YIELD AND PLASTIC FLOW

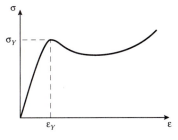

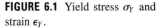

**FIGURE 6.1** Yield stress $\sigma_Y$ and strain $\epsilon_Y$.

## CHAPTER 6

In our overview of the tensile stress-strain curve in Chapter 1, we described yield as a permanent molecular rearrangement that begins at a sufficiently high stress, denoted $\sigma_Y$ in Fig. 6.1. The yielding process is very material-dependent, being related directly to molecular mobility. It is often possible to control the yielding process by optimizing the materials processing in a way that influences mobility. General-purpose polystyrene, for instance, is a weak and brittle plastic often credited with giving plastics a reputation for shoddiness that plagued the industry for years. The reason is that polystyrene at room temperature has so little molecular mobility that it experiences brittle fracture at stresses less than those needed to induce yield with its associated ductile flow. When that same material is blended with rubber particles of suitable size and composition, however, it becomes so tough that it is used for batting helmets and ultradurable children's toys. This "magic" is done by control of the yielding process. Yield control to balance strength against toughness is one of the most important aspects of materials engineering for structural applications, and all engineers should be aware of the possibilities.

Another important reason for understanding yield is more prosaic. If the material is not allowed to yield, it is not likely to fail. This is not true of brittle materials such as ceramics that fracture before they yield, but in most of the tougher structural materials no damage occurs before yield. It is common design practice to size the structure so as to keep the stresses in the elastic range, short of yield by a suitable safety factor. We therefore need to be able to predict when yielding will occur in general multidimensional stress states, given an experimental value of $\sigma_Y$.

Fracture is driven by *normal* stresses, acting to separate one atomic plane from another. Yield, conversely, is driven by *shearing* stresses, sliding one plane along another. These two distinct mechanisms are illustrated in Fig. 6.2. Of course, bonds must be broken during the sliding associated with yield, but unlike those broken in fracture, they are allowed to reform in new positions. This process can generate substantial change in the material and even lead eventually to fracture (as in bending a metal rod back and forth repeatedly to break it). The "plastic" deformation that underlies yielding is essentially a viscous flow process and follows kinetic laws quite similar to liquids. Like flow in liquids, plastic flow usually takes place without change in volume, corresponding to Poisson's ratio $\nu = \frac{1}{2}$.

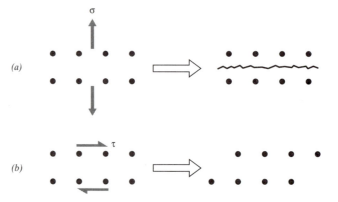

**FIGURE 6.2** Cracking is caused by normal stresses (*a*); sliding is
caused by shear stresses (*b*).

# 6.1 PHENOMENOLOGICAL ASPECTS OF YIELD

### 6.1.1 Drawability and Strengthening

The design engineer can choose from materials with a wonderfully diverse collection of yield behaviors. Some materials (such as porcelain) fracture before yielding; others (such as aluminum) yield but fail shortly thereafter; still others (such as polyethylene) yield and then go on to stretch hundreds or even thousands of percent before fracturing. A graphical method known as the *Considère construction* helps quantify the differences in drawability from material to material. This method replots the tensile stress-strain curve with "true" stress $\sigma_t$ as the ordinate and extension ratio $\lambda = L/L_0$ as the abscissa. The true stress is defined as load divided by the *current* area $A$, as opposed to the *original* area $A_0$.

These "true" stress-strain curves do not show the characteristic drop in stress at the yield point seen in conventional curves. The drop occurs because the specimen narrows at the location of plastic flow, and the reduced section does not transmit as much load. (This discussion assumes that the tensile testing is being done in displacement control; if the testing machine is in load control, it will "run away" at the yield point as the control system tries to bring the load transmitted by the specimen back up to the programmed level.) If the actual reduced area is used in normalizing the load to plot true stress, the effect of area is compensated and the curve continues to rise throughout the test.

If we assume constant volume, which is true at yield, the original and current specimen areas and lengths are related by $A_0L_0 = AL$. The "engineering" stress $\sigma_e$, the load divided by the original area, is then related to the true stress by

$$\sigma_e = \frac{P}{A_0} = \frac{PL_0}{AL} = \frac{\sigma_t}{\lambda}$$

Hence, the engineering stress $\sigma_e$ corresponding to any value of true stress $\sigma_t$ is the slope of a secant line drawn from origin ($\lambda = 0$, not $\lambda = 1$) to intersect the $\sigma_t$-$\lambda$ curve at $\sigma_t$.

Among the many possible shapes the true stress-strain curves could assume, let us consider the concave up, concave down, and sigmoidal shapes shown in Fig. 6.3. These differ in the number of tangent points that can be found for the secant line, and produce the following yield characteristics:

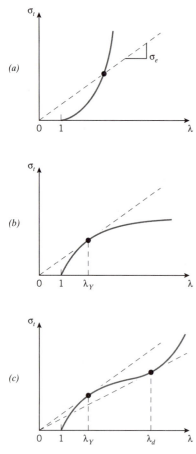

**FIGURE 6.3** True stress-strain curves
with (*a*) no tangents, (*b*) one tangent,
and (*c*) two tangents.

1. *No tangents:* Here the curve is always concave upward, as in Fig. 6.3a, so the slope of the secant line rises continuously. Therefore the engineering stress rises as well, without showing a yield drop. Eventually fracture intercedes, so a true stress-strain curve of this shape identifies a material that fractures before it yields.

2. *One tangent:* The curve is concave downward as in Fig. 6.3b, so a secant line reaches a tangent point at $\lambda = \lambda_Y$. The slope of the secant line, and therefore the engineering stress as well, begins to fall at this point. This is then the yield stress $\sigma_Y$ seen as a maximum in stress on a conventional stress-strain curve, and $\lambda_Y$ is the extension ratio at yield. The yielding process begins at some adventitious location in the gage length of the specimen, and it continues at that location rather than being initiated elsewhere because the secant modulus has been reduced at the first location. The specimen is now flowing at a single location with decreasing resistance, leading eventually to failure. Ductile metals such as aluminum fail in this way, showing a marked reduction in cross-sectional area at the position of yield and eventual fracture.

3. *Two tangents:* For sigmoidal stress-strain curves as in Fig. 6.3c, the engineering stress begins to fall at an extension ratio $\lambda_Y$ and then rises again at $\lambda_d$. As in the previous one-tangent case, material begins to yield at a single position when $\lambda = \lambda_Y$, producing a neck, implying in turn a nonuniform distribution of strain along the gage length. Material at the neck location then stretches to $\lambda_d$, after which the engineering stress there would have to rise to stretch it further. However, this stress is greater than that needed to stretch material at the edge of the neck from $\lambda_Y$ to $\lambda_d$, so material already in the neck stops stretching and the neck propagates outward from the initial yield location. Only material within the neck shoulders is being stretched during propagation, while the material inside the necked-down region holds constant at $\lambda_d$ (the material's *natural draw ratio*) and the material outside holds at $\lambda_Y$. When all the material has been drawn into the necked region, the stress begins to rise uniformly in the specimen until, eventually, fracture occurs.

The Considère construction provides useful insight, but it doesn't really explain *why* different materials fall into one or another of these categories. What governs the shape of the true stress-strain curves? The principal factor here is the balance of *strain softening* versus *strain hardening* developed in the material by yielding flow. "Softening" indicates a reduction in the tangent modulus; that is an increasing strain increment for equal increments of increased stress. We visualize the microstructure "breaking loose" under increasing stress, after which it flows more easily. This obviously leads to an instability at the point of initial yielding. Flow there makes continued flow easier, and so on until the specimen fails. This is the nature of the concave-downward, single-tangent case in the foregoing list.

On the other hand, the process of yield flow can harden the material rather than softening it. This makes the deformed material better able to withstand the increased true stress in the necked region, so that rather than stretching further, the material outside the neck shoulders stretches instead, and the neck propagates rather than thinning more and more until failure occurs. For this to happen, the true stress needed to increase the strain of the already-necked material has to be greater than that needed to draw the original material by more than a factor of $A_0/A = \lambda_d$. Then the stress needed to neck more material is less than that needed to stretch the already-necked material further, so the neck propagates.

Many metals harden during plastic flow by processes of dislocation multiplication and entanglement, to be discussed later. However, this type of hardening is not usually sufficient to permit continued drawing. The strain hardening needed to sustain the drawing process requires a dramatic transformation in the material; that is just what happens in such materials as the polyethylene used in six-pack holders. These materials are semicrystalline, with the crystals initially being flat lamellar plates, perhaps 10 nm thick, arranged radially outward like spokes, in spherical domains called *spherulites*. As the induced strain increases, these spherulites are first deformed in the straining direction. As the strain increases further, the spherulites are broken apart and the lamellar fragments rearrange with a predominantly axial molecular orientation to become what is known as a *fibrillar* microstructure. With the strong covalent bonds now dominantly lined up in the load-bearing direction, the material exhibits markedly greater strengths and stiffnesses—by perhaps an order of magnitude—than in the original material. These structural changes and their influence on the stress-strain curve are illustrated in Fig. 6.4.

Many of nature's polymers, such as spider silks, human tendons, and wood cellulose, are fibrillar and oriented with amazing specificity to optimize themselves for the loads they will be required to bear. Our "advanced" composite materials use this same method, orienting either molecular chains or reinforcing fibers to obtain the desired strength. We are getting pretty good at this, but nature has us hopelessly outclassed.

As might be expected, the strengthening that is obtained in the stretching direction increases as the perfection of molecular alignment increases. The alignment tends to increase with the draw ratio but is also influenced by such processing factors as temperature. At temperatures close to the melting point the polymer stretches easily but also has enough mobility to retract back to the unoriented, entropically favored state. The key to strengthening is the molecular orientation, not the stretch ratio itself, because the same orientation can be achieved with a large number of draw-temperature combinations. But when the orientation parameter $f$ of polypropylene fibers is plotted against strength (see Fig. 6.5), a smooth correlation is noted.[1] Here $f$ is a microstructural orientation factor, being 1 at perfect axial orientation and 0 at random orientation. The break in the curve at $f \approx 0.7$ corresponds to the spherulite-to-fibril transformation.

At a given temperature, the maximum draw ratio is determined by the inflections in the true stress-strain curve. Given that sufficient strengthening is needed to draw a material, what determines the natural draw ratio $\lambda_d$? In searching for more and more perfect orientation, we naturally want $\lambda_d$ to be as large as possible. Drawing a

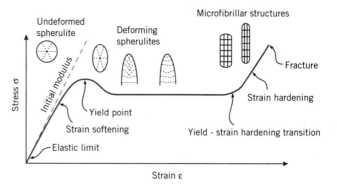

**FIGURE 6.4** Stress-strain curve of a drawable polymer. (From R. J. Samuels, *Structured Polymer Properties,* © 1974, Wiley. Reprinted by permission of John Wiley & Sons, Inc.)

---

[1]R.J. Samuels, *Structured Polymer Properties,* John Wiley & Sons, New York, 1974. This is a classic materials process-structure-property study.

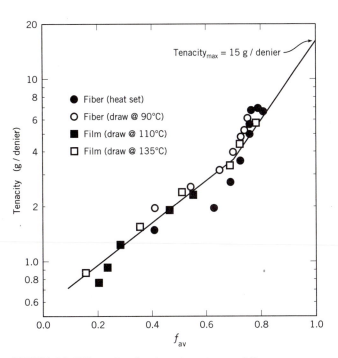

**FIGURE 6.5** Effect of molecular orientation on failure stress. (From R. J. Samuels, *Structured Polymer Properties,* © 1974, Wiley. Reprinted by permission of John Wiley & Sons, Inc.)

polymer is much like trying to straighten out an entangled garden hose. Long before the hose is straight, segments caught between entanglement points pull tight, and the hose will break if pulled further. The draw ratio $\lambda_d$ in polymers is similarly related to the molecular entanglement density, and the key to stretching a polymer to ultrahigh orientations involves reducing the entanglement density.

This is done commercially in two very different ways. The *aramid* fibers (such as DuPont Kevlar™) are polymers with such a rigid molecular structure that their molecular chains exist as rigid rods rather than entangled random coils. These rods align themselves as liquid crystals which can be oriented along a common direction during spinning. Another method known as *gel spinning* reduces entanglements by diluting the molten polymer with an organic diluent, such as hexane or decalin, and spinning from that state. This is the method used by Allied Signal to make Spectra™ polyethylene fibers. Both of these fibers have very high strengths and stiffnesses along their axis, comparable with steel but at a fraction of the weight, albeit with problematically poor strength in the lateral direction.

## 6.1.2 Multiaxial Stress States

The yield stress $\sigma_Y$ is usually determined in a tensile test, where a single uniaxial stress acts. However, the engineer must be able to predict when yield will occur in more complicated real-life situations involving multiaxial stresses. This is done by use of a *yield criterion,* an observation derived from experimental evidence as to just what it is about the stress state that causes yield. One of the simplest of these criteria, known as the maximum shear stress or *Tresca* criterion, states that yield occurs when the maximum shear stress reaches a critical value $\tau_{\max} = k$. The numerical value of $k$ for a given material could be determined directly in a pure-shear test, such as torsion of a circular shaft, but it can also be found indirectly from the tension test as well. As shown in Fig. 6.6, Mohr's circle shows that the maximum shear stress acts on a

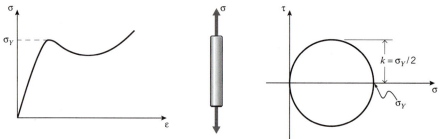

**FIGURE 6.6** Mohr's circle construction for yield in uniaxial tension.

plane 45° away from the tensile axis and is half the tensile stress in magnitude; then $k = \sigma_Y/2$.

In cases of plane stress, Mohr's circle gives the maximum shear stress *in that plane* as half the difference of the principal stresses:

$$\tau_{max} = \frac{\sigma_{p1} - \sigma_{p2}}{2} \tag{6.1}$$

## EXAMPLE 6.1

Using $\sigma_{p1} = \sigma_\theta = pr/b$ and $\sigma_{p2} = \sigma_z = pr/2b$ in Eq. 6.1, the shear stress in a cylindrical pressure vessel with closed ends is

$$\tau_{max,\theta z} = \frac{1}{2}\left(\frac{pr}{b} - \frac{pr}{2b}\right) = \frac{pr}{4b}$$

where the $\theta z$ subscript indicates a shear stress in a plane tangential to the vessel wall. Based on this, we might expect the pressure vessel to yield when

$$\tau_{max,\theta z} = k = \frac{\sigma_Y}{2}$$

which would occur at a pressure of

$$p_Y = \frac{4b\tau_{max,\theta z}}{r} \stackrel{?}{=} \frac{2b\sigma_Y}{r}$$

However, this analysis is in error, as can be seen by drawing Mohr's circles not only for the $\theta z$ plane but for the $\theta r$ and $rz$ planes as well, as shown in Fig. 6.7.

The shear stresses in the $\theta r$ plane are seen to be twice those in the $\theta z$ plane, because in the $\theta r$ plane the second principal stress is zero:

$$\tau_{max,\theta r} = \frac{1}{2}\left(\frac{pr}{b} - 0\right) = \frac{pr}{2b}$$

Yield will therefore occur in the $\theta r$ plane at a pressure of $b\sigma_Y/r$, half the value needed to cause yield in the $\theta z$ plane. Failing to consider the shear stresses acting in this third direction would lead to a seriously underdesigned vessel.

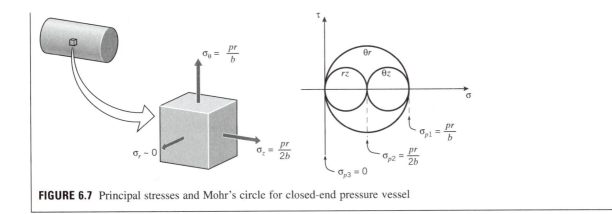

FIGURE 6.7 Principal stresses and Mohr's circle for closed-end pressure vessel

Situations similar to this example occur in plane stress whenever the principal stresses in the $xy$ plane are of the same sign (both tensile or both compressive). The maximum shear stress, which controls yield, is half the *difference* between the principal stresses; if they are both of the same sign, an even larger shear stress will occur on the perpendicular plane containing the larger of the principal stresses in the $xy$ plane.

This concept can be used to draw a *yield locus,* as shown in Fig. 6.8, an envelope in $\sigma_1$-$\sigma_2$ coordinates outside of which yield is predicted. This locus obviously crosses the coordinate axes at values corresponding to the tensile yield stress $\sigma_Y$. In quadrants I and III the principal stresses are of the same sign, so according to the maximum shear stress criterion, yield is determined by the difference between the larger principal stress and zero. In quadrants II and IV the locus is given by $\tau_{max} = |\sigma_1 - \sigma_2|/2 = \sigma_Y/2$, so $\sigma_1 - \sigma_2 = $ constant; this gives straight diagonal lines running from $\sigma_Y$ on one axis to $\sigma_Y$ on the other.

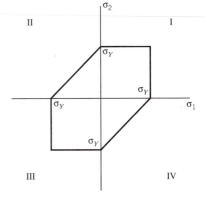

FIGURE 6.8 Yield locus for the maximum shear stress criterion.

## EXAMPLE 6.2

A circular shaft is subjected to a torque of half that needed to cause yielding, as shown in Fig. 6.9; we now ask what tensile stress could be applied simultaneously without causing yield.

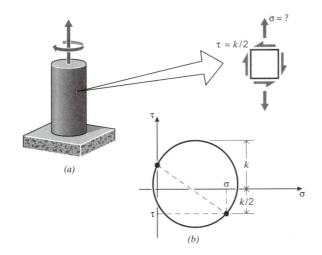

FIGURE 6.9 (*a*) Circular shaft subjected to simultaneous twisting and tension. (*b*) Mohr's circle construction.

A Mohr's circle is drawn with shear stress $\tau = k/2$ and unknown tensile stress $\sigma$. Using the Tresca maximum-shear yield criterion, yield will occur when $\sigma$ is such that

$$\tau_{max} = k = \sqrt{\left(\frac{\sigma}{2}\right)^2 + \left(\frac{k}{2}\right)^2}$$

$$\sigma = \sqrt{3}\, k$$

The Tresca criterion is convenient to use in practice, but a somewhat better fit to experimental data can often be obtained from the *von Mises*[2] *criterion,* in which the driving force for yield is the strain energy associated with the deviatoric components of stress. The *von Mises stress* (also called the *equivalent* or *effective* stress) is defined as

$$\sigma_M = \sqrt{\tfrac{1}{2}\left[(\sigma_x - \sigma_y)^2 + (\sigma_x - \sigma_z)^2 + (\sigma_y - \sigma_z)^2 + 6(\tau_{xy} + \tau_{yz} + \tau_{xz})\right]}$$

In terms of the principal stresses this is

$$\sigma_M = \sqrt{\tfrac{1}{2}\left[(\sigma_1 - \sigma_2)^2 + (\sigma_1 - \sigma_3)^2 + (\sigma_2 - \sigma_3)^2\right]}$$

where the stress differences in parentheses are proportional to the maximum shear stresses on the three principal planes.[3] (Since the quantities are squared, the order of stresses inside the parentheses is unimportant.) The von Mises stress can also be written in compact form in terms of the second invariant of the deviatoric stress tensor $\Sigma_{ij}$:

$$\sigma_M = \sqrt{3\Sigma_{ij}\Sigma_{ij}/2} \qquad (6.2)$$

It can be shown that this is proportional to the total distortional strain energy in the material and also to the shear stress $\tau_{oct}$ on the "octahedral" plane oriented equally to the 1-2-3 axes. The von Mises stress is the driving force for damage in many ductile engineering materials and is routinely computed by most commercial finite element stress analysis codes.

The value of the von Mises stress $\sigma_{M,Y}$ needed to cause yield can be determined from the tensile yield stress $\sigma_Y$, since in tension at the yield point we have $\sigma_1 = \sigma_Y$, $\sigma_2 = \sigma_3 = 0$. Then

$$\sigma_{M,Y} = \sqrt{\tfrac{1}{2}\left[(\sigma_Y - 0)^2 + (\sigma_Y - 0)^2 + (0 - 0)^2\right]} = \sigma_Y$$

Hence the value of von Mises stress needed to cause yield is the same as the simple tensile yield stress.

---

[2]Richard von Mises (1883–1953), Austrian-born American mathematician and aerodynamicist.

[3]Some authors use a factor other than 1/2 within the radical. This is immaterial, because it will be absorbed by the calculation of the critical value of $\sigma_M$.

The shear yield stress $k$ can similarly be found by inserting the principal stresses corresponding to a state of pure shear in the Mises equation. Using $k = \sigma_1 = -\sigma_3$ and $\sigma_2 = 0$, we have

$$\sqrt{\frac{1}{2}\left[(k-0)^2 + (k+k)^2 + (0-k)^2\right]} = \sqrt{\frac{6k^2}{2}} = \sigma_Y$$

$$k = \frac{\sigma_Y}{\sqrt{3}}$$

Note that this result is different than the Tresca case, in which we had $k = \sigma_Y/2$.

The von Mises criterion can be plotted as a yield locus as well. Like the Tresca criterion, it must pass through $\sigma_Y$ on each axis. However, it plots as an ellipse rather than the prismatic shape of the Tresca criterion (see Fig. 6.10).

### 6.1.3 Effect of Hydrostatic Pressure

Up to this point in the discussion, yield has been governed only by shear stress, so it has not mattered whether a uniaxial stress is compressive or tensile; yield occurs when $\sigma = \pm\sigma_Y$. This corresponds to the hydrostatic component of the stress $-p = (\sigma_x + \sigma_y + \sigma_z)/3$ having no influence on yield, which is observed experimentally to be valid for slip in metallic systems. Polymers, however, are much more resistant to yielding in compressive stress states than in tension. The atomic motions underlying slip in polymers can be viewed as requiring "free volume" as the molecular segments move, and this free volume is diminished by compressive stresses. It is thus difficult to form solid polymers by deformation processing such as stamping and forging in the same way steel can be shaped; this is one reason the vast majority of automobile body panels continue to be made of steel rather than plastic.

This dependency on hydrostatic stress can be modeled by modifying the yield criterion to state that yield occurs when

$$\tau_{\max} \text{ or } \sigma_M \geq \tau_0 + Ap \tag{6.3}$$

where $\tau_0$ and $A$ are constants. As $p$ increases (the hydrostatic component of stress becomes more positive) the shear stress needed for yield becomes greater as well, since there is less free volume and more hindrance to molecular motion. The effect of this modification is to slide the von Mises ellipse to extend less into quadrant I and more into quadrant III, as shown in Fig. 6.11. This shows graphically that greater stresses are needed for yield in compression, and lesser stresses in tension.

Several amorphous glassy polymers—notably polystyrene, polymethylmethacrylate, and polycarbonate—are subject to a yield mechanism termed *crazing,* in which elongated voids are created within the material by a tensile cavitation process. Figure 6.12 shows a craze in polystyrene, grown in plasticizing fluid near $T_g$. The voids, or crazes, are approximately 1000 Å thick and several microns or more in length, and appear visually to be much like conventional cracks. They differ from cracks, however, in that the broad faces of the crazes are spanned by a great many elongated fibrils that have been drawn from the polymer as the craze opens. The fibril formation requires shear flow, but the process is also very dependent on free volume. A successful multiaxial stress criterion for crazing has been proposed,[4] incorporating

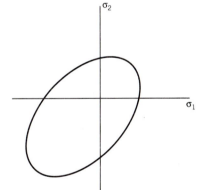

**FIGURE 6.10** Yield locus for the von Mises criterion.

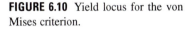

**FIGURE 6.11** Effect of pressure on the von Mises yield envelope.

---

[4]S. Sternstein and L. Ongchin, *Polymer Preprints,* vol. 10, p. 1117, 1969.

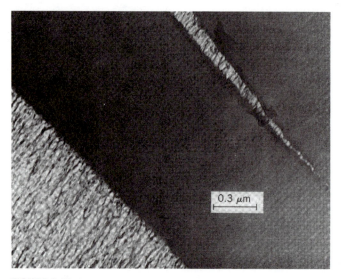

**FIGURE 6.12** A craze in polystyrene (from *Polymers: An Encyclopedic Sourcebook of Engineering Properties,* Wiley, 1987. Reprinted by permission of John Wiley & Sons, Inc.)

both these features. It has the form

$$\sigma_1 - \sigma_2 = A(T) + \frac{B(T)}{\sigma_1 + \sigma_2}$$

The left-hand side of this relation is proportional to the shear stress, and the denominator in the second term on the right-hand side is related to the hydrostatic component of the stress. As the hydrostatic tension increases, the shear needed to cause crazing decreases. The parameters $A$ and $B$ are adjustable, and both depend on temperature. This relation plots as a "bat wing" on the yield locus diagram, as seen in Fig. 6.13, approaching a 45° diagonal drawn through quadrants II and IV. Crazing occurs to the right of the curve; note that crazing never occurs in compressive stress fields.

Crazing is a yield mechanism, but it also precipitates brittle fracture as the craze height increases and the fibrils are brought to rupture. The point where the craze locus crosses the shear yielding locus is therefore a type of mechanically induced ductile-brittle transition, as the failure mode switches from shear yielding to craze embrittlement. Environmental agents, such as acetone, that expand the free volume in these polymers greatly exacerbate the tendency for craze brittleness. Conversely, modifications, such as rubber particle inclusions, that stabilize the crazes and prevent them from becoming true cracks can provide remarkable toughness. Rubber particles not only stabilize crazes; they also cause a great increase in the number of crazes, so the energy absorption of craze formation can add to the toughness as well. This is the basis of the "high-impact polystyrene" (HIPS) mentioned at the outset of this chapter.

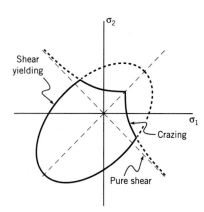

**FIGURE 6.13** The Sternstein envelopes for crazing and pressure-inhibited shear yielding.

### 6.1.4 Effect of Rate and Temperature

The yield process can be viewed as competing with fracture, and whichever process has the lowest stress requirements will dominate. As the material is made less and less mobile—for instance, by lowering the temperature or increasing the number and tightness of chemical bonds—yielding becomes more and more difficult. The fracture process is usually much less dependent on mobility. Both yield and fracture stresses usually increase with decreasing temperature, but yield is more temperature-dependent (see Fig. 6.14). This implies that below a critical temperature (called the *ductile-brittle transition temperature, $T_{DB}$*) the material will fracture before it yields. Several notable failures in ships and pipelines have occurred during winter temperatures when the steels used in their manufacture were stressed below their $T_{DB}$ and were thus unable to resist catastrophic crack growth. In polymers the ductile-brittle transition temperature is often coincident with the glass transition temperature. Clearly, we need an engineering model capable of showing how yield depends on temperature. One popular approach is outlined in the following paragraphs.

Yield processes are thermally activated, stress-driven motions, much like the flow of viscous liquids. Even without going into much detail as to the specifics of the motions, it is possible to write down quite effective expressions for the dependency of these motions on strain rate and temperature. In the *Eyring* view of thermally activated processes, an energy barrier $E_Y^*$ must be overcome for the motion to proceed. (We shall use the asterisk superscript to indicate activation parameters, and the $Y$ subscript here indicates the yield process.) A stress acts to lower the barrier when it acts in the direction of flow and to raise the barrier when the stress opposes the flow.

Consider now a constant strain rate test ($\dot{\epsilon}$ = const), in which the stress rises until yield occurs at $\sigma = \sigma_Y$. At the yield point we have $d\sigma/d\epsilon = 0$, so a fluidlike state is achieved in which an increment of strain can occur without a corresponding incremental increase in stress. Analogously with rate theories for viscous flow, an Eyring rate equation can be written for the yielding process as

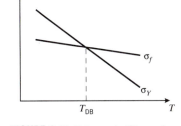

**FIGURE 6.14** Schematic illustration of the temperature dependence of yield and fracture stress.

$$\dot{\epsilon} = \dot{\epsilon}_0 \exp \frac{-(E_Y^* - \sigma_Y V^*)}{kT} \qquad (6.4)$$

Here $k$ is Boltzmann's constant, and $V^*$ is a factor governing the effectiveness of the stress in reducing the activation barrier. It must have units of volume for the product $\sigma_Y V^*$ to have units of energy; thus it is called the *activation volume* of the process. Taking logs and rearranging,

$$\frac{\sigma_Y}{T} = \frac{E_Y^*}{V^* T} + \left(\frac{k}{V^*}\right) \ln\left(\frac{\dot{\epsilon}}{\dot{\epsilon}_0}\right)$$

Hence plots of $\sigma_Y/T$ versus $\ln \dot{\epsilon}$ should be linear with a slope $k/V^*$, as seen in Fig. 6.15, from which the activation volume may be computed. The horizontal spacing between two lines at differing temperatures $T_1$ and $T_2$ gives the activation energy:

$$E_Y^* = \frac{k(\ln \dot{\epsilon}^{T_2} - \ln \dot{\epsilon}^{T_1})}{\left(\frac{1}{T_1} - \frac{1}{T_2}\right)}$$

Apparent activation volumes in polymers are on the order of 5000 Å³, much larger than a single repeat unit. This is taken to indicate that yield in polymers involves the cooperative motion of several hundred repeat units.

---

## EXAMPLE 6.3

The yield stress for polycarbonate is reported at 60 MPa at room temperature (23°C = 296 K), and we wish to know its value at 0°C (273 K), keeping the strain rate the same.

This can be accomplished by writing Eq. 6.4 out twice (once for each temperature) and then dividing one by the other. The parameters $\dot{\epsilon}$ and $\dot{\epsilon}_0$ cancel, leaving

$$1 = \exp\left(\frac{E_Y^* - \sigma_Y^{273} V^*}{R(273)} - \frac{E_Y^* - \sigma_Y^{296} V^*}{R(296)}\right)$$

From the data in Fig. 6.15, the yield activation parameters are $E_Y^* = 309$ kJ/mol, $V^* = 3.9 \times 10^{-3} \text{m}^3/\text{mol}$. Using these along with $R = 8.314$ J/mol and $\sigma_Y^{296} = 60 \times 10^6$ N/m², we have

$$\sigma_Y^{273} = 61.5 \text{ MPa}$$

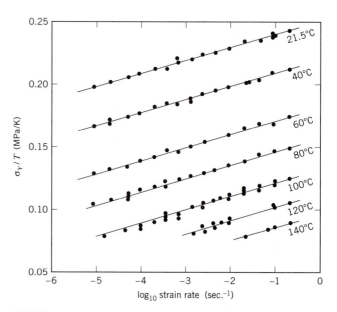

**FIGURE 6.15** Eyring plot showing dependence of yield strength on temperature and strain rate in polycarbonate (from N. G. Mc-Crum, C. P. Buckley, and C. B. Bucknall, *Principles of Polymer Engineering,* Oxford University Press, 1988).

## 6.2 THE DISLOCATION BASIS OF YIELD IN CRYSTALLINE MATERIALS

Phenomenological treatments such as the Eyring model are very useful for engineering predictions, but they provide only limited insight to the molecular mechanisms that underlie yield. Molecular understanding is a higher level of insight and also guides processing adjustments that can optimize the material. As we saw in Chapter 1, the high level of order present in crystalline materials led to good atomistic models for the stiffness. Early workers naturally sought an atomistic treatment of the yield process as well. This turned out to be a much more subtle problem than had been anticipated; it required researchers to hypothesize a type of crystalline defect, the *dislocation,* to explain the experimentally observed results. Dislocation theory permits a valuable intuitive understanding of yielding in metals (and even some ceramics and

### 6.2.1 Theoretical Yield Strength

In yield, atoms slide tangentially from one equilibrium position to another. The forces required to bring this about are given by the bond energy function, which is the anharmonic curve resulting from the balance of attractive and repulsive atomic forces described in Chapter 1. The force needed to displace the atom from equilibrium is the derivative of the energy function, being zero at the equilibrium position (see Fig. 6.16). As a simplifying assumption, let us approximate the force function with a harmonic expression, and write

$$\tau = \tau_{max} \sin\left(2\pi\frac{x}{a}\right)$$

where $a$ is the interatomic spacing. The stress reaches a maximum a quarter of the distance between the two positions, dropping to zero at the metastable position midway between them. After that the stress changes sign, meaning that force is required to hold the atom back as it tries to fall toward the new equilibrium position. Using $\gamma = x/a$ as the shear strain, the maximum shear stress $\tau_{max}$ can be related to the shear modulus $G$ as

$$\frac{d\tau}{d\gamma} = \frac{d\tau}{dx}\frac{dx}{d\gamma} = a\frac{d\tau}{dx} = a\tau_{max}\frac{2\pi}{a}\cos\frac{2\pi x}{a}$$

$$G = \frac{d\tau}{d\gamma}\Big|_{\gamma\to 0} = \tau_{max}\cdot 2\pi$$

This implies a shear stress at yield of $\tau_{max} = G/2\pi \approx G/10$, which would be on the order of 10 GPa. Measured values are 10 to 100 MPa, so the theoretical value is 2 to 3 orders of magnitude too large. More elaborate derivations give a somewhat smaller value for the theoretical yield stress, but still much larger than what is observed experimentally.

### 6.2.2 Edge, Screw, and Mixed Dislocations

A rationale for the low experimental values for the yield strengths of crystalline materials was proposed independently by Taylor, Polyani, and Orowan in 1934. These workers realized that it was not necessary to slip entire planes of atoms past one another to deform the material plastically, a process that would require breaking all the bonds connecting the planes simultaneously. The stress needed to do this would be very high, on the order of $G/10$ as described in the foregoing section. However, it isn't necessary to move all the atoms at once; only a few at a time need to move, requiring a much smaller stress. Analogously to the way an inchworm moves, only those atoms lying in a plane above a single line might be displaced one atomic spacing. This would force the plane of atoms previously there into a midway position as shown in Fig. 6.17, creating an "extra" plane of atoms halfway between the normal equilibrium positions. The termination of this plane then constitutes a line defect in the crystal, known as a *dislocation*.

Viewed end-on, as seen in Fig. 6.18, it can be appreciated that the extra plane of atoms creates a region of compression near the plane but above the dislocation line, and a tensile region below it. In a "soft" crystal, whose interatomic bonds are relatively

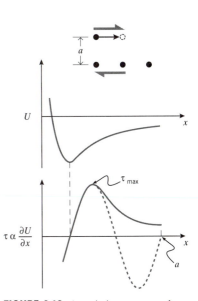

**FIGURE 6.16** Atomistic energy and stress functions.

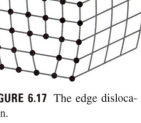

**FIGURE 6.17** The edge dislocation.

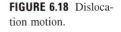

**FIGURE 6.18** Dislocation motion.

compliant, the distortion extends an appreciable distance from the dislocation. Conversely, in "hard" crystals, with stiffer bonds, the distortion is confined to a smaller region near the dislocation. Face-centered cubic (fcc) metals such as copper and gold have close-packed planes (those with (111) Miller indices[5]), which corresponds to large distances between those planes. This gives rise to relatively soft interplanar bonds, so the dislocation width is large. The dislocation width is substantially smaller in body-centered cubic (bcc) metals such as iron and steel, smaller still in ionically bonded ceramics, and even smaller in covalently bonded ceramics.

The dislocation associated with this extra plane of atoms can be moved easily, since only a small adjustment in position is required to break the bonds on the next plane over and allow them to form on the originally "extra" plane. Now the third plane is the extra one, and the dislocation will have moved by one atomic position. Slip is obviously made much easier if dislocation motion is available to the material. In fact, it first appears that the dislocation concept does *too* good a job in explaining crystal plasticity, since the dislocation is in a balanced metastable position and should be capable of being moved either left or right with a vanishingly small force. If this were true, the crystal would have essentially *no* shear strength.

However, as the dislocation moves, it drags with it the regions of compressive and tensile distortion in the lattice around it. This is accompanied by a sort of frictional drag, giving rise to a resistance to dislocation motion known as the *Peierls force.* This force is dependent on such factors as the crystal type and the temperature and plays an important role in determining the material's yield stress. As seen in Table 6.1, materials with wide dislocations have low Peierls forces, because the distortion is spread out over a large volume and is much less intense at its core.

Table 6.1 also indicates that the effect of temperature on the Peierls force is low for fcc materials having wide dislocations, and thus the yield stress has only a small temperature dependency. Conversely, materials with narrow and intense dislocation fields have high Peierls forces and therefore a large temperature sensitivity of the yield stress; higher temperatures facilitate dislocation mobility and thus reduce the yield strength. Among the important consequences of these factors is the dangerous tendency of steel to become brittle at low temperatures; as the temperature is lowered, the yield stress can rise to such high levels that brittle fracture intervenes.

Dislocations can have geometries other than the simple edge dislocation shown in Fig. 6.18. A more general view is provided by considering displacing a portion of the atoms in a "slip plane" *acfg* a distance $\bar{b}$, as shown in Fig. 6.19. The vector $\bar{b}$ is also a measure of the magnitude and direction of the crystal dislocation, and is known as *Burgers' vector.* The boundary between slipped and unslipped atoms on the slip plane is the dislocation line, shown as a dotted line. At position *e* the dislocation line

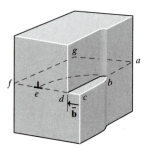

**FIGURE 6.19** The mixed dislocation.

**TABLE 6.1** Relationship between dislocation width and yield strength temperature sensitivity

| Material | Crystal type | Dislocation width | Peierls stress | Yield strength temperature sensitivity |
|----------|--------------|-------------------|----------------|----------------------------------------|
| Metal | fcc | Wide | Very small | Negligible |
| Metal | bcc | Narrow | Moderate | Strong |
| Ceramic | Ionic | Narrow | Large | Strong |
| Ceramic | Covalent | Very narrow | Very large | Strong |

From R. W. Hertzberg, *Deformation and Fracture Mechanics of Engineering Materials,* © 1976, John Wiley & Sons. Reprinted by permission of John Wiley & Sons, Inc.

[5]See Appendix E for a summary of crystallographic notation.

is perpendicular to Burgers' vector, so these two quantities lie in the slip plane. A dislocation so situated is called an *edge* dislocation and is constrained to move only in the slip plane defined by the dislocation line and Burgers' vector.

At position $b$ a spiral-like defect is formed such that a circular transit around the dislocation line ends on a plane a distance $\bar{b}$ from the starting point. Now the defect is known as a *screw* dislocation. The dislocation line is now parallel to Burgers' vector, so these two quantities do not define a unique slip plane the way an edge dislocation does. A screw dislocation can therefore *cross-slip* to another easy-glide plane passing through the dislocation line, and this mechanism enables screw dislocations to maneuver around obstacles that might otherwise impede their motion. Edge dislocations are more easily pinned, because they must "climb" to surmount obstacles by diffusion of vacancies, as illustrated in Fig. 6.20.

As the curved dislocation line is traversed from point $b$ to point $e$, the dislocation changes gradually from screw to edge character. At intermediate points the dislocation has both edge and screw character and is known as a *mixed* dislocation.

### 6.2.3 Dislocation-Controlled Yield

Single crystals tend to slip on their most closely packed planes and in directions of minimum atomic separation distance. The distance between planes is maximum for the close-packed planes, so these are the most loosely bonded. Slip in close-packed directions minimizes the distance over which the stresses need to displace the slipping atoms. Both of these effects act to minimize the energy needed for slip in the close-packed directions. There are 12 such slip systems in the face-centered cubic (fcc) systems; using Miller indices (see Appendix E) these are the {111} planes and the ⟨110⟩ directions. There are four independent nonparallel (111) planes, and three independent [110] directions in each plane.

Slip occurs when the shear stress on the slip plane and in the slip direction reaches a value $\tau_{\mathrm{crss}}$, the *critical resolved shear stress;* experimental values for $\tau_{\mathrm{crss}}$ are listed in Table 6.2 for a number of single-crystal materials. The *resolved shear stress* corresponding to an arbitrary stress state can be computed using the transformation relations of Chapter 3. In a simple tension test it can be written by inspection of Fig. 6.21 as

$$\tau_{\mathrm{rss}} = \frac{P \cos \theta}{A_s} = \frac{F \cos \theta}{A_0/\cos \phi} = \sigma(\cos \theta \cos \phi) \equiv \frac{\sigma}{m}$$

where $m$ is a structure factor dependent on the orientation of the slip system relative to the applied tensile stress. For single crystals of arbitrary alignment, the yield stress

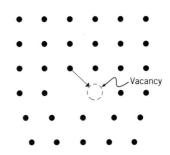

**FIGURE 6.20** Dislocation climb by vacancy diffusion.

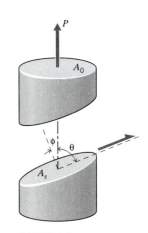

**FIGURE 6.21** Critical resolved shear stress.

**TABLE 6.2** Critical resolved shear stress for single crystals of various materials

| Material | Crystal type | Slip system | $\tau_{\mathrm{crss}}$ MPa |
|---|---|---|---|
| Nickel | fcc | {111} ⟨110⟩ | 5.7 |
| Copper | fcc | {111} ⟨110⟩ | 0.98 |
| Gold | fcc | {111} ⟨110⟩ | 0.90 |
| Silver | fcc | {111} ⟨110⟩ | 0.60 |
| Magnesium | hcp | {1101} ⟨001⟩ | 0.81 |
| NaCl | Cubic | {110} ⟨110⟩ | 0.75 |

will then be of the form

$$\sigma_Y = \tau_{\text{crss}} \cdot m \tag{6.5}$$

This is known as *Schmid's law,* and $m$ is the *Schmid factor.*

The yield stress will generally be higher in polycrystalline materials because many of the grains will be oriented unfavorably (have high Schmid factors). Equation 6.5 can be modified for polycrystalline systems as

$$\sigma_Y = \tau_{\text{crss}} \cdot \overline{m}$$

where $\overline{m}$ is an equivalent Schmid factor that is generally somewhat higher than a simple average over all the individual grains; for fcc and bcc systems $\overline{m} \approx 3$.

### 6.2.4  Strain Energy in Dislocations

Many calculations in dislocation mechanics are done more easily with energy concepts than with Newtonian force-displacement approaches. As seen in Fig. 6.22, the shear strain associated with a screw dislocation is the deflection $\overline{b}$ divided by the circumference of a circular path around the dislocation core:

$$\gamma = \frac{\overline{b}}{2\pi r} \tag{6.6}$$

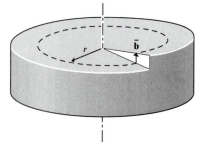

**FIGURE 6.22** Shear strain associated with a screw dislocation.

where $r$ is the distance from the dislocation core. Assuming Hookean elasticity, the corresponding strain energy per unit volume is

$$U = \int \tau \, d\gamma = \frac{1}{2}\tau\gamma = \frac{G\gamma^2}{2} = \frac{G\overline{b}^2}{8\pi^2 r^2}$$

The total strain energy associated with the screw dislocation is now obtained by integrating this over the volume around the dislocation:

$$U_{\text{screw}} = \int U \, dV = l \cdot \int_{r_0}^{r} \frac{G\overline{b}^2}{8\pi^2 r^2} 2\pi r \, dr$$

where here $l$ is the length of the dislocation line and $r_0$ is the radius of the dislocation "core" inside which the energy is neglected. (Mathematically, the energy density increases without bound inside the core; however, its volume becomes very small.) Taking $l = 1$ to obtain energy per unit length and carrying out the integration,

$$\boxed{U_{\text{screw}} = \frac{G\overline{b}^2}{4\pi} \ln \frac{r}{r_0} \approx G\overline{b}^2} \tag{6.7}$$

This last approximation sign should be read "scales as," because the limiting value $r$ is arbitrarily selected so that $\ln(r/r_0) \approx 4\pi$. The important conclusion is that the dislocation energy increases linearly with the shear modulus $G$ and quadratically with the magnitude of Burgers' vector $\bar{b}$. A similar expression can be obtained for the strain energy per unit length of edge dislocation; it can be shown that

$$U_{\text{edge}} = \frac{G\bar{b}^2}{1 - \nu} \qquad (6.8)$$

where $\nu$ is Poisson's ratio.

The dislocation energy represents an increase in the total energy of the system, which the material will try to eliminate if possible. For instance, two dislocations of opposite sign will be attracted to one another, because their strain fields will tend to cancel and lower the energy. Conversely, two dislocations of same sign will repel one another. The force of this attraction or repulsion will scale as

$$F\,dr = dU \Rightarrow F_{\text{screw}} \approx \frac{G\bar{b}^2}{r}$$

where here $r$ is the distance between dislocations.

## 6.2.5 Dislocation Motion and Hardening

The ductility of crystalline materials is determined by dislocation mobility, and factors that impede dislocation motion can produce dramatic increases in the material's yield strength. This increased resistance to plastic flow also raises the indentation hardness of the material, so strengthening of this sort is known as *hardening*. Alloying elements, grain boundaries, and even dislocations themselves can provide this impediment, and these provide the means by which the materials technologist controls yield. The following paragraphs will provide a brief outline of some of these mechanisms.

When one dislocation, moving on its slip plane under the influence of a driving shear stress, passes through another, a "jog" will be created in the second dislocation, as shown in Fig. 6.23. The portion of the dislocation line in the jog is now no longer on its original glide plane and is "pinned" in position. If the dislocation concentration is large, these jogs become a powerful impediment to plastic flow by dislocation motion. Paradoxically, the very dislocations that permit plastic flow in the first place can impede it if they become too numerous.

When a moving dislocation becomes pinned by jogs or other impediments, the shear stress $\tau$ that had been driving the dislocation now causes the line segment between the obstacles to bow forward, as shown in Fig. 6.24, with an angle $\phi$ between adjacent segments. The extra length of the bowed line represents an increase in the strain energy of the dislocation; if the shear stress were not present, the line would straighten out to reduce this energy. The line acts similarly to an elastic band, with a "line tension" $T$ that acts to return the line to a straight shortest-distance path between pinning points. The units of dislocation energy per unit length (N-m/m) are the same as simple tension, and we can write

$$T = \frac{\partial E}{\partial l} \approx Gb^2$$

As shown in Fig. 6.25, a free-body diagram of the line segment between two pinning points gives a force balance of the form

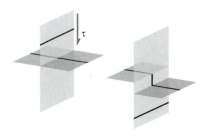

FIGURE 6.23 A dislocation jog.

FIGURE 6.24 Dislocation bowing.

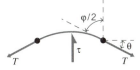

FIGURE 6.25 Force balance on dislocation segment.

$$2T \sin \frac{d\theta}{2} = \tau \overline{b} \cdot r \, d\theta$$

where here $r$ is the radius of curvature of the line (not the distance from the dislocation, as in Eq. 6.6). Rearranging and canceling the $d\theta$ factor, we obtain

$$\tau = \frac{G\overline{b}}{r} \tag{6.9}$$

This relation gives the curvature of the dislocation in terms of the shear stress acting on it. The maximum shear stress is that needed to bend the dislocation into a semi-circle (smallest $r$), after which the dislocation expands spontaneously. When the loops meet, annihilation occurs at that point, spawning a new dislocation line embedded in a circular loop. The process can be repeated with the new dislocation as well, and by this mechanism a large number of dislocations can be spawned as shown in Fig. 6.26. This is the *Frank-Read source* and is an important means by which dislocations can multiply during plastic deformation. The increasing number of dislocations leads to more and more entanglements, with jogs acting as pinning points.

Equation 6.9 also provides an estimate of the influence of dislocation density on yield strength. If the obstacles pinning dislocation motion are "soft," the dislocation will be able to overcome them at relatively low driving stress, corresponding to a low critical angle $\phi_c$. As the obstacle becomes "harder" (provides more resistance to dislocation motion), the angle approaches zero and the radius of curvature becomes on the order of the obstacle spacing $L$. The shear stress needed to overcome such obstacles is then

$$\tau \approx \frac{G\overline{b}}{L}$$

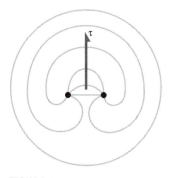

**FIGURE 6.26** The Frank-Read dislocation source.

When the hard obstacles arise from jogs created by intersections with other dislocations, the obstacle spacing $L$ can be written in terms of the dislocation density. If the number of dislocations passing through a unit area is $\rho$, the number of dislocations encountered in moving along a straight line will be proportional to $\sqrt{\rho}$. The spacing between them is proportional to the reciprocal of this, so $\tau \propto G\overline{b}\sqrt{\rho}$. The yield stress is then the stress $\tau_0$ needed to move dislocations in the absence of interfering dislocations, plus that needed to break through the obstacles; this can be written as

$$\boxed{\tau_Y = \tau_0 + AG\overline{b}\sqrt{\rho}} \tag{6.10}$$

where $A$ is a constant that has been found to vary between 0.3 and 0.6 for a number of fcc, bcc, and polycrystalline metals as well as some ionic crystals. Experimental corroboration of this relation is provided in Fig. 6.27.

The action of plastic flow therefore creates new dislocations by Frank-Read and other sources, making the material harder and harder, that is, increasingly resistant to further plastic flow. Eventually the yield stress for continued deformation becomes larger than the fracture stress, and the material will now break before it deforms further. If continued working of the material is desired, the number of dislocations must be reduced, for instance by thermal *annealing,* which can produce *recovery* (dislocation climb around obstacles by vacancy diffusion) or recrystallization of new dislocation-free grains.

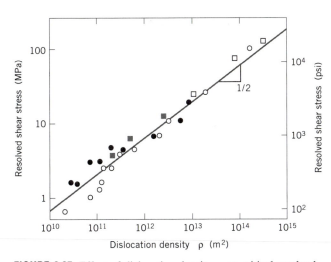

**FIGURE 6.27** Effect of dislocation density $\rho$ on critical resolved shear stress (for copper single and polycrystals, from T. H. Courtney, *Mechanical Behavior of Materials,* McGraw-Hill, 1990).

Grain boundaries act to impede dislocation motion, since the slip systems in adjoining grains will usually not line up; increases in yield strength arising from this mechanism are called *boundary strengthening*. Fine-grained metals have increased grain boundary area and thus have higher yield strengths than coarse-grained ones. The influence of grain size can often be described by the *Hall-Petch* formula

$$\sigma_Y = \sigma_0 + k_Y d^{-1/2} \tag{6.11}$$

where $\sigma_0$ is the lattice friction stress needed to move dislocations and $K$ is a constant. This relation is essentially empirical, but it can be rationalized by viewing the second term as being related to the stress needed to activate a new mobile dislocation in the unfavorably oriented grain.

As dislocations pile up against the boundary in the originally deforming grain, they act much like a crack whose length scales with the grain size $d$, as shown in Fig. 6.28: the larger the grain, the more dislocations in the pileup, and the larger the virtual crack. We saw in Chapter 5 that the stress in front of a sharp crack of length $a$ scales as $\sqrt{a}$. Hence, the stress in front of the crack containing the dislocation pileup is increased by a factor that scales with $\sqrt{d}$; when this stress exceeds that needed to generate a new dislocation, the unfavorably oriented grain begins to deform by dislocation motion. This stress diminishes according to $d^{-1/2}$ as the size of the original grain is scaled down, thus strengthening the metal according to the Hall-Petch relationship. Grain size is determined by the balance between nucleation and growth rates as the metal is solidified, and these are in turn controllable by the cooling rates imposed. This is an important example of processing-structure-property control available to the materials technologist.

A related phenomenon accounts for the very high strengths ($\approx$ 4 GPa, or 600 kpsi) of piano wire, a eutectoid steel that has been drawn through a sequence of reducing dies to obtain a small final diameter. The "pearlitic" structure obtained on cooling this steel through the eutectoid temperature is a two-phase mixture of $Fe_3C$ ("cementite") in bcc iron ("ferrite"). As the diameter is reduced during drawing, the ferrite cells are reduced as well, forming a structure analogous to a fine-grained metal. The cell boundaries restrict dislocation motion, leading to the very high yield strengths.

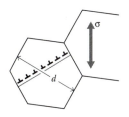

**FIGURE 6.28** Dislocation pileup at a grain boundary.

Impurity atoms in solid solution can also serve to harden a crystalline material by impeding dislocation motion; this is called *solution strengthening.* An impurity atom smaller than the atoms of the host lattice will create an approximately spherical tensile field around itself, which will attract the compressive regions around mobile dislocations; a larger impurity atom will tend to trap the tensile region of nearby dislocations. On average, the population of dislocations will maneuver so as to lower their strain energies by associating with the nonuniform strain fields around impurities. This association impedes dislocation motion, inhibiting plastic flow and increasing the yield stress.

Solution hardening is not usually an especially effective strengthening mechanism in commercial materials, largely because the solubility of impurity atoms is not sufficient to generate an appreciable number of obstacles. One important exception to this is the iron-carbon, or steel, system. If steel at approximately the eutectoid carbon composition (0.8% C) is cooled rapidly from above the eutectoid temperature of 723°C, the carbon atoms become trapped in the iron lattice at much higher concentrations than bcc iron's equilibrium carbon solubility would normally allow. To accommodate these metastable impurity atoms, the iron lattice transforms to a body-centered tetragonal form named *martensite* (see Fig. 6.29), with a strong nonspherical strain field around the carbon atoms. These tetragonal distortions are very effective impediments to dislocation motion, making martensite an extremely hard phase. The periodic water quenches a blacksmith uses during metalworking serve to tailor the material's hardness by developing martensitic inclusions in the steel.

Martensite is so hard and brittle that the rapidly quenched steel must usually be *tempered* by heating it to approximately 400°C for an hour or so. This allows diffusion of carbon to take place, creating a dispersion of cementite inclusions; it also permits recovery of the dislocations present in the martensite. The resulting material is much tougher than the as-formed martensitic steel, but it still retains a high strength level because of the strengthening effect of the carbide inclusions.

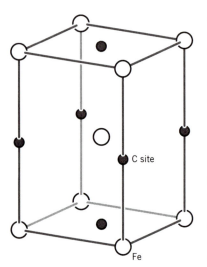

**FIGURE 6.29** The body-centered tetragonal structure of martensite.

## 6.3 CREEP

We have described creep in Chapter 1 as the continuing increase in strain that materials may exhibit under constant stress. This phenomenon can take place at stress levels less than the yield stress, but it involves material mobility and is thus related to the concepts already discussed. In the sections to follow, we will extend our earlier discussions to include the effect of temperature on creep rates and the role of dislocation and vacancy motion in creep of crystalline materials.

### 6.3.1 Thermorheologically Simple Materials

We focus our attention on creep in polymers initially, although the discussion in this section might be applied to other material types as well. In our earlier discussions of creep, we took time as the independent variable. We noted that because of the very wide range of times observed for various polymer relaxations, logarithmic plotting must usually be used to cover the complete relaxation. Experimentalists must work in straight rather than logarithmic time, however, so real-time observation of phenomena spanning ten or twenty orders of magnitude of time is a bit impractical.

Fortunately, polymer creep and stress relaxation times are highly temperature-dependent, so the experimenter can manipulate the temperature to make life easier. If a viscoelastic transition occurs too quickly at room temperature for easy measurement, the experimenter simply lowers the temperature to slow things down. In many materials, especially "simple" materials such as polyisobutylene and other amorphous thermoplastics that have few complicating features in their microstructure, the relation between time and temperature can be described by correspondingly simple models. Such materials are termed *thermorheologically simple*.

Consider for simplicity a creep response that can be modeled with a single relaxation time $\tau$, as in Eq. 1.29, replotted for convenience in Fig. 6.30:

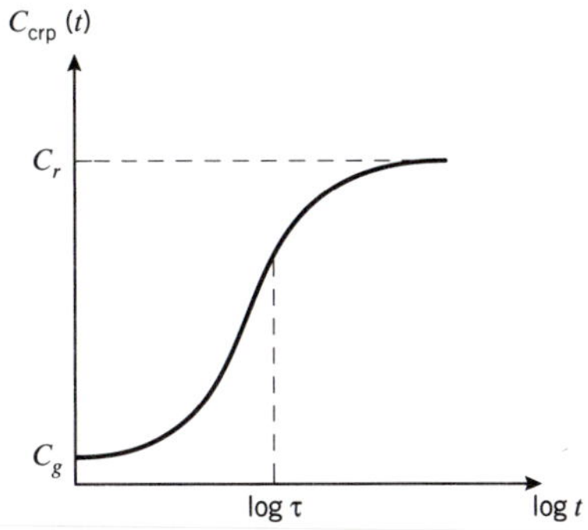

$$C_{\text{crp}} = \frac{\epsilon(t)}{\sigma_0} = C_g + (C_r - C_g)\left(1 - e^{t/\tau}\right) \tag{6.12}$$

where $C_g$ and $C_r$ are the glassy and rubbery compliances and $\tau$ is the *retardation time*. For thermally activated segmental motions, the retardation time can be described by an inverse Eyring relation of the form

$$\tau = \tau_0 \exp\left(\frac{E^*}{RT}\right)$$

**FIGURE 6.30** Creep compliance.

where $E^*$ is the apparent activation energy of the process, typically $\approx 40$ to $400$ kJ/mole for polymer viscoelastic transitions. This relation can be linearized by taking logs:

$$\log \tau = \log \tau_0 + \frac{E^*}{2.303RT} \tag{6.13}$$

Here the factor 2.303 is the conversion between natural and base 10 logarithms, which are commonly used to facilitate graphical plotting using log paper. Now suppose we know $\tau$ at a reference temperature $T_{\text{ref}}$ and wish to compute it for another temperature $T$. We evaluate Eq. 6.13 at the reference temperature:

$$\log \tau_{\text{ref}} = \log \tau_0 + \frac{E^*}{2.303RT_{\text{ref}}} \tag{6.14}$$

Subtracting Eq. 6.14 from Eq. 6.13, we obtain

$$\boxed{\log \tau - \log \tau_{\text{ref}} \equiv \log a_T = \frac{E^*}{2.303R}\left(\frac{1}{T} - \frac{1}{T_{\text{ref}}}\right)} \tag{6.15}$$

where $\log a_T$ is a "time-temperature shift factor" that gives the shift in logarithmic time of a creep transition in terms of a reference value at $T_{\text{ref}}$. If creep curves are measured at two different temperatures, as shown in Fig. 6.31, then $\log a_T$ is the horizontal displacement between them along the $\log t$ axis. We also have

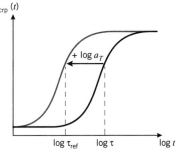

$$\log a_T = \log(\tau/\tau_{\text{ref}})$$
$$a_T = (\tau/\tau_{\text{ref}})$$

**FIGURE 6.31** The time-temperature shifting factor.

so $a_T$ can be found by ratio rather than horizontal shifting.

A series of creep data taken over a range of temperatures can be converted to a single "master curve" via this horizontal shifting. A particular curve is chosen as reference, and then the other curves are shifted horizontally to obtain a single curve spanning a wide range of log time, as shown in Fig. 6.32. Curves representing data

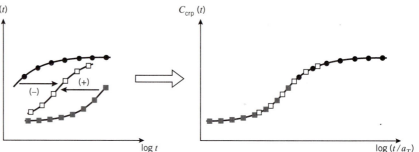

**FIGURE 6.32** Time-temperature superposition.

obtained at temperatures lower than the reference temperature appear at longer times, to the right of the reference curve, so they will have to shift left; this is a *positive* shift as we have defined the shift factor in Eq. 6.15. Each curve produces its own value of $a_T$, so $a_T$ becomes a tabulated function of temperature. The master curve is valid only at the reference temperature, but it can be used at other temperatures by shifting it by the appropriate value of log $a_T$.

The labeling of the abscissa as $\log(t/a_T) = \log t - \log a_T$ in Fig. 6.32 merits some discussion. Rather than shifting the master curve to the right for temperatures less than the reference temperature, or to the left for higher temperatures, it is easier simply to renumber the axis, increasing the numbers for low temperatures and decreasing them for high. The label therefore indicates that the numerical values on the horizontal axis have been adjusted for temperature by subtracting the log of the shift factor. Because lower temperatures have positive shift factors, the numbers are smaller than they need to be and have to be *increased* by the appropriate shift factor. Labeling axes this way is admittedly ambiguous and tends to be confusing, but the correct adjustment is easily made by remembering that lower temperatures slow the creep rate, so times have to be made longer by increasing the numbers on the axis. Conversely, for higher temperatures the numbers must be made smaller.

## EXAMPLE 6.4

We wish to find the extent of creep in a two-temperature cycle that consists of $t_1 = 10$ hours at 20°C followed by $t_2 = 5$ minutes at 50°C. The log shift factor for 50°C, relative to a reference temperature of 20°C, is known to be $-2.2$.

Using the given shift factor, we can adjust the time of the second temperature at 50°C to an equivalent time $t_2'$ at 20°C as follows:

$$t_2' = \frac{t_2}{a_T} = \frac{5 \text{ min}}{10^{-2.2}} = 792 \text{ min} = 13.2 \text{ h}$$

Hence 5 min at 50°C is equivalent to over 13 h at 20°C. The total effective time is then the sum of the two temperature steps:

$$t' = t_1 + t_2' = 10 + 13.2 = 23.2 \text{ h}$$

The total creep can now be evaluated by using this effective time in Eq. 6.12.

The effective-time approach to response at varying temperatures can be extended to an arbitrary number of temperature steps:

$$t' = \sum_j t'_j = \sum_j \left( \frac{t_j}{a_T(T_j)} \right) \longrightarrow \int \frac{dt}{a_T(T)}$$

This approach, while perhaps seeming a bit abstract, is of considerable use in modeling time-dependent materials response. Factors such as damage due to applied stress or environmental exposure can accelerate or retard the rate of a given response, and this change in rate can be described by a time-expansion factor similar to $a_T$ but dependent on other factors in addition to temperature.

If the kinetic treatment used to develop Eq. 6.15 is in fact applicable, then a plot of $\log a_T$ versus $1/T$ should produce a straight line whose slope is $E^*/2.303R$ as in Fig. 6.33. The degree to which the plot is actually straight is a measure (but not a guarantee) of the applicability of the model.

While the Eyring-Arrhenius kinetic treatment is usually applicable to secondary polymer transitions, many workers believe that the glass-rubber primary transition appears governed by other principles. A popular alternative is to use the *WLF equation*[6] at temperatures near or above the glass temperature:

$$\log a_T = \frac{-C_1(T - T_{\text{ref}})}{C_2 + (T - T_{\text{ref}})} \tag{6.16}$$

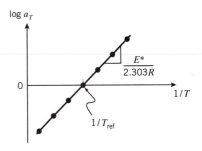

**FIGURE 6.33** Arrhenius plot of shift factors.

Here $C_1$ and $C_2$ are arbitrary material constants whose values depend on the choice of reference temperature $T_{\text{ref}}$. It has been found that if $T_{\text{ref}}$ is chosen to be $T_g$, then $C_1$ and $C_2$ assume "universal" values applicable to a wide range of polymers:

$$\boxed{\log a_T = \frac{-17.44(T - T_{\text{ref}})}{51.6 + (T - T_{\text{ref}})}} \tag{6.17}$$

where $T$ is in Celsius. Note that some caution must be exercised when using WLF equations in automatic computation: There is a singularity at $T = T_g - 51.6$, and values of $\log a_T$ become meaningless there.

The original WLF paper developed this relation empirically but rationalized it in terms of free-volume concepts. To outline this approach, take the fractional free volume $f$ available for molecular motions to be governed by simple thermal expansion

$$f = f_g + \alpha(T - T_g)$$

where $f_g$ is the available volume at the glass transition temperature, and $\alpha$ is the coefficient of volumetric thermal expansion above $T_g$. The creep retardation time $\tau$ is now taken to vary exponentially with the free volume:

$$\tau = \tau_0 \exp(\beta/f)$$

The underlying concept here is that as the volume increases, the number of molecular segments having enough "elbow room" to act cooperatively in allowing macroscopic

[6]M.L. Williams, R.F. Landel, and J.D. Ferry, *J. Am. Chem. Soc.,* Vol. 77, No. 14, pp. 3701–3707, 1955.

creep also increases. The number of possible molecular combinations rises factorially with the number of available segments, much as it is easier to find new partners on a relatively uncrowded dance floor than when everyone is packed together cheek to jowl. We now proceed as with the Eyring approach, obtaining the shift factor by comparing the creep time at an arbitrary temperature with that at a reference temperature. Taking $T_{\text{ref}} = T_g$:

$$\log a_T = \log \tau - \log \tau_g$$

$$= \beta \left[ \frac{1}{f_g + \alpha(T - T_g)} - \frac{1}{f_g} \right] = \frac{-(\beta/f_g)(T - T_g)}{(f_g/\alpha) + (T - T_g)}$$

where the 2.303 factor needed to convert from natural to base 10 logarithms has been incorporated into the constant factor $\beta$. This is clearly of the same form as Eq. 6.17, where $(f_g/\alpha) = 51.6$. It has been noted that if we take $\alpha \approx 5 \times 10^{-4}/°C$ as a value approximately correct for many amorphous polymers, we can calculate that $f_g = 51.6\alpha \approx 0.025$ for these materials as well. In this simple view, the glass temperature is simply that temperature above which the free volume is sufficient for large-scale cooperative chain motion, and that critical free volume has a constant value of 2.5 percent.

The free-volume viewpoint is simplistic, but it provides a valuable generalized insight to rate processes. We now argue that anything that expands internal volume, such as swelling of polymers by organic liquids or the presence of tensile hydrostatic stress components, produces very strong increases in rate similar to those caused by increases in temperature. A similar line of thought was used in our discussion of the effect of pressure on yield, in which compressive hydrostatic stress components impede yield by reducing the free volume and thus suppressing molecular motions.

### 6.3.2 Kinetics of Creep in Crystalline Materials

Creep in polycrystalline metallic and ceramic systems is usually important only well above room temperature, at approximately half the absolute melting temperature. High-temperature creep is of concern in such applications as jet engines or nuclear reactors. The more complicated microstructures and mechanisms in these materials result in a loss of thermorheological simplicity, and more general models are necessary. The creep response often consists of three distinct regimes, as seen in Fig. 6.34: *primary* creep, in which the material appears to harden so the creep rate diminishes with time; *secondary* or steady-state creep, in which hardening and softening mechanisms appear to balance to produce a constant creep rate $\dot{\epsilon}_{\text{II}}$; and *tertiary* creep, in which the material softens until creep rupture occurs. The entire creep curve reflects a competition between hardening mechanisms, such as dislocation pileup, and mechanisms such as dislocation climb and cross-slip, which are termed *recovery* and augment dislocation mobility.

In most applications the secondary regime consumes most of the time to failure, so much of the modeling effort has been directed to this stage. The secondary creep rate $\dot{\epsilon}_{\text{II}}$ can often be described by a general nonlinear expression of the form

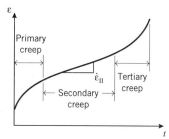

**FIGURE 6.34** The three stages of creep.

$$\dot{\epsilon}_{\text{II}} = A\sigma^m \exp \frac{-E_c^*}{RT} \qquad (6.18)$$

where $A$ and $m$ are adjustable constants, $E_c^*$ is an apparent activation energy for creep, $\sigma$ is the stress, $R$ is the gas constant (to be replaced by Boltzmann's constant if a molar basis is not used) and $T$ is the absolute temperature. This is known as the *Weertman-Dorn equation.*

The plastic flow rate is related directly to dislocation velocity, which can be visualized by considering a section of material of height $h$ and width $L$ as shown in Fig. 6.35. A single dislocation, having traveled in the width direction for the full distance $L$, will produce a transverse deformation of $\delta_i = \bar{b}$. If the dislocation has propagated through the crystal only a fraction $x_i/L$ of the width, the deformation can be reduced by this same fraction: $\delta_i = \bar{b}(x_i/L)$. The total deformation in the crystal is then the sum of the deformations contributed by each dislocation:

$$\delta = \sum_i \delta_i = \sum_i \bar{b}(x_i/L)$$

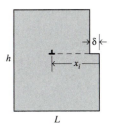

**FIGURE 6.35** Dislocation motion and creep rate.

The shear strain is the ratio of the transverse deformation to the height over which it is distributed:

$$\gamma = \frac{\delta}{h} = \frac{\bar{b}}{Lh} \sum_i x_i$$

The value $\sum_i x_i$ can be replaced by the quantity $N\bar{x}$, where $N$ is the number of dislocations in the crystal segment and $\bar{x}$ is the average propagation distance. We can then write

$$\gamma = \rho\bar{b}\bar{x}$$

where $\rho = N/Lh$ is the dislocation density in the crystal. The shear strain *rate* $\dot{\gamma}$ is then obtained by differentiation:

$$\boxed{\dot{\gamma} = \rho\bar{b}v} \tag{6.19}$$

where $v = \dot{\bar{x}}$ is the average dislocation velocity. Hence the creep rate scales directly with the dislocation velocity.

To investigate the temperature and stress dependence of this velocity, we consider the rate at which dislocations can overcome obstacles to be yet another example of a thermally activated, stress-aided rate process and write an equation analogous to Eq. 6.4 for the creep rate:

$$\dot{\epsilon} \propto v \propto \exp\frac{-(E_d^* - \sigma V^*)}{kT} - \exp\frac{-(E_d^* + \sigma V^*)}{kT}$$

where $V^*$ is an apparent activation volume. The second term here indicates that the activation barrier for motion in the direction of stress is raised by the stress and that the barrier is lowered for motions in the opposite direction. When we discussed yielding, the stress was sufficiently high that motion in the direction opposing flow could be neglected. Here we are interested in creep taking place at relatively low stresses and at high temperature, so reverse flow can be appreciable. Factoring,

$$\dot{\epsilon} \propto \exp\frac{-E_d^*}{RT}\left(\exp\frac{+\sigma V^*}{RT} - \exp\frac{-\sigma V^*}{RT}\right)$$

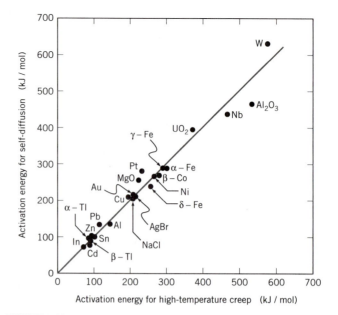

**FIGURE 6.36** Correlation of activation energies for diffusion and creep.

Since $\sigma V^* \ll RT$, we can neglect quadratic and higher-order terms in the series expansion $e^x = 1 + x + (x^2/2!) + (x^3/3!) + \cdots$ to give

$$\dot{\epsilon} = A\left(\frac{\sigma V^*}{RT}\right)\exp\frac{-(E_d^*)}{RT}$$

If now we neglect the temperature dependence in the preexponential factor in comparison with the much stronger temperature dependence of the exponential itself, this model predicts a creep rate in agreement with the Weertman-Dorn equation with $m = 1$.

Creep by dislocation glide occurs over the full range of temperatures from absolute zero to the melting temperature, although the specific equation developed here contains approximations valid only at higher temperature. The stresses needed to drive dislocation glide are on the order of one-tenth of the theoretical shear strength of $G/10$. At lower stresses the creep rate is lower and becomes limited by the rate at which dislocations can climb over obstacles by vacancy diffusion, as is hinted at in the similarity of the activation energies for creep and self-diffusion as shown in Fig. 6.36. (Note that these values also correlate with the tightness of the bond energy functions, as discussed in Chapter 1; diffusion is impeded in more tightly bonded lattices.) Vacancy diffusion is another stress-aided, thermally activated rate process, again leading to models in agreement with the Weertman-Dorn equation.

## 6.4 CONTINUUM PLASTICITY

Plasticity theory, which seeks to determine stresses and displacements in structures all or part of which have been stressed beyond the yield point, is an important aspect of solid mechanics. The situation is both materially and geometrically nonlinear, so it is not a trivial undertaking. However, in such areas as metal forming, plasticity theory has provided valuable insight. In the following paragraphs we will only introduce some of the fundamental concepts, which the reader can extend in future study.

## 6.4.1 Plastic Deformation

Although permanent flow can take place as described earlier at stresses below the yield stress, a useful idealization is to consider the material linearly elastic up to the yield point as shown in Fig. 6.37, and "perfectly plastic" at strains beyond yield. Strains up to yield (the line between points *a* and *b*) are recoverable, and the material unloads along the same elastic line it followed during loading; this is conventional elastic response. But if the material is strained beyond yield (point *b*), the plastic straining beyond *b* takes place at constant stress and is unrecoverable. If the material is strained to point *c* and then unloaded, it follows the path *cd* (a line parallel to the original elastic line *ab*) rather than returning along *cba*. When the stress has been brought to zero (point *d*), the plastic strain *ad* remains as a *residual* strain.

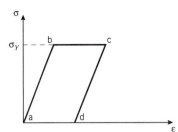

Plastic deformation can generate beneficial or damaging residual stresses in structures. To illustrate this, consider two rods having different stress-strain curves, connected in parallel (so that their strains are always equal) as shown in Fig. 6.38. When the rods are strained up to the yield point of rod *B* (point *a* on the strain axis), rod *A* will have experienced an amount of permanent plastic deformation $\epsilon^p$. When the applied load is removed, rod *B* unloads along its original stress-strain curve, but rod *A* follows a path parallel to its original elastic line. When rod *A* reaches zero stress (point *b*), rod *B* will still be in tension (point *c*). In order for the load transmitted by the rods together to come to zero, rod *B* will pull rod *A* into compression until $-\sigma_B = \sigma_A$ as indicated by points *d* and *e*. Residual stresses are left in the rods, and the assembly as a whole is left with a residual tensile strain.

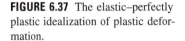

**FIGURE 6.37** The elastic–perfectly plastic idealization of plastic deformation.

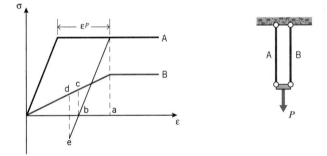

**FIGURE 6.38** Plastic deformation of two-bar assembly.

Compressive residual stress can be valuable if the structure must bear tensile loads. Similarly to the way rapid quenching can be used to make safety glass by putting the surfaces in compression, plastic deformation can be used to create favorable compressive stresses. One famous such technique is called *autofrettage;* this is a method used to strengthen cannon barrels against bursting by pressurizing them from the inside so as to bring the inner portion of the barrel into the plastic range. When the pressure is removed, the inner portions are left with a compressive residual stress just as with bar *A* in the foregoing example.

## 6.4.2 Wire Drawing

To quantify the plastic flow process in more detail, consider next the *drawing* of wire,[7] in which wire is pulled through a reducing die so as to reduce its cross-sectional area from $A_0$ to $A$ as shown in Fig. 6.39. Because volume is conserved during plastic

---

[7]G.W. Rowe, *Elements of Metalworking Theory,* Edward Arnold, London, 1979.

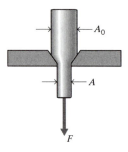

**FIGURE 6.39** Wire drawing.

deformation, this corresponds to an axial elongation of $L/L_0 = A_0/A$. Considering the stress state to be simple uniaxial tension, we have

$$\sigma_1 = \sigma_Y, \qquad \sigma_2 = \sigma_3 = 0$$

where 1 denotes the direction along the wire and 2 and 3 are the transverse directions. The work done in stretching the wire by an increment of length $dL$, per unit volume of material, is

$$dU = \frac{dW}{AL} = \frac{\sigma_Y A \, dL}{AL}$$

Integrating this from $L_0$ to $L$, we obtain the total work as

$$U = \int_{L_0}^{L} dU = \frac{F \, dL}{AL} = \sigma_Y \ln \frac{L}{L_0}$$

The quantity $\ln(L/L_0)$ is known as the *logarithmic strain* $\epsilon_T$.

---

## EXAMPLE 6.5

The logarithmic strain can be written in terms of either length increase or area reduction, because volume remains constant during plastic deformation: $\epsilon_T = \ln(L/L_0) = \ln(A_0/A)$. In terms of diameter reduction, the relation $A = \pi d^2/4$ leads to

$$\epsilon_T = \ln\left(\frac{\pi d^2/4}{\pi d_0^2/4}\right) = 2\ln\left(\frac{d}{d_0}\right)$$

Taking the pearlite cell size to shrink commensurately with the diameter, we expect the wire strength $\sigma_f$ to vary according to the Hall-Petch relation with $1/\sqrt{d}$. The relation between wire strength and logarithmic drawing strain is then

$$\sigma_f \propto \frac{\exp(\epsilon_T/4)}{\sqrt{d_0}}$$

---

The work done by the constant pulling force $F$ in drawing an initial length $L_0$ of wire to a new length $L$ is $W = FL$. This must equal the work per unit volume done in the die, multiplied by the total volume of wire:

$$FL = (AL)\sigma_Y \ln \frac{L}{L_0}$$

Written in terms of area reduction, this is

$$F = A\sigma_Y \ln \frac{A_0}{A}$$

This simple result is useful in estimating the requirements of wire drawing, even though it neglects the actual complicated flow field within the die and the influence of friction at the die walls. Both friction at the surface and constraints to flow within the field raise the force needed in drawing, but the present analysis serves to establish a lower-limit approximation. It is often written in terms of the drawing stress $\sigma_1 = F/A$ and the area reduction ratio $r = (A_0 - A)/A_0 = 1 - (A/A_0)$:

$$\sigma_1 = \sigma_Y \ln \frac{1}{1-r}$$

Note that the draw stress for a small area reduction is less than the tensile yield stress. In fact, the maximum area reduction that can be achieved in a single pass can be estimated by solving for the value of $r$ that brings the draw stress up to the value of the yield stress, which it obviously cannot exceed. This calculation gives

$$\ln \frac{1}{1-r_{\max}} = 1 \Rightarrow r_{\max} = 1 - \frac{1}{e} = 0.63$$

Hence the maximum area reduction is approximately 63 percent, assuming perfect lubrication at the die. This lower-bound treatment gives an optimistic result, but is not far from the approximately 50 percent reduction often used as a practical limit. If the material hardens during drawing, the maximum reduction can be slightly greater.

### 6.4.3  Slip-Line Fields

In cases of plane strain, a graphical technique called *slip-line theory*[8] permits a more detailed examination of plastic flow fields and the loads needed to create them. Friction and internal flow constraints can be included, so upper-bound approximations are obtained that provide more conservative estimates of the forces needed in deformation. Considerable experience is needed to become proficient in this method, but the following will outline some of the basic ideas.

Consider plane strain in the 1-3 plane, with no strain in the 2-direction. There is a Poisson stress in the 2-direction, given by

$$\epsilon_2 = 0 = \frac{1}{E}[\sigma_2 - \nu(\sigma_1 + \sigma_3)]$$

Because $\nu = \frac{1}{2}$ in plastic flow,

$$\sigma_2 = \tfrac{1}{2}(\sigma_1 + \sigma_3)$$

The hydrostatic component of stress is then

$$p = \tfrac{1}{3}(\sigma_1 + \sigma_2 + \sigma_3) = \tfrac{1}{2}(\sigma_1 + \sigma_3) = \sigma_2$$

Hence, the Poisson stress $\sigma_2$ in the zero-strain direction is the average of the other two stresses $\sigma_1$ and $\sigma_2$ and is also equal to the hydrostatic component of stress. The stress state can be specified in terms of the maximum shear stress, which is just $k$ during plastic flow, and the superimposed hydrostatic pressure $p$:

$$\sigma_1 = -p + k \qquad \sigma_2 = -p \qquad \sigma_3 = -p - k$$

[8]W. Johnson and P.B. Mellor, *Plasticity for Mechanical Engineers,* Van Nostrand Co., New York, 1962.

The shear stress is equal to $k$ everywhere, so the problem is one of determining the directions of $k$ (the direction of maximum shear, along which slip occurs) and the magnitude of $p$.

The graphical technique involves sketching lines that lie along the directions of $k$. Since maximum shear stresses act on two orthogonal planes, there will be two sets of these lines, always perpendicular to one another and referred to as $\alpha$ lines and $\beta$ lines. The direction of these lines is specified by an inclination angle $\phi$. Any convenient inclination can be used for the $\phi = 0$ datum, and the identification of $\alpha$ versus $\beta$ lines is such as to make the shear stress positive according to the usual convention. As the pressure $p$ varies from point to point, there is a corresponding variation of the angle $\phi$, given by the *Hencky equations* as

$$p + 2k\phi = C_1 = \text{constant, along an } \alpha \text{ line}$$

$$p - 2k\phi = C_2 = \text{constant, along a } \beta \text{ line}$$

Hence the pressure can be determined from the curvature of the slip lines, once the constant is known.

The slip-line field must obey certain constraints at boundaries:

1. *Free surfaces:* Since there can be no stress normal to a free surface, we can put $\sigma_3 = 0$ there and then

$$p = k, \qquad \sigma_1 = -p - k = -2k$$

   Hence the pressure is known to be just the shear yield strength at a free surface. Furthermore, because the directions normal and tangential to the surface are principal directions, the directions of maximum shear must be inclined at 45° to the surface.

2. *Frictionless surface:* The shear stress must be zero tangential to a frictionless surface, which again means that the tangential and normal directions must be principal directions. Hence the slip lines must meet the surface at 45°. However, there will in general be a stress acting normal to the surface, so $\sigma_3 \neq 0$, and thus $p$ will not be equal to $k$.

3. *Perfectly rough surface:* If the friction is so high as to prevent any tangential motion at the surface, the shearing must be maximum in a direction that is also tangential to the surface. One set of slip lines must then be tangential to the surface, and the other set must be normal to it.

Consider a flat indentor of width $b$ being pressed into a semi-infinite block, with negligible friction (see Fig. 6.40). Since the slip lines must meet the indentor surface at 45°, we can draw a triangular flow field $ABC$. All lines in this region are straight, so there can be no variation in the pressure $p$, and the field is one of "constant state." This cannot be the full extent of the field, however, because it would be constrained both vertically and laterally by rigid metal. The field must extend to the free surfaces adjacent to the punch, so that downward motion under the punch can be compensated by upward flow adjacent to it. Two more triangular regions $ADF$ and $BEG$ are added that satisfy the boundary conditions at free surfaces, and these are connected to the central triangular regions by "fans" $AFC$ and $BCG$. Fans are very useful in slip line constructions; they are typically centered on singularities such as points $A$ and $B$, where there is no defined normal to the surface.

The pressure on the punch needed to establish this field can be determined from the slip lines, and this is one of their principal uses. Since $BE$ is a free surface, $\sigma_3 = 0$

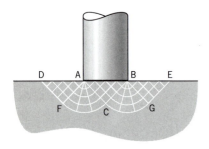

**FIGURE 6.40** Slip-line construction for a flat indentation.

there and $p = k$. The pressure remains constant along line $EG$ because $\phi$ is unchanging, but as $\phi$ decreases along the curve $GC$ (the line curves clockwise), the pressure must increase according to the Hencky equation. At point $C$ it has rotated through $-\pi/2$, so the pressure there is

$$p_C + 2k\phi = p_C + 2k\left(-\frac{\pi}{2}\right) = \text{constant} = p_G = k$$

$$p_C = k(1 + \pi)$$

The pressure remains unchanged along lines $CA$ and $CB$, so the pressure along the punch face is also $k(1 + \pi)$. The total stress acting upward on the punch face is therefore

$$\sigma_1 = p + k = 2k\left(1 + \frac{\pi}{2}\right)$$

The ratio of punch pressure to the tensile yield strength $2k$ is

$$\frac{\sigma_1}{2k} = 1 + \frac{\pi}{2} = 2.571$$

The factor 2.571 represents the increase over the tensile yield strength caused by the geometrical constraints on the flow field under the punch.

The *Brinell*[9] *hardness test* is similar to the punch yielding scenario just described, but it uses a hard steel sphere instead of a flat indentor. The Brinell hardness $H$ is calculated as the load applied to the punch divided by the projected area of the indentation. Analysis of the Brinell test differs somewhat in geometry but produces a result not much different from that of the flat punch:

$$\frac{H}{\sigma_Y} \approx 2.8\text{–}2.9$$

This relation is very useful in estimating the yield strength of metals by simple nondestructive indentation hardness tests.

# 6.5 PROBLEMS

**1.** Show that an ideal rubber whose engineering stress-strain law is $\sigma_e = G[\lambda - (1/\lambda^2)]$ will not form a neck according to the Considère construction.

**2.** The open-ended pressure vessel shown in Fig. P.6.2 is constructed of aluminum, with diameter 0.3 m and wall thickness 3 mm. ("Open-ended" in this context means that both ends of the vessel are connected to other structural parts able to sustain pressure, as in a hose connected between two reservoirs.) Determine the internal pressure at which the vessel will yield according to the (a) Tresca and (b) von Mises criterion.

**3.** Repeat Problem 2, but with the pressure vessel now being closed-ended as shown in Fig. P.6.3.

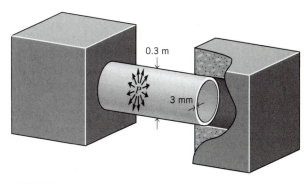

**FIGURE P.6.2**

---

[9]Johann August Brinell (1849–1925), Swedish engineer.

**FIGURE P.6.3**

**4.** A steel plate is clad with a thin layer of aluminum on both sides at room temperature as shown in Fig. P.6.4, and the temperature is then raised. At what temperature increase $\Delta T$ will the aluminum yield?

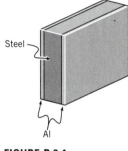

Steel

Al

**FIGURE P.6.4**

**5.** If the temperature in Problem 4 is raised 40°C beyond the value at which yielding occurs, and the plate is thereafter cooled back to room temperature, what is the residual stress left in the aluminum?

**6.** Copper alloy is subjected to the stress state $\sigma_x = 100$, $\sigma_y = -200$, $\tau_{xy} = 100$ (all in MPa). Determine whether yield will occur according to the (a) Tresca and (b) von Mises criterion.

**7.** Repeat Problem 6 but with the stress state $\sigma_x = 190$, $\sigma_y = 90$, $\tau_{xy} = 120$ (all in MPa).

**8.** The thin-walled tube shown in Fig. P.6.8 is placed in simultaneous tension and torsion, causing a stress state as shown here. Construct a plot of $\tau/\sigma_Y$ versus $\sigma/\sigma_Y$ at which yield will occur according to the (a) Tresca and (b) von Mises criterion.

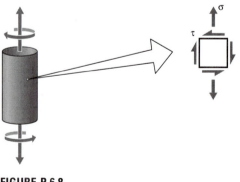

**FIGURE P.6.8**

**9.** A solid circular steel shaft is loaded by belt pulleys at both ends as shown in Fig. P.6.9. Determine the diameter of the shaft required

to avoid yield according to the von Mises criterion, with a factor of safety of 2.

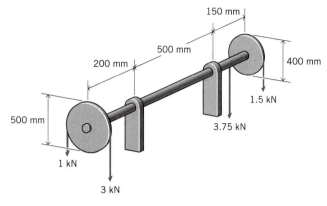

**FIGURE P.6.9**

**10.** For polycarbonate, the kinetic parameters in Eq. 6.4 are found to be $\dot{\epsilon}_0 = 448$ s$^{-1}$, $E_Y^* = 309$ kJ/mol, and $V^* = 3.9 \times 10^{-3}$ m$^3$/mol. Find the yield stress $\sigma_Y$ at a strain rate of $\dot{\epsilon} = 10^2$ s$^{-1}$ and temperature 40°C.

**11.** Yield stresses (in MPa) have been measured at various strain rates and temperatures as follows:

|  | $\dot{\epsilon} = 10^{-3}$ s$^{-1}$ | $\dot{\epsilon} = 10^{-1}$ s$^{-1}$ |
|---|---|---|
| $T = 0°C$ | 54.1 | 62.7 |
| $T = 40°C$ | 42.3 | 52.1 |

Determine the activation volume for the yield process. What physical significance might this parameter have?

**12.** The yield stress of a polymer is measured to be 20 MPa at a temperature of 300 K and a strain rate of $10^{-3}$ s$^{-1}$. When the strain rate is doubled from this value, the yield stress is observed to increase by 10 percent. What is the apparent activation volume for yield in this case?

**13.** Show that the von Mises stress can be written in index notation as $\sigma_M = \sqrt{3\Sigma_{ij}\Sigma_{ij}/2}$

**14.** A sample of linear polyethylene was tested in uniaxial loading at $T = 23°C$ and $\dot{\epsilon} = 10^{-3}$ s$^{-1}$. The yield stress $\sigma_Y$ was found to be 30.0 MPa in tension and 31.5 MPa in compression. Determine the pressure-dependency constant $A$ in Eq. 6.3.

**15.** A thermorheologically simple material with $T_g = 40°C$ is subjected to a creep test lasting 1000 hours at a temperature of 50°C, after which the creep strain is measured. If a new creep test were conducted at a temperature of 60°C, how long would it take to reach this same strain?

**16.** A circular shaft of radius $R$ is subjected to a torque $T$.
(a) What value of $T$ will be just large enough to induce yielding at the outer surface?

(b) As the value of $T$ is increased beyond the level found in part (a), determine the radius $r_e$ within which the material is still in the elastic range.

(c) What value of $T$ will make the shaft fully plastic, that is, $r_e = 0$?

**17.** The yield stresses $\sigma_Y$ have been measured using steel and aluminum specimens of various grain sizes, as follows:

| Material | $d, \mu$ | $\sigma_Y,$ MPa |
|---|---|---|
| Steel | 60.5 | 160 |
|  | 136 | 130 |
| Aluminum | 11.1 | 235 |
|  | 100 | 225 |

(a) Determine the coefficients $\sigma_0$ and $k_Y$ in the Hall-Petch relation (Eq. 6.11) for these two materials.

(b) Determine the yield stress in each material for a grain size of $d = 30 \mu$.

**18.** A two-element truss frame is constructed of steel with the geometry shown in Fig. P.6.18. What load $P$ can the frame support without yielding in either element?

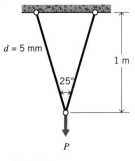

$d = 5$ mm

1 m

25°

$P$

**FIGURE P.6.18**

**19.** A three-element truss frame is constructed of steel with the geometry shown in Fig. P.6.19. What load $P$ can the frame support before all three elements have yielded?

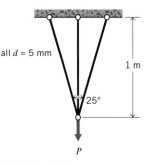

all $d = 5$ mm

1 m

25°

$P$

**FIGURE P.6.19**

**20.** If the frame of Problem 19 is loaded until all three members have yielded and the load is then reduced to zero, find the residual stress in the central element.

**21.** A rigid beam is hinged at one end and supported by two vertical rods as shown in Figure P.6.21.

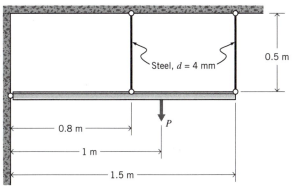

Steel, $d = 4$ mm

0.5 m

0.8 m

1 m

$P$

1.5 m

**FIGURE P.6.21**

(a) What load $P$ can the structure support before both vertical rods have yielded?

(b) What is the residual stress in the vertical rods after the load has been reduced to zero?

**22.** Estimate the drawing force required to reduce the diameter of a 0.125″ aluminum rod by 50 percent in a wire-drawing operation.

# 7 FRACTURE AND FATIGUE

In 1983, the National Bureau of Standards (now the National Institute for Science and Technology) and Battelle Memorial Institute[1] estimated the costs for failure due to fracture to be $119 billion per year in 1982 dollars. The dollars are important, but the cost of many failures in human life and injury is infinitely more so.

Failures have occurred for many reasons, including uncertainties in the loading or environment, defects in the materials, inadequacies in design, and deficiencies in construction or maintenance. Design against fracture has a technology of its own, and this is a very active area of current research. This chapter will provide an introduction to this field because without an understanding of fracture the methods in stress analysis discussed previously would be of little use. We will focus on fractures due to simple tensile overstress, but the designer is cautioned again about the need to consider absolutely as many factors as possible that might lead to failure, especially when life is at risk. The 1981 tragedy at the Kansas City Hyatt Regency hotel, outlined in Example 7.1, is a grim reminder of how a design can be correct at first glance but yet catastrophically flawed.

## EXAMPLE 7.1

In 1981 two walkways spanning an atrium at the Kansas City Hyatt-Regency hotel collapsed under the weight of persons dancing on them. Over a hundred people were killed, making this the most catastrophic structural failure in the history of the United States. A Pulitzer Prize–winning investigation by *The Kansas City Star* and its engineer consultants showed that the failure was due largely to a design change in how the walkways were supported.

As seen in Fig. 7.1*a*, the original design called for each walkway beam to hang from a long rod that dropped from the ceiling of the atrium to the floor. Each beam would then be simply supported and would have to bear only the bending moment and shear force imposed on it by the persons standing on it and the weight of the walkway itself. Each walkway would be loaded independently of the others. The tension in the vertical rod, of course, would increase from a minimum at the bottom to a maximum at the ceiling, where it would have to bear the full weight of all the walkways below it.

---

[1]R.P. Reed et al., *NBS Special Publication 647-1,* Washington, DC, 1983.

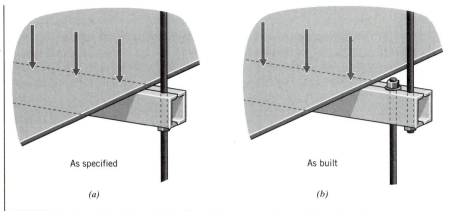

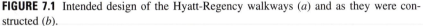

As specified

(a)

As built

(b)

**FIGURE 7.1** Intended design of the Hyatt-Regency walkways (*a*) and as they were constructed (*b*).

The walkway beams were supposed to sit on nuts that were threaded onto the vertical rod. But how can a design like this be constructed? In order for the rod to have threads at the location of the beams, the entire rod would have to be threaded; it's impossible to put threads on only a small region somewhere along the rod. To avoid threading the entire rod or finding some other alternative, the design was changed to that shown in Fig. 7.1*b*. This is certainly easier to construct, but it should be clear that this causes a huge change in the stresses at the beam supports. Now the walkways are no longer independent, and the attachments of an upper walkway carry the weight of all walkways below it. Further, they are loaded with an intense shearing couple that was not present in the original design. A beam sized to support the loads of the original design is not guaranteed to be able to support the altered, unanalyzed design, and tragedy followed.

This painful story is another example of the multifaceted nature of engineering design. If a design isn't manufacturable, it isn't much good. And as was true in the terrible Hyatt-Regency disaster, the work-around of an unmanufacturable design can easily be unsafe. When that happens, none of the careful mechanics of materials work that was present in the original design is relevant.

## 7.1 STATISTICS OF FRACTURE

One particularly troublesome aspect of fracture, especially in high-strength and brittle materials, is its *variability*. The designer must be able to cope with this and limit stresses to those that reduce the *probability* of failure to an acceptably low level. Selection of an acceptable level of risk is a difficult design decision itself; obviously it must be as close to zero as possible in cases where human safety is involved, but it can be higher in doorknobs and other inexpensive items where failure is not too much more than a nuisance. The following sections will not replace a thorough study of statistics, but they will introduce at least some of the basic aspects of statistical theory needed in design against fracture. The text by Collins[2] includes an extended treatment of statistical analysis of fracture and fatigue data and is recommended for further reading.

---

[2]J.A. Collins, *Failure of Materials in Mechanical Design,* Wiley, 1993.

## 7.1.1 Basic Statistical Measures

The value of tensile strength $\sigma_f$ cited in materials property handbooks is usually the *arithmetic mean,* simply the sum of a number of individual strength measurements divided by the number of specimens tested:

$$\overline{\sigma_f} = \frac{1}{N} \sum_{i=1}^{N} \sigma_{f,i} \qquad (7.1)$$

where the overline denotes the mean and $\sigma_{f,i}$ is the measured strength of the *i*th (out of *N*) individual specimen. Of course, not all specimens have strengths exactly equal to the mean; some are weaker, some are stronger. There are several measures of how widely scattered the distribution of strengths is, one important one being the sample *standard deviation,* a sort of root mean square average of the individual deviations from the mean:

$$s = \sqrt{\frac{1}{N-1} \sum_{i=1}^{N} \left(\overline{\sigma_f} - \sigma_{f,i}\right)^2} \qquad (7.2)$$

The significance of *s* to the designer is usually in relation to how large it is compared to the mean, so the *coefficient of variation,* or C.V., is commonly used:

$$\text{C.V.} = \frac{s}{\overline{\sigma_f}}$$

This is often expressed as a percentage. Coefficients of variation for tensile strength are commonly in the range of 1 to 10 percent, with values much over that indicating substantial inconsistency in the specimen preparation or experimental error.

### EXAMPLE 7.2

In order to illustrate the statistical methods to be outlined in this section, we will use a sequence of thirty measurements of the room-temperature tensile strength of a graphite/epoxy composite.[3] These data (in kpsi) are

| | | |
|---|---|---|
| 72.5 | 67.95 | 72.85 |
| 73.8 | 82.84 | 77.81 |
| 68.1 | 79.83 | 75.33 |
| 77.9 | 80.52 | 71.75 |
| 65.5 | 70.65 | 72.28 |
| 73.23 | 72.85 | 79.08 |
| 71.17 | 77.81 | 71.04 |
| 79.92 | 72.29 | 67.84 |
| 65.67 | 75.78 | 69.2 |
| 74.28 | 67.03 | 71.53 |

---

[3] P. Shyprykevich, *ASTM STP 1003,* pp. 111–135, 1989.

Another thirty measurements from the same source, but taken at 93°C, are given in Problem 2 in this chapter and can be subjected to the same treatments as homework.

There are several computer packages available for doing statistical calculations, and most of the procedures to be outlined here can be done with spreadsheets. The Microsoft Excel™ functions for mean and standard deviation are `average()` and `stdev()`, where the arguments are the range of cells containing the data. These give for the foregoing data

$$\overline{\sigma_f} = 73.28, \qquad s = 4.63 \text{ (kpsi)}$$

The coefficient of variation is C.V. $= (4.63/73.28) \times 100\% = 6.32\%$.

## 7.1.2 The Normal Distribution

A more complete picture of strength variability is obtained if the number of individual specimen strengths falling in a discrete strength interval $\Delta\sigma_f$ is plotted versus $\sigma_f$ in a *histogram* as shown in Fig. 7.2; the maximum in the histogram will be near the mean strength, and its width will be related to the standard deviation.

As the number of specimens increases, the histogram can be drawn with increasingly finer $\Delta\sigma_f$ increments, eventually forming a smooth *probability distribution function* (pdf). The mathematical form of this function is up to the material (and also the test method in some cases) to decide, but many phenomena in nature can be described satisfactorily by the *normal,* or *Gaussian,* function:

$$f(X) = \frac{1}{\sqrt{2\pi}} \exp\frac{-X^2}{2}, \qquad X = \frac{\sigma_f - \overline{\sigma_f}}{s} \tag{7.3}$$

Here $X$ is the *standard normal variable* and is simply how many standard deviations an individual specimen strength is away from the mean. The factor $1/\sqrt{2\pi}$ normalizes the function so that its integral is unity, which is necessary if the specimen is to have a 100 percent chance of having *some* value of tensile stress. In this expression we have assumed that the measure of standard deviation determined from Eq. 7.2 based on a discrete number of specimens is acceptably close to the "true" value that would be obtained if every piece of the material in the universe could somehow be tested.

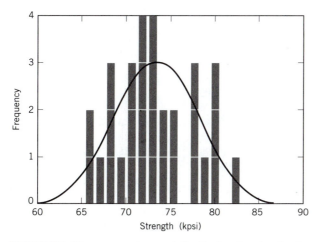

**FIGURE 7.2** Histogram and normal distribution function for the strength data of Example 7.2.

The normal distribution function $f(X)$ plots as the "bell curve" familiar to all grade-conscious students. Its integral, known as the *cumulative distribution function* or $P_f(X)$, is also used commonly; its ordinate is the probability of fracture, also the fraction of specimens having a strength lower than the associated abscissal value. Since the normal pdf has been normalized, the cumulative function rises with an S (sigmoidal) shape to approach unity at large values of $X$. The two functions $f(X)$ and $F(X)$ are plotted in Fig. 7.3 and tabulated in Tables 1 and 2 of Appendix H. (Often the probability of survival $P_s = 1 - P_f$ is used as well; this curve begins at near unity and falls in a sigmoidal shape toward zero as the applied stress increases.)

One convenient means of determining whether or not a particular set of measurements is normally distributed involves using special graph paper (a copy is included in Appendix H) whose ordinate has been distorted to make the sigmoidal cumulative distribution $P_f$ plot as a straight line. (Sometimes it is easier to work with straight lines on curvy paper than with curvy lines on straight paper.) Experimental data are ranked from lowest to highest, and each assigned a rank based on the fraction of strengths having higher values. If the ranks are assigned as $i/(N + 1)$, where $i$ is the position of a datum in the ordered list and $N$ is the number of specimens, the ranks are always greater than zero and less than one; this facilitates plotting.

The degree to which these rank-strength data plot as straight lines on normal probability paper is then a visual measure of how well the data are described by a normal distribution. The best-fit straight line through the points passes the 50 percent cumulative fraction line at the sample mean, and its slope gives the standard distribution. Plotting several of these lines, for instance for different processing conditions of a given material, is a convenient way to characterize the strength differences arising from the two conditions (see Problem 2 in this chapter).

## EXAMPLE 7.3

For our 30-specimen test population, the ordered and ranked data are

| $i$ | $\sigma_{f,i}$ | $P_f = \frac{i}{N+1}$ | $i$ | $\sigma_{f,i}$ | $P_f = \frac{i}{N+1}$ |
|---|---|---|---|---|---|
| 1 | 65.50 | 0.0323 | 16 | 72.85 | 0.5161 |
| 2 | 65.67 | 0.0645 | 17 | 72.85 | 0.5484 |
| 3 | 67.03 | 0.0968 | 18 | 73.23 | 0.5806 |
| 4 | 67.84 | 0.1290 | 19 | 73.80 | 0.6129 |
| 5 | 67.95 | 0.1613 | 20 | 74.28 | 0.6452 |
| 6 | 68.10 | 0.1935 | 21 | 75.33 | 0.6774 |
| 7 | 69.20 | 0.2258 | 22 | 75.78 | 0.7097 |
| 8 | 70.65 | 0.2581 | 23 | 77.81 | 0.7419 |
| 9 | 71.04 | 0.2903 | 24 | 77.81 | 0.7742 |
| 10 | 71.17 | 0.3226 | 25 | 77.90 | 0.8065 |
| 11 | 71.53 | 0.3548 | 26 | 79.08 | 0.8387 |
| 12 | 71.75 | 0.3871 | 27 | 79.83 | 0.8710 |
| 13 | 72.28 | 0.4194 | 28 | 79.92 | 0.9032 |
| 14 | 72.29 | 0.4516 | 29 | 80.52 | 0.9355 |
| 15 | 72.50 | 0.4839 | 30 | 82.84 | 0.9677 |

When these are plotted using probability scaling on the ordinate, the graph in Fig. 7.4 is obtained.

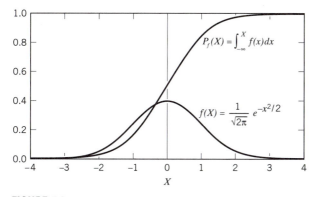

**FIGURE 7.3** Differential $f(X)$ and cumulative $P_f(X)$ normal probability functions.

The normal distribution function has been characterized thoroughly, and it is possible to infer a great deal of information from it for strength distributions that are close to normal. For instance, Table 2 of Appendix H shows that 68.3 percent of all members of a normal distribution lie within $\pm 1s$ of the mean, 95 percent lie within $\pm 1.96s$, and 99.7 percent lie within $\pm 3s$. It is common practice in much aircraft design to take $\overline{\sigma_f} - 3s$ as the safe fracture strength; then 99.7 percent of all specimens will have a strength at least this high. This doesn't really mean that three out of every thousand airplane wings are unsafe; within the accuracy of the theory, 0.3 percent is a negligible number, and the $3s$ tolerance includes essentially the entire population. Having to reduce the average strength by $3s$ in design can be a real penalty for advanced materials such as composites that have high strengths but also high variability because their processing methods are relatively undeveloped. This is a major factor that limits the market share of these advanced materials.

Beyond the visual check of the linearity of the probability plot, several "goodness-of-fit" tests are available to assess the degree to which the population can reasonably be defined by the normal (or some other) distribution function. The *chi-square* test is often used for this purpose. Here a test statistic measuring how far the observed

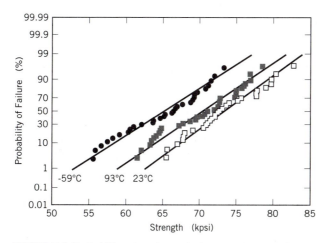

**FIGURE 7.4** Probability plot of cumulative probability of failure for the strength data of Example 7.2. Also shown are test data taken at higher and lower temperatures.

data deviate from those expected from a normal distribution, or any other proposed distribution, is

$$\chi^2 = \sum \frac{(\text{expected} - \text{observed})^2}{\text{observed}}$$

$$= \sum_{i=1}^{N} \frac{(Np_i - n_i)^2}{n_i}$$

where $N$ is the total number of specimens, $n_i$ is the number of specimens actually failing in a strength increment $\Delta\sigma_{f,i}$, and $p_i$ is the probability expected from the assumed distribution of a specimen having a strength in that increment.

## EXAMPLE 7.4

To apply the chi-square test to our 30-test population, we begin by counting the number of strengths falling in selected strength increments, much as if we were constructing a histogram. We choose five increments to obtain reasonable counts in each increment. The number *expected* to fall in each increment is determined from the normal pdf table, and the square of the difference calculated.

| Lower limit | Upper limit | Observed frequency | Expected frequency | $\chi^2$ |
|---|---|---|---|---|
| 0 | 69.33 | 7 | 5.9 | 0.198 |
| 69.33 | 72.00 | 5 | 5.8 | 0.116 |
| 72.00 | 74.67 | 8 | 6.8 | 0.214 |
| 74.67 | 77.33 | 2 | 5.7 | 2.441 |
| 77.33 | $\infty$ | 8 | 5.7 | 0.909 |
| | | | | $\chi^2 = 3.878$ |

The number of degrees of freedom for this chi-square test is 4; this is the number of increments less one, since we have the constraint that $n_1 + n_2 + n_3 + n_4 + n_5 = 30$. From Table 3 in Appendix H we read that $\alpha = 0.05$ for $\chi^2 = 9.488$, where $\alpha$ is the fraction of the $\chi^2$ population with values of $\chi^2$ greater than 9.488. Equivalently, values of $\chi^2$ above 9.488 would imply that there is less than a 5 percent chance that a population described by a normal distribution would have the computed $\chi^2$ value. Our value of 3.878 is substantially less than this, and we are justified in claiming our data to be normally distributed.

Several governmental and voluntary standards-making organizations have worked to develop standardized procedures for generating statistically allowable property values for design of critical structures.[4] One such procedure defines the *B-allowable* strength as that level for which we have 95 percent confidence that 90 percent of all specimens will have at least that strength. (The use of *two* percentages here may be confusing. We mean that if we were to measure the strengths of 100 groups each containing 10 specimens, at least 95 of these groups would each have at least 9

---

[4]*Military Handbook 17B*, Army Materials Technology Laboratory, Part I, Vol. 1, 1987.

tion function is found to provide a suitable description of the population, the *B-basis value* can be computed from the mean and standard deviation using the formula

$$B = \overline{\sigma_f} - k_B s$$

where $k_B$ is given by the formula

$$k_B = 1.282 + \exp(0.958 - 0.520 \ln N + 3.19/N)$$

## EXAMPLE 7.5

In the case of the previous 30-test example, $k_B$ is computed to be 1.78, so this is less conservative than the $3s$ guide. The B-basis value is then

$$B = 73.28 - (1.78)(4.632) = 65.05$$

Having a distribution function available lets us say something about the confidence we can have in how reliably we have measured the mean strength, based on a necessarily limited number of individual strength tests. A famous and extremely useful result in mathematical statistics states that, if the mean of a distribution is measured $N$ times, the *distribution of the means* will have its own standard deviation $s_m$ that is related to the mean of the underlying distribution $s$ and *the number of determinations, N*, as

$$s_m = \frac{s}{\sqrt{N}} \tag{7.4}$$

This result can be used to establish *confidence limits.* Since 95 percent of all measurements of a normally distributed population lie within 1.96 standard deviations from the mean, the ratio $\pm 1.96 s/\sqrt{N}$ is the range over which we can expect 95 out of 100 measurements of the mean to fall. So, even in the presence of substantial variability, we can obtain measures of mean strength to any desired level of confidence; we simply make more measurements to increase the value of $N$ in Eq. 7.4. The "error bars" often seen in graphs of experimental data are not always labeled, and the reader must be somewhat cautious: They are *usually* standard deviations, but they may indicate maximum and minimum values, and occasionally they are 95 percent confidence limits. The significance of these three is obviously quite different.

## EXAMPLE 7.6

Equation 7.4 tells us that were we to repeat the 30-test sequence of the previous example over and over (obviously with new specimens each time), 95 percent of the measured sample means would lie within the interval

$$73.278 - \frac{(1.96)(4.632)}{\sqrt{30}}, \quad 73.278 + \frac{(1.96)(4.632)}{\sqrt{30}} = 71.62, 74.94$$

## 7.1.3 The *t* Distribution

The *t* distribution, tabulated in Table 4 of Appendix H, is similar to the normal distribution, plotting as a bell-shaped curve centered on the mean. It has several useful applications to strength data. When there are few specimens in the sample, the *t* distribution should be used in preference to the normal distribution in computing confidence limits. As seen in the table, the *t*-value for the 95th percentile and the 29 degrees of freedom of our 30-test sample in Example 7.6 is 2.045. (The number of degrees of freedom is one less than the total specimen count, because the sum of the number of specimens having each recorded strength is constrained to be the total number of specimens.) The 2.045 factor replaces 1.96 in this example without much change in the computed confidence limits. As the number of specimens increases, the *t*-value approaches 1.96. For fewer specimens the factor deviates substantially from 1.96; it is 2.571 for $n = 5$ and 3.182 for $n = 3$.

The *t* distribution is also useful in deciding whether two test samplings indicate significant differences in the populations they are drawn from, or whether any difference in, say, the means of the two samplings can be ascribed to expected statistical variation in what are two essentially identical populations. For instance, Fig. 7.4 shows the cumulative failure probability for graphite-epoxy specimens tested at three different temperatures, and it appears that the mean strength is reduced somewhat by high temperatures and even more by low temperatures. But are these differences real, or are they merely statistical scatter?

This question can be answered by computing a value for *t* using the means and standard deviations of any two of the samples, according to the formula

$$t = \frac{\left| \overline{\sigma}_{f1} - \overline{\sigma}_{f2} \right|}{\sqrt{\dfrac{s_1^2}{n_1 - 1} + \dfrac{s_2^2}{n_2 - 1}}} \tag{7.5}$$

This statistic is known to have a *t* distribution if the deviations $s_1$ and $s_2$ are not too different. The mean and standard deviation of the $-59°C$ data shown in Fig. 7.4 are 65.03 and 5.24, respectively. Using Eq. 7.5 to compare the room-temperature (23°C) and $-59°C$ data, the *t*-statistic is

$$t = \frac{(73.28 - 65.03)}{\sqrt{\dfrac{(4.63)^2}{29} + \dfrac{(5.24)^2}{29}}} = 6.354$$

From Table 4 in Appendix H we now look up the value of *t* for 29 degrees of freedom corresponding to 95 percent (or some other value, as desired) of the population. We do this by scanning the column for $F(t) = 0.975$ rather than 0.95, since the *t* distribution is symmetric and another 0.025 fraction of the population lies beyond $t = -0.975$. The *t* value for 95 percent ($F(t) = 0.975$) and 29 degrees of freedom is 2.045.

This result means that were we to select repeatedly any two arbitrary 30-specimen samples from a single population, 95 percent of these selections would have *t*-statistics, as computed with Eq. 7.5, less than 2.045; only 5 percent would produce larger values of *t*. Since the 6.354 *t*-statistic for the $-59°C$ and 23°C samplings is much greater than 2.045, we can conclude that it is very unlikely that the two sets of data are from the same population. Conversely, we conclude that the two datasets are in fact statistically independent and that temperature has a statistically significant effect on the strength.

## 7.1.4 The Weibull Distribution

We mentioned in Chapter 1 that large specimens tend to have lower average strengths than small ones, simply because large ones are more likely to contain a flaw large enough to induce fracture at a given applied stress. This effect can be measured directly, for instance by plotting the strengths of fibers versus the fiber circumference, as in Fig. 7.5. For similar reasons, brittle materials tend to have higher strengths when tested in flexure than in tension, since in flexure the stresses are concentrated in a smaller region near the outer surfaces.

The hypothesis of the size effect led to substantial development effort in the statistical analysis community of the 1930s and 40s, with perhaps the most noted contribution being that of W. Weibull[5] in 1939. Weibull postulated that the probability of survival at a stress $\sigma$ (that is, the probability that the specimen volume does not contain a flaw large enough to fail under the stress $\sigma$) could be written in the form

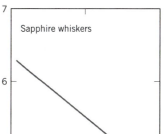

**FIGURE 7.5** Effect of circumference $c$ on fracture strength $\sigma_f$ for sapphire whiskers. (From L. J. Broutman and R. H. Krock, *Modern Composite Materials,* © 1967 Addison-Wesley Publishing Company, Inc. Reprinted by permission of Addison-Wesley.)

$$P_s(\sigma) = \exp\left[-\left(\frac{\sigma}{\sigma_0}\right)^m\right] \tag{7.6}$$

Weibull selected the form of this expression for its mathematical convenience rather than some fundamental understanding, but it has been found over many trials to describe fracture statistics well. The parameters $\sigma_0$ and $m$ are adjustable constants; Fig. 7.6 shows the form of the Weibull function for two values of the parameter $m$. Materials with greater variability have smaller values of $m$; steels have $m \approx 100$, whereas ceramics have $m \approx 10$. This parameter can be related to the coefficient of variation; to a reasonable approximation, $m \approx 1.2/C.V.$

A variation on the normal probability paper graphical method outlined earlier can be developed by taking logarithms of Eq. 7.6:

$$\ln P_s = -\left(\frac{\sigma}{\sigma_0}\right)^m$$

$$\ln(\ln P_s) = -m \ln\left(\frac{\sigma}{\sigma_0}\right)$$

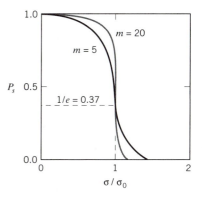

**FIGURE 7.6** The Weibull function.

Hence the double logarithm of the probability of exceeding a particular strength $\sigma$ versus the logarithm of the strength should plot as a straight line with slope $m$.

---

## EXAMPLE 7.7

Again using the 30-test sample of the previous examples, an estimate of the $\sigma_0$ parameter can be obtained by plotting the survival probability (1 minus the rank) and noting the value of $\sigma_f$ at which $P_s$ drops to $1/e = 0.37$; this gives $\sigma_0 \approx 74$. (A more accurate regression method gives 75.46.) A tabulation of the double logarithm of $P_s$ against the logarithm of $\sigma_f/\sigma_0$ is then

---

[5] See B. Epstein, *J. Appl. Phys.,* Vol. 19, p. 140, 1948 for a useful review of the statistical treatment of the size effect in fracture and for a summary of extreme-value statistics as applied to fracture problems.

| $i$ | $\sigma_{f,i}$ | $\ln\ln(1/P_s)$ | $\ln(\sigma_{f,i}/\sigma_0)$ |
|---|---|---|---|
| 1 | 65.50 | −3.4176 | −0.1416 |
| 2 | 65.67 | −2.7077 | −0.1390 |
| 3 | 67.03 | −2.2849 | −0.1185 |
| 4 | 67.84 | −1.9794 | −0.1065 |
| 5 | 67.95 | −1.7379 | −0.1048 |
| 6 | 68.10 | −1.5366 | −0.1026 |
| 7 | 69.20 | −1.3628 | −0.0866 |
| 8 | 70.65 | −1.2090 | −0.0659 |
| 9 | 71.04 | −1.0702 | −0.0604 |
| 10 | 71.17 | −0.9430 | −0.0585 |
| 11 | 71.53 | −0.8250 | −0.0535 |
| 12 | 71.75 | −0.7143 | −0.0504 |
| 13 | 72.28 | −0.6095 | −0.0431 |
| 14 | 72.29 | −0.5095 | −0.0429 |
| 15 | 72.50 | −0.4134 | −0.0400 |
| 16 | 72.85 | −0.3203 | −0.0352 |
| 17 | 72.85 | −0.2295 | −0.0352 |
| 18 | 73.23 | −0.1404 | −0.0300 |
| 19 | 73.80 | −0.0523 | −0.0222 |
| 20 | 74.28 | 0.0355 | −0.0158 |
| 21 | 75.33 | 0.1235 | −0.0017 |
| 22 | 75.78 | 0.2125 | 0.0042 |
| 23 | 77.81 | 0.3035 | 0.0307 |
| 24 | 77.81 | 0.3975 | 0.0307 |
| 25 | 77.90 | 0.4961 | 0.0318 |
| 26 | 79.08 | 0.6013 | 0.0469 |
| 27 | 79.83 | 0.7167 | 0.0563 |
| 28 | 79.92 | 0.8482 | 0.0574 |
| 29 | 80.52 | 1.0083 | 0.0649 |
| 30 | 82.84 | 1.2337 | 0.0933 |

The Weibull plot of these data is shown in Fig. 7.7; the regression slope is 17.4.

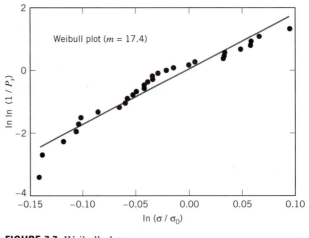

**FIGURE 7.7** Weibull plot.

Similarly to the B-basis design allowable for the normal distribution, the B-allowable can also be computed from the Weibull parameters $m$ and $\sigma_0$. The procedure is[6]

$$B = Q \exp\left[\frac{-V}{m\sqrt{N}}\right]$$

where $Q$ and $V$ are

$$Q = \sigma_0 (0.10536)^{1/m}$$

$$V = 3.803 + \exp\left[1.79 - 0.516\ln(N) + \frac{5.1}{N}\right]$$

## EXAMPLE 7.8

The B-allowable is computed for the 30-test population as

$$Q = 75.46 (0.10536)^{1/17.4} = 66.30$$

$$V = 3.803 + \exp\left[1.79 - 0.516\ln(30) + \frac{5.1}{30}\right] = 5.03$$

$$B = 66.30 \exp\left[\frac{-5.03}{17.4\sqrt{30}}\right] = 62.89$$

This value is somewhat lower than the 65.05 obtained as the normal-distribution B-allowable, so in this case the Weibull method is a bit more lenient.

The Weibull equation can be used to predict the magnitude of the size effect. If, for instance, we take a reference volume $V_0$ and express the volume of an arbitrary specimen as $V = nV_0$, then the probability of failure at volume $V$ is found by multiplying $P_s(V_0)$ by itself $n$ times:

$$P_s(V) = [P_s(V_0)]^n = [P_s(V_0)]^{V/V_0}$$

$$\boxed{P_s(V) = \exp\left[-\frac{V}{V_0}\left(\frac{\sigma}{\sigma_0}\right)^m\right]} \tag{7.7}$$

Hence, the probability of failure increases exponentially with the specimen volume. This is another danger in simple scaling, beyond the area vs. volume argument we outlined in Chapter 1.

---

[6]S.W. Rust, et al., *ASTM STP 1003*, p. 136, 1989. (Also *Military Handbook 17*.)

EXAMPLE 7.9

Solving Eq. 7.7, the stress for a given probability of survival is

$$\sigma = \left[ \frac{-\ln(P_s)}{(V/V_0)} \right]^{1/m} \sigma_0$$

Using $\sigma_0 = 75.46$ and $m = 17.4$ for the 30-specimen population, the stress for $P_s = 0.5$ and $V/V_0 = 1$ is $\sigma = 73.9$ kpsi. If now the specimen size is doubled, so that $V/V_0 = 2$, the probability of survival at this stress as given by Eq. 7.7 drops to $P_s = 0.25$. If, on the other hand, the specimen size is halved ($V/V_0 = 0.5$), the probability of survival rises to $P_s = 0.71$.

A final note of caution somewhat recalls the famous Mark Twain aphorism about there being "lies, damned lies, and statistics": It is often true that populations of simple tensile or other laboratory specimens can be well described by classical statistical distributions. This should not be taken to imply that more complicated structures such as bridges and airplanes can be so neatly described. For instance, one aircraft study cited by Gordon[7] found failures to occur randomly and uniformly over a wide range extending both above and below the statistically based design safe load. Any real design, especially for structures that put human life at risk, must be checked in every reasonable way the engineer can imagine. This will include proof testing to failure, consideration of the worst possible environmental factors, consideration of construction errors resulting from difficult-to-manufacture designs, and so on almost without limit. Experience, caution, and common sense will usually be at least as important as elaborate numerical calculations.

## 7.2 ATOMISTIC FRACTURE MODELS

In seeking to understand fracture processes in solids, it is natural to begin by examining the inherent strength of the chemical bonds within the material. In our earlier development of a theoretical yield strength, we used a harmonic approximation to the bond force-distance function to estimate the stress needed to shear an atom from one equilibrium position to the next; this gave a maximum of $\tau_{max} \approx G/2\pi$. An analogous development, but with the atoms being displaced in the direction normal to a given fracture plane, gives a similar result: $\sigma_{max} \approx E/2\pi$ (see Problem 7 in this chapter). Just as with yield strengths, this value is some two orders of magnitude higher than those observed experimentally, and it is necessary to postulate the existence of some sort of strength-reducing flaw in the microstructure. The dislocation played this role in the case of shear yield and flow, and we shall see that in the case of fracture the culprit is a crack within the material or at its surface. When special care is made to avoid these cracks, as in the case of "whiskers" that can be made having essentially perfect microstructures, observed fracture strengths do indeed approach the theoretical maximum.

---

[7]J.E. Gordon, *Structures, or Why Things Don't Fall Down*, Plenum Press, New York, 1978.

An alternative view, which provides an estimate of the influence of time and temperature in the fracture process, takes damage accumulation to be a first-order process in which the rate of bond rupture is proportional to the fraction $n$ of bonds remaining unbroken at any time:

$$\frac{dn}{dt} = -kn$$

where $k$ is a rate constant, also the probability of a single bond suffering scission in a unit time. Separating variables and integrating, we obtain

$$\frac{dn}{n} = -k\,dt$$

$$n = n_0 e^{-kt}$$

where $n_0 = 1$ is the fraction of unbroken bonds at the beginning of the process. The *average* time $\bar{t}$ for a bond to rupture is

$$\bar{t} = \frac{\int_0^\infty t \cdot n\,dt}{\int_0^\infty n\,dt} = \frac{\int_0^\infty t \cdot n_0 e^{-kt}\,dt}{\int_0^\infty n_0 e^{-kt}\,dt} = \frac{1}{k}$$

(See Fig. 7.8.) Hence, the average bond rupture lifetime is just the inverse of the rate constant.

Next, the rate "constant" is taken to vary with stress and temperature as a thermally activated, stress-aided process similar to the model used for thermally activated yield processes in Section 6.1.4:

$$k = k_0 \exp \frac{-(E^* - V^*\psi)}{RT}$$

where $E^*$ and $V^*$ are the activation energy and volume of the process, and $\psi$ is the stress on the bond. Finally, we take the creep rupture lifetime $t_f$—the time to failure under a constant stress—to scale with the average bond lifetime, and the molecular stress $\psi$ to scale with the macroscopic applied stress $\sigma$. If these admittedly questionable assertions are accepted, we obtain *Zhurkov's equation:*[8]

$$\boxed{t_f = t_0 \exp \frac{(E^* - V^*\sigma)}{RT}} \qquad (7.8)$$

where $t_0$ is a constant that scales with $1/k_0$. For constant temperature this reduces to

$$t_f = \alpha \exp(-\beta\sigma) \qquad (7.9)$$

where

$$\alpha = t_0 \exp(E^*/RT) \qquad (7.10)$$

$$\beta = V^*/RT \qquad (7.11)$$

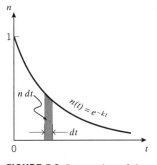

**FIGURE 7.8** Integration of the bond survival function.

---

[8]S.N. Zhurkov, *Int. J. Fracture Mech.*, Vol. 1, p. 311, 1964.

EXAMPLE 7.10

There are usually a number of ways to fit theoretical models to experimental data. To illustrate one such method for the three-parameter Zhurkov equation, consider the following table of creep rupture data for drawn polyamide (nylon) fibers, at three different temperatures:

| $T$, °C | $\sigma$, MPa | $t_f$, s |
|---------|---------------|----------|
| −10     | 875           | 987      |
|         | 900           | 239      |
|         | 925           | 57.8     |
|         | 950           | 14.0     |
|         | 975           | 3.39     |
| 23      | 800           | 714      |
|         | 825           | 202      |
|         | 850           | 57.4     |
|         | 875           | 16.3     |
|         | 900           | 4.61     |
| 50      | 750           | 340      |
|         | 775           | 107      |
|         | 800           | 33.7     |
|         | 825           | 10.6     |
|         | 850           | 3.35     |

(These data are fictitious; in reality, substantially more testing would be required to establish statistically valid results.)

Equations 7.9 and 7.10 can be linearized by taking logs; Eq. 7.9 gives $\ln t_f = \ln \alpha - \beta \sigma$, so plotting $\ln t_f$ versus $\sigma$ for the three temperatures (see Fig. 7.9a) gives three straight lines whose slopes are $-\beta(T)$ and whose intercepts are $\ln \alpha(T)$:

| $T$, K | $\ln \alpha$ | $\beta$, m²/N |
|--------|--------------|---------------|
| 263    | 56.5         | $5.67 \times 10^{-8}$ |
| 296    | 46.9         | $5.04 \times 10^{-8}$ |
| 323    | 40.5         | $4.62 \times 10^{-8}$ |

Now, from Eq. 7.10 a plot of $\ln \alpha$ versus $1/T$ (Fig. 7.9b) gives a straight line whose slope is $E^*/R$ and whose intercept is $\ln t_0$. Plotting $\beta$ versus $1/T$ (Fig. 7.9c) gives a straight line through the origin whose slope is $V^*/R$. Substituting the known value of $R$, the three kinetic parameters can be computed as

$$E^* = 45 \text{ kcal/mol}$$

$$t_0 = 1 \times 10^{-13} \text{ s}$$

$$V^* = 2.05 \times 10^{-28} \text{ m}^3 = 205 \text{ Å}^3$$

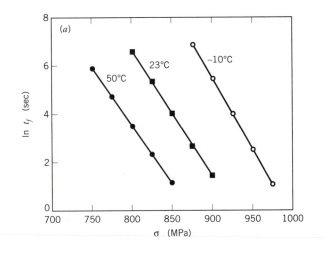

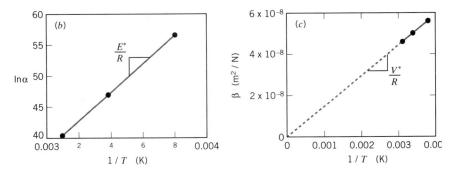

**FIGURE 7.9**  Plots to obtain Zhurkov fracture parameters.

The Zhurkov equation seems to provide a workable predictive model for creep rupture lifetimes in a wide range of ionic, metallic, and covalent materials. Table 7.1 lists numerical parameters for three of these. The reasonable fit to experimental data may be due more to the very adaptable form of the equation than to its accurate portrayal of the underlying physics, but it does provide a useful, albeit simplistic, mental picture of what is usually a rather complicated event. When the adjustable constants $t_0$, $E^*$, and $V^*$ are selected to fit experimental data, the activation energy is found to be perhaps half the calorimetric dissociation energy of a chemical bond, and the activation volume is on the order of a thousand times larger than the volume associated with a single bond. Both of these are related to the reasons for the much-too-high estimate of $E/2\pi$ obtained from a simple analysis of the bond energy function. Evidently, the stress is *not* distributed uniformly over all the bonds, and of course we would be

**TABLE 7.1**  Numerical parameters in the Zhurkov equation

| Material | $t_0$, s | $E^*$, kJ/mol | $V^*$, m³/mol |
| --- | --- | --- | --- |
| Silver chloride | $1 \times 10^{-13}$ | 126 | $3.50 \times 10^{-3}$ |
| Aluminum | $1 \times 10^{-13}$ | 222 | $2.59 \times 10^{-3}$ |
| Polymethyl methacrylate | $1 \times 10^{-13}$ | 226 | $2.41 \times 10^{-3}$ |

astounded if they were. Some bonds—those near crack tips, we shall argue—are more critically stressed, and it is these that rupture during the fracture process.

The Zhurkov approach can be generalized to provide a cumulative damage model in which stress and temperature can vary with time. In a time increment $dt$ we take the fraction of lifetime consumed to be

$$\frac{dt}{t_f} = \frac{dt}{t_0 \exp(E^* - V^*\sigma)/RT}$$

If we now assume that the damage accumulates linearly, fracture can be predicted when the total fraction of lifetime consumed adds to unity:

$$\int_0^{t_f} \frac{dt}{t_f} = \int_0^{t_f} \frac{dt}{t_0 \exp(E^* - V^*\sigma)/RT} = 1 \qquad (7.12)$$

---

## EXAMPLE 7.11

We wish to estimate the time to failure for oriented nylon fibers in a constant-stress-rate experiment at a constant temperature of 23°C (296 K). Here the stress is a function of $t$ only:

$$\sigma(t) = R_\sigma t$$

where $R_\sigma = 1$ MPa/s is the stress rate for this example. Using the definitions of $\alpha$ and $\beta$ in Eqs. 7.10 and 7.11, Eq. 7.12 gives

$$\int_0^{t_f} \frac{dt}{\alpha \exp(-\beta R_\sigma t)} = \frac{\exp(\beta R_\sigma t_f) - 1}{\alpha \beta R_\sigma} = 1$$

$$t_f = \frac{\ln(1 + \alpha \beta R_\sigma)}{\beta R_\sigma}$$

Now using the values $\alpha = 2.33 \times 10^{20}$ s and $\beta = 5.04 \times 10^{-7}$ m$^2$/N from the previous example, this gives

$$t_f = 871 \text{ s}$$

In practice, it is difficult to corroborate the applicability of treatments such as this, since damage accumulation depends so strongly on stress. Having stress inside an exponential causes the time to failure to be dominated by the last moments of the loading, with very little sensitivity to damage induced early in the process.

---

## 7.3 FRACTURE MECHANICS

The story of the DeHavilland Comet aircraft of the early 1950s, in which at least two aircraft disintegrated in flight, provides a tragic but fascinating insight into the importance of fracture theory. It is an eerie story as well, having been all but predicted in a 1948 novel by Nevil Shute named *No Highway*. The book later became a movie starring James Stewart as a perseverant metallurgist convinced that his company's new aircraft (the "Reindeer") was fatally prone to metal fatigue. When just a few years later the Comet (see Fig. 7.10) was determined to have almost exactly this problem, both the book and the movie became rather famous in the materials engineering community.

The postmortem study of the Comet's problems was one of the most extensive in engineering history.[9] It required salvaging almost the entire aircraft from scattered wreckage on the ocean floor and also involved full-scale pressurization of an aircraft in a giant water tank. Although valuable lessons were learned, it is hard to overstate the damage done to the DeHavilland Company and to the British aircraft industry in general. It is sometimes argued that the long predominance of the United States in commercial aircraft is due at least in part to the Comet's misfortune.

The central difficulty in designing against fracture in high-strength materials is that the presence of cracks can modify the local stresses to such an extent that the elastic stress analyses done so carefully by the designers are insufficient. When a crack reaches a certain critical length, it can propagate catastrophically through the structure, *even though the gross stress is much less than would normally cause yield or failure in a tensile specimen.* The term *fracture mechanics* refers to a vital specialization within solid mechanics in which the presence of a crack is assumed and we wish to find quantitative relations between the crack length, the material's inherent resistance to crack growth, and the stress at which the crack propagates at high speed to cause structural failure.

### 7.3.1 The Energy-Balance Approach

When A. A. Griffith (1893–1963) began his pioneering studies of fracture in glass in the years just prior to 1920, he was aware of the work of Inglis in calculating the stress concentrations around elliptical holes, and naturally he considered how it might be

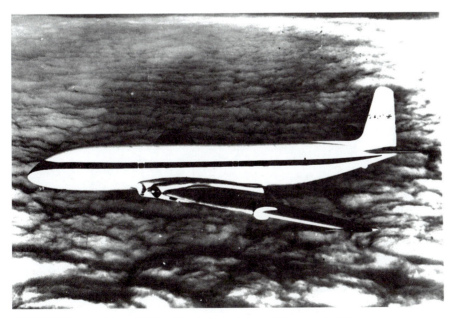

**FIGURE 7.10**  The ill-fated DeHavilland Comet jetliner. (Bettman Archive.)

---

[9]T. Bishop, *Metal Progress,* Vol. 67, pp. 79–85, May 1955.

used in developing a fundamental approach to predicting fracture strengths. However, the Inglis solution poses a mathematical difficulty. If the Inglis ellipse is flattened to a sharp crack ($\rho$ approaches zero in Eq. 5.35), the stresses approach infinity. This is obviously nonphysical (actually the material generally undergoes some local yielding to blunt the crack tip), and using such a result would predict that materials would have near-*zero* strength: Even for very small applied loads, the stresses near crack tips would become infinite, and the bonds there would rupture. Rather than focusing on the crack tip stresses directly, Griffith employed an energy-balance approach that has become one of the most famous developments in materials science.[10]

The strain energy per unit volume of stressed material is

$$U^* = \frac{1}{V} \int f \, dx = \int \frac{f}{A} \frac{dx}{L} = \int \sigma \, d\epsilon$$

If the material is linear ($\sigma = E\epsilon$), then the strain energy per unit volume is

$$U^* = \frac{E\epsilon^2}{2} = \frac{\sigma^2}{2E}$$

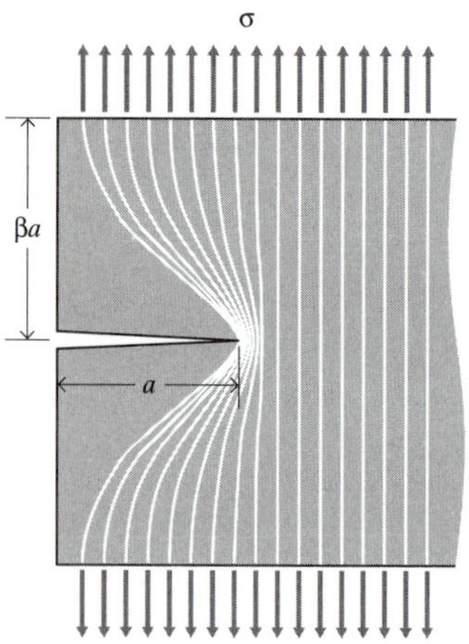

When a crack has grown into a solid to a depth $a$, a region of material adjacent to the free surfaces is unloaded, and its strain energy is released. Using the Inglis solution, Griffith was able to compute just how much energy this is.

A simple way of visualizing this energy release, illustrated in Fig. 7.11, is to regard two triangular regions near the crack flanks, of width $a$ and height $\beta a$, as being completely unloaded, while the remaining material continues to bear the full stress $\sigma$. The parameter $\beta$ can be selected so as to agree with the Inglis solution, and it turns out that for plane stress loading $\beta = \pi$. The total strain energy $U$ released is then the strain energy per unit volume times the volume in both triangular regions:

$$U = -\frac{\sigma^2}{2E} \cdot \pi a^2$$

Here the dimension normal to the x-y plane is taken to be unity, so $U$ is the strain energy released per unit thickness of specimen. This strain energy is *liberated* by crack growth. However, to form the crack, bonds must be broken, and the requisite bond energy is in effect *absorbed* by the material. The surface energy $S$ associated with a crack of length $a$ (and unit depth) is

$$S = 2\gamma a$$

where $\gamma$ is the surface energy (joules/meter$^2$), and the factor 2 is needed because two free surfaces have been formed. As shown in Fig. 7.12, the total energy associated with the crack is then the sum of the (positive) energy absorbed to create the new surfaces, plus the (negative) strain energy liberated by allowing the regions near the crack flanks to become unloaded.

As the crack grows longer ($a$ increases), the quadratic dependence of strain energy on $a$ eventually dominates the surface energy, and beyond a critical crack length

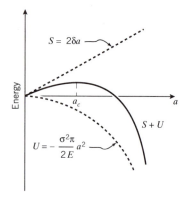

**FIGURE 7.11** Idealization of unloaded region near crack flanks.

**FIGURE 7.12** The fracture energy balance.

---

[10] A.A. Griffith, *Philosophical Transactions,* Series A, Vol. 221, pp. 163–198, 1920. The importance of Griffith's work in fracture was unrecognized until the 1950s. See J. E. Gordon, *The Science of Structures and Materials,* Scientific American Library, 1988, for a personal account of the Griffith story.

$a_c$ the system can lower its energy by letting the crack grow still longer. Up to the point where $a = a_c$, the crack will grow only if the stress is increased. Beyond that point, crack growth is spontaneous and catastrophic.

The value of the critical crack length can be found by setting the derivative of the total energy $S + U$ to zero:

$$\frac{\partial(S + U)}{\partial a} = 2\gamma - \frac{\sigma_f^2}{E}\pi a = 0$$

Since fast fracture is imminent when this condition is satisfied, we write the stress as $\sigma_f$. Solving, we obtain

$$\sigma_f = \sqrt{\frac{2E\gamma}{\pi a}}$$

Griffith's original work dealt with very brittle materials, specifically glass rods. When the material exhibits more ductility, consideration of the surface energy alone fails to provide an accurate model for fracture. This deficiency was later remedied, at least in part, independently by Irwin[11] and Orowan.[12] They suggested that in a ductile material a good deal—in fact the vast majority—of the released strain energy was absorbed not by creating new surfaces but by energy dissipation due to plastic flow in the material near the crack tip. They suggested that catastrophic fracture occurs when the strain energy is released at a rate sufficient to satisfy the needs of all these energy "sinks," and denoted this *critical strain energy release rate* by the parameter $\mathcal{G}_c$; the Griffith equation can then be rewritten in the form

$$\sigma_f = \sqrt{\frac{E\mathcal{G}_c}{\pi a}} \tag{7.13}$$

This expression describes, in a very succinct way, the interrelation between three important aspects of the fracture process: the *material,* as evidenced in the critical strain energy release rate, $\mathcal{G}_c$; the *stress level,* $\sigma_f$; and the *size, a,* of the flaw. In a design situation we might choose a value of $a$ based on the smallest crack that could be easily detected. Then for a given material with its associated value of $\mathcal{G}_c$, the safe level of stress $\sigma_f$ could be determined. The structure would then be sized so as to keep the working stress comfortably below this critical value.

It is important to realize that the critical crack length is an absolute number, not depending on the size of the structure containing it. Each time the crack jumps ahead, say by a small increment $\delta a$, an additional quantity of strain energy is released from the newly unloaded material near the crack. Again using our simplistic picture of a triangular-shaped region that is at zero stress while the rest of the structure continues to bear the overall applied stress, it is easy to see in Fig. 7.13 that much more energy is released due to the jump at position 2 than at position 1. This is yet another reason why small things tend to be stronger: They simply aren't large enough to contain a critical-length crack.

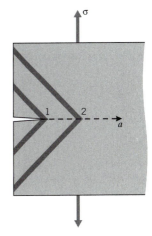

**FIGURE 7.13** Energy released during an increment of crack growth, for two different crack lengths.

---

[11] G.R. Irwin, "Fracture Dynamics," *Fracturing of Metals,* American Society for Metals, Cleveland, 1948.

[12] E. Orowan, "Fracture and Strength of Solids," *Report of Progress in Physics,* Vol. 12, 1949.

Gordon[13] tells of a ship's cook who one day noticed a crack in the steel deck of
his galley. His superiors assured him that it was nothing to worry about—the crack
was certainly small compared with the vast bulk of the ship—but the cook began
painting dates on the floor to mark the new length of the crack each time a bout of
rough weather caused it to grow longer. With each advance of the crack, additional
decking material was unloaded, and the strain energy formerly contained in it was
released. The amount of energy released grows *quadratically* with the crack length,
so eventually enough was available to keep the crack growing even with no further
increase in the gross load. When this happened, the ship broke into two pieces. This
story seems amazing, but there are more than a few such occurrences that are very
well documented. As it happened, the part of the ship with the marks showing the
crack's growth was salvaged, and this has become one of the very best documented
examples of slow crack growth followed by final catastrophic fracture.

## EXAMPLE 7.12

The Comet aircraft had a fuselage of clad aluminum, with $\mathcal{G}_c \approx 300$ in-psi. The
hoop stress due to relative cabin pressurization was 20,000 psi, and at that stress the
length of crack that will propagate catastrophically is

$$a = \frac{\mathcal{G}_c E}{\pi \sigma^2} = \frac{(300)(11 \times 10^6)}{\pi (20 \times 10^3)^2} = 2.62''$$

A crack would presumably be detected in routine inspection long before it could
grow to this length. In the Comet, however, the cracks were propagating from rivet
holes near the cabin windows. When the crack reached the window, the size of the
window opening was effectively added to the crack length, leading to disaster.

Modern aircraft are built with this failure mode in mind and have "tear strips"
that are supposedly able to stop any rapidly growing crack. However, as the Aloha
Airlines tragedy demonstrated, this remedy is not always effective. That aircraft
had stress-corrosion damage at a number of rivets in the fuselage lap splices, and
this permitted multiple small cracks to link up to form a large crack. A great deal
of attention is currently being directed to protection against this sort of "multisite
damage."

### 7.3.2 Compliance Calibration

A number of means are available by which the material property $\mathcal{G}_c$ can be measured.
One of these, known as *compliance calibration,* employs the concept of compliance
as a ratio of deformation to applied load: $C = \delta/P$. The total strain energy $U$ can be
written in terms of this compliance as

$$U = \tfrac{1}{2}P\delta = \tfrac{1}{2}CP^2$$

The compliance of a suitable specimen, for instance a cantilevered beam, could be
measured experimentally as a function of the length $a$ of a crack that is grown into

---

[13]J.E. Gordon, *Structures, or Why Things Don't Fall Down,* Plenum, New York, 1978.

the specimen (see Fig. 7.14). The strain energy release rate can then be determined by differentiating the curve of compliance versus length:

$$\mathcal{G} = \frac{\partial U}{\partial a} = \frac{1}{2}P^2\frac{\partial C}{\partial a} \tag{7.14}$$

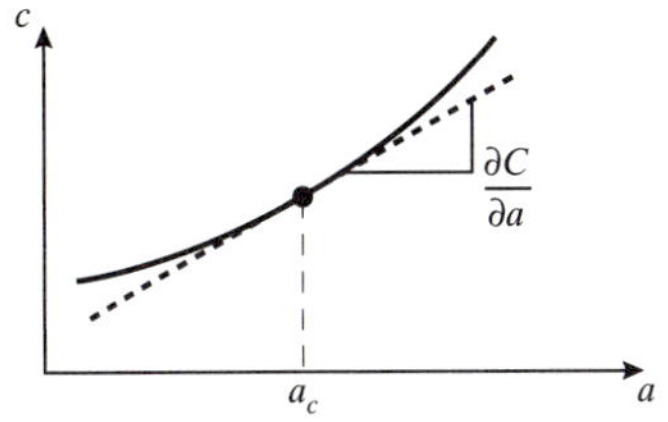

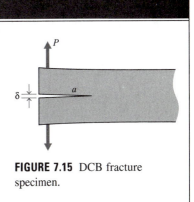

**FIGURE 7.14** Compliance as a function of crack length.

The *critical* value of $\mathcal{G}$, $\mathcal{G}_c$, is then found by measuring the critical load $P_c$ needed to fracture a specimen containing a crack of length $a_c$ and using the slope of the compliance curve at that same value of $a$:

$$\boxed{\mathcal{G}_c = \frac{1}{2}P_c^2\frac{\partial C}{\partial a}\bigg|_{a=a_c}} \tag{7.15}$$

## EXAMPLE 7.13

For a double-cantilever beam (DCB) specimen such as that shown in Fig. 7.15, beam theory gives the deflection as

$$\frac{\delta}{2} = \frac{Pa^3}{3EI}$$

where $I = bh^3/12$. The elastic compliance is then

$$C = \frac{\delta}{P} = \frac{2a^3}{3EI}$$

If the crack is observed to jump forward when $P = P_c$, Eq. 7.15 can be used to compute the critical strain energy release rate as

$$\mathcal{G}_c = \frac{1}{2}P_c^2 \cdot \frac{2a^2}{EI} = \frac{12P_c^2a^2}{b^2h^3E}$$

**FIGURE 7.15** DCB fracture specimen.

### 7.3.3 The Stress Intensity Approach

While the energy-balance approach provides a great deal of insight into the fracture process, an alternative method that examines the stress state near the tip of a sharp crack directly has proven more useful in engineering practice. The literature treats three types of cracks, termed mode I, II, and III, as illustrated in Fig. 7.16. Mode I is a normal-opening mode and is the one we shall emphasize here, whereas modes II and III are shear sliding modes. As was outlined in Chapter 5, the semi-inverse method developed by Westergaard shows the opening-mode stresses to be

$$\begin{aligned}
\sigma_x &= \frac{K_\mathrm{I}}{\sqrt{2\pi r}}\cos\frac{\theta}{2}\left(1 - \sin\frac{\theta}{2}\sin\frac{3\theta}{2}\right) + \cdots \\
\sigma_y &= \frac{K_\mathrm{I}}{\sqrt{2\pi r}}\cos\frac{\theta}{2}\left(1 + \sin\frac{\theta}{2}\sin\frac{3\theta}{2}\right) + \cdots \\
\tau_{xy} &= \frac{K_\mathrm{I}}{\sqrt{2\pi r}}\cos\frac{\theta}{2}\cos\frac{3\theta}{2}\sin\frac{\theta}{2}\cdots
\end{aligned} \tag{7.16}$$

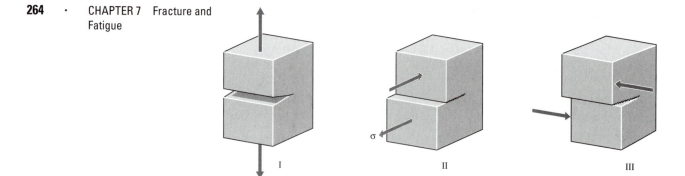

**FIGURE 7.16** Fracture modes.

For distances close to the crack tip ($r \le 0.1a$), the second and higher-order terms, indicated by dots, may be neglected. At large distances from the crack tip these relations cease to apply and the stresses approach their far-field values, which would obtain were the crack not present.

The $K_I$ in Eqs. 7.16 is a very important parameter known as the *stress intensity factor*. The I subscript is used to denote the crack opening mode, but similar relations apply in modes II and III. The equations show three factors that, taken together, depict the stress state near the crack tip:

1. The denominator factor $(2\pi r)^{-1/2}$ shows the singular nature of the stress distribution; $\sigma$ approaches infinity as the crack tip is approached, with an $r^{-1/2}$ dependency.

2. The angular dependence is separable as another factor; for example, $f_x = \cos\theta/2 \cdot (1 - \sin\theta/2 \sin 3\theta/2) + \cdots$.

3. The factor $K_I$ contains the dependence on applied stress $\sigma_\infty$, the crack length $a$, and the specimen geometry. The $K_I$ factor gives the overall intensity of the stress distribution, hence its name.

For the specific case of a central crack of width $2a$ or an edge crack of length $2a$ in a large sheet, $K_I = \sigma_\infty \sqrt{\pi a}$, and $K_I = 1.12\sigma_\infty \sqrt{\pi a}$ for an edge crack of length $a$ in the edge of a large sheet. (The factor $\pi$ could obviously be canceled with the $\pi$ in the denominator of Eq. 7.16, but it is commonly retained for consistency with earlier work.) Expressions for $K_I$ for some additional geometries are given in Table 7.2. The literature contains expressions for $K$ for a large number of crack and loading geometries, and both numerical and experimental procedures exist for determining the stress intensity factor in specific actual geometries.

These stress intensity factors are used in design and analysis by arguing that the material can withstand crack tip stresses up to a critical value of stress intensity, termed $K_{Ic}$, beyond which the crack propagates rapidly. This *critical stress intensity factor* is then a measure of material toughness. The failure stress $\sigma_f$ is then related to the crack length $a$ and the fracture toughness by

$$\boxed{\sigma_f = \frac{K_{Ic}}{\alpha \sqrt{\pi a}}} \qquad (7.17)$$

where $\alpha$ is a geometrical parameter equal to 1 for edge cracks and generally on the order of unity for other situations. Expressions for $\alpha$ are tabulated for a wide variety

**TABLE 7.2**  Stress intensity factors for several common geometries

| Type of crack | Stress intensity factor, $K_I$ |
|---|---|
| Center crack, length $2a$, in an infinite plate | $\sigma_\infty \sqrt{\pi a}$ |
| Edge crack, length $a$, in a semi-infinite plate | $1.12\sigma_\infty \sqrt{\pi a}$ |
| Central penny-shaped crack, radius $a$, in an infinite body | $2\sigma_\infty \sqrt{\dfrac{a}{\pi}}$ |
| Center crack, length $2a$ in plate of width $W$ | $\sigma_\infty \sqrt{W \tan\left(\dfrac{\pi a}{W}\right)}$ |
| Two symmetrical edge cracks, each length $a$, in plate of total width $W$ | $\sigma_\infty \sqrt{W\left[\tan\left(\dfrac{\pi a}{W}\right) + 0.1\sin\left(\dfrac{2\pi a}{W}\right)\right]}$ |

of specimen and crack geometries, and specialty finite element methods are available to compute it for new situations.

The stress intensity and energy viewpoints are interrelated, as can be seen by comparing Eqs. 7.13 and 7.17 (with $\alpha = 1$):

$$\sigma_f = \sqrt{\frac{E\mathcal{G}_c}{\pi a}} = \frac{K_{Ic}}{\sqrt{\pi a}} \rightarrow K_{Ic}^2 = E\mathcal{G}_c$$

This relation applies in plane stress; it is slightly different in plane strain:

$$K_{Ic}^2 = E\mathcal{G}_c(1 - \nu^2)$$

For metals with $\nu = 0.3$, $(1 - \nu^2) = 0.91$. This is not a big change; however, the numerical values of $\mathcal{G}_c$ or $K_{Ic}$ are very different in plane stress or plane strain situations, as will be described subsequently.

Typical values of $G_{Ic}$ and $K_{Ic}$ for various materials are listed in Table 7.3, and it is seen that they vary over a very wide range from material to material. Some polymers

**TABLE 7.3**  Fracture toughness of materials

| Material | $G_{Ic}$ (kJ-m$^{-2}$) | $K_{Ic}$ (MN-m$^2$) | $E$ (GPa) |
|---|---|---|---|
| Steel alloy | 107 | 150 | 210 |
| Aluminum alloy | 20 | 37 | 69 |
| Polyethylene | 20 ($J_{Ic}$) | — | 0.15 |
| High-impact polystyrene | 15.8 ($J_{Ic}$) | — | 2.1 |
| Steel (mild) | 12 | 50 | 210 |
| Rubber | 13 | — | 0.001 |
| Glass-reinforced thermoset | 7 | 7 | 7 |
| Rubber-toughened epoxy | 2 | 2.2 | 2.4 |
| PMMA | 0.5 | 1.1 | 2.5 |
| Polystyrene | 0.4 | 1.1 | 3 |
| Wood | 0.12 | 0.5 | 2.1 |
| Glass | 0.007 | 0.7 | 70 |

can be very tough, especially when rated on a per-pound basis, but steel alloys are hard to beat in terms of absolute resistance to crack propagation.

---

## EXAMPLE 7.14

Equation 7.17 provides a design relation among the applied stress $\sigma$, the material's toughness $K_{Ic}$, and the crack length $a$. Any one of these parameters can be calculated once the other two are known. To illustrate one application of the process, suppose we wish to determine the safe operating pressure in an aluminum pressure vessel 0.25 m in diameter and with a 5 mm wall thickness. First, assuming failure by yield when the hoop stress reaches the yield stress (330 MPa) and using a safety factor of 0.75, we can compute the maximum pressure as

$$p = \frac{0.75\sigma t}{r} = \frac{0.75 \times 330 \times 10^6}{0.25/2} = 9.9 \text{ MPa} = 1400 \text{ psi}$$

To insure against failure by rapid crack growth, we now calculate the maximum crack length permissible at the operating stress, using a toughness value of $K_{Ic} = 41 \text{ MPa}\sqrt{\text{m}}$:

$$a = \frac{K_{Ic}^2}{\pi\sigma^2} = \frac{(41 \times 10^6)^2}{\pi(0.75 \times 330 \times 10^6)^2} = 0.01 \text{ m} = 0.4 \text{ in}$$

Here an edge crack with $\alpha = 1$ has been assumed. An inspection schedule must be implemented that is capable of detecting cracks before they reach this size.

---

### 7.3.4 Effect of Specimen Geometry

The toughness, or resistance to crack growth, of a material is governed by the energy absorbed as the crack moves forward. In an extremely brittle material such as window glass this energy is primarily just that of rupturing the chemical bonds across the crack plane. As already mentioned, in tougher materials bond rupture plays a relatively small role in resisting crack growth; by far the largest part of the fracture energy is associated with plastic flow near the crack tip. A "plastic zone" is present near the crack tip; within that zone the stresses as predicted by Eq. 7.16 would be above the material's yield stress $\sigma_Y$. Since the stress cannot rise above $\sigma_Y$, the stress in this zone is $\sigma_Y$ rather than that given by Eq. 7.16. To a first approximation, the distance $r_p$ this zone extends along the $x$-axis can be found by using Eq. 7.16 with $\theta = 0$ to find the distance at which the crack tip stress reduces to $\sigma_Y$:

$$\sigma_y = \sigma_Y = \frac{K_I}{\sqrt{2\pi r_p}}$$

$$r_p = \frac{K_I^2}{2\pi\sigma_Y^2}$$

(7.18)

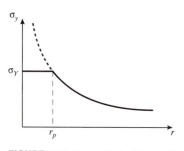

**FIGURE 7.17** Stress limited by yield within zone $r_p$.

This relation is illustrated in Fig. 7.17. As the stress intensity is increased, either by raising the imposed stress or by crack lengthening, the plastic zone size will increase as well. However, the extent of plastic flow is ultimately limited by the material's molecular or microstructural mobility, and the zone can become only so large. When the zone can grow no larger, the crack can no longer be constrained, and unstable propagation ensues. The value of $K_I$ at which this occurs can then be considered a materials property, named $K_{Ic}$.

In order for the measured value of $K_{Ic}$ to be valid, the plastic zone size should not be so large as to interact with the specimen's free boundaries or to destroy the basic nature of the singular stress distribution. The ASTM specification for fracture toughness testing[14] specifies the specimen geometry to ensure that the specimen is large compared to the crack length and the plastic zone size (see Fig. 7.18):

$$a, B, (W - a) \geq 2.5 \left( \frac{K_I}{\sigma_Y} \right)^2$$

A great deal of attention has been paid to the important case in which enough ductility exists that the foregoing criterion cannot be satisfied. In these cases the stress intensity view must be abandoned and alternative techniques, such as the $J$-integral or the crack tip opening displacement method, used instead. The reader is referred to the references listed at the end of the chapter for discussion of these approaches.

The fracture toughness as measured by $K_c$ or $\mathcal{G}_c$ is essentially a measure of the extent of plastic deformation associated with crack extension. The quantity of plastic flow would be expected to scale linearly with the specimen thickness, because reducing the thickness by half would naturally cut the volume of plastically deformed material approximately in half as well. The toughness therefore rises linearly, at least initially, with the specimen thickness, as seen in Fig. 7.19. Eventually, however, the toughness is observed to go through a maximum and fall thereafter to a lower value. This loss of toughness beyond a certain critical thickness $t^*$ is extremely important in design against fracture, because using too thin a specimen in measuring toughness will yield an unrealistically optimistic value for $\mathcal{G}_C$. The specimen size requirements for valid fracture toughness testing are such that the most conservative value is measured.

The *critical thickness* causes the specimen to be dominated by a state of *plane strain,* as opposed to plane *stress.* The stress in the through-thickness $z$ direction must become zero at the sides of the specimen, because no traction is applied there, and in a thin specimen the stress will not have room to rise to appreciable values within the material. The *strain* in the $z$ direction is not zero, of course, and the specimen will experience a Poisson contraction given by $\epsilon_z = \nu(\sigma_x + \sigma_y)$. When the specimen is thicker, however, material near the center will be unable to contract laterally, being constrained by adjacent material. Now the $z$-direction strain is zero, so a tensile stress will arise as the material tries to contract but is prevented from doing so. The value of $\sigma_z$ rises from zero at the outer surface and approaches a maximum value, given by $\sigma_z \approx \nu(\sigma_x + \sigma_y)$, in a distance $t^*$, as seen in Fig. 7.20. To guarantee that plane strain conditions dominate, the specimen thickness $t$ must be such that $t \gg 2t^*$.

The triaxial stress state set up near the center of a thick specimen near the crack tip *reduces* the maximum shear stress available to drive plastic flow, since the maximum shear stress is equal to one-half the difference of the largest and smallest principal stresses, and the smallest is now greater than zero. Or equivalently, we can state that the mobility of the material is constrained by the inability to contract laterally. From either a stress or a strain viewpoint, the extent of available plasticity is reduced by making the specimen thick.

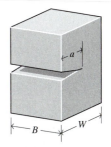

**FIGURE 7.18**
Dimensions of fracture toughness specimen.

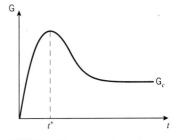

**FIGURE 7.19** Effect of specimen thickness on toughness.

---

[14]E 399-83, "Standard Test Method for Plane-Strain Fracture Toughness of Metallic Materials," ASTM, 1983.

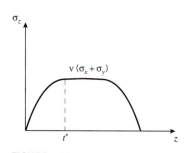

**FIGURE 7.20** Transverse stress at crack tip.

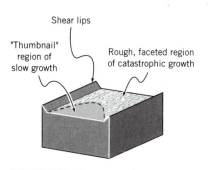

**FIGURE 7.21** Fracture surface topography.

Even in a thick specimen the $z$-direction stress must approach zero at the side surfaces. Regions near the surface are therefore free of the triaxial stress constraint and exhibit greater shear-driven plastic flow. After a cracked specimen has been tested to failure, a flat "thumbnail" pattern will often be visible, as illustrated in Fig. 7.21. This is the region of slow crack growth, where the crack is able to maintain its preferred orientation transverse to the $y$-direction stress. The crack growth near the edges is retarded by the additional plastic flow there, so the crack line bows inward. When the stress is increased enough to cause the crack to grow catastrophically, it typically does so at speeds high enough that the transverse orientation is not always maintained. The region of rapid fracture is thus faceted and rough, leading some backyard mechanics to claim that the material failed because it "crystallized."

Along the edges of the specimen, "shear lips" can often be found, on which the crack has developed by shear flow and with intensive plastic deformation. The lips will be near a 45° angle, the orientation of the maximum shear planes.

### 7.3.5 Grain Size and Temperature

Steel is such an important and widely used structural material that it is easy to forget that steel is a fairly recent technological innovation. Well into the nineteenth century, wood was the dominant material for many bridges, buildings, and ships. As the use of iron and steel became more widespread in the latter part of that century and the first part of the present one, a number of disasters took place that can be traced to the then-incomplete state of understanding of these materials, especially concerning their tendency to become brittle at low temperatures. Many of these failures have been described and analyzed in a fascinating book by Parker.[15]

One of these brittle failures is perhaps the most famous disaster of the last several centuries: the sinking of the transatlantic ocean liner *Titanic* on April 15, 1912, with a loss of some 1500 people and only 705 survivors (see Fig. 7.22). Until very recently, the tragedy was thought to be caused by a long gash torn through the ship's hull by an iceberg. However, when the wreckage of the ship was finally discovered in 1985 using undersea robots, no evidence of such a gash was found. Further, the robots were later able to return samples of the ship's steel, whose analysis has given rise to an alternative explanation.

It is now well known that lesser grades of steel, especially those having large concentrations of impurities such as interstitial carbon inclusions, are subject to embrittlement at low temperatures. William Garzke, a naval architect with the New York firm of Gibbs & Cox, and his colleagues have argued that the steel in the *Titanic* was indeed brittle in the 31°F waters of the Atlantic that night and that the 22-knot collision with the iceberg generated not a gash but extensive cracking through which water could enter the hull. Had the steel remained tough at this temperature, these authors believe, the cracking might have been much less extensive. This would have slowed the flooding and allowed more time for rescue vessels to reach the scene, which could have greatly increased the number of survivors.

In the bcc transition metals such as iron and carbon steel, brittle failure can be initiated by dislocation glide within a crystalline grain. The slip takes place at the yield stress $\sigma_Y$, which varies with grain size according to the Hall-Petch law, as described in Chapter 6:

$$\sigma_Y = \sigma_0 + k_Y d^{-1/2}$$

---

[15]E.R. Parker, *Brittle Behavior of Engineering Structures,* John Wiley & Sons, 1957.

**FIGURE 7.22** The fate of the R.M.S. *Titanic* might have been less catastrophic if the hull steel had better low-temperature toughness. (Bettman Archive.)

Dislocations are not able to propagate beyond the boundaries of the grain, because adjoining grains will not in general have their slip planes suitably oriented. The dislocations then "pile up" against the grain boundaries, as illustrated in Fig. 7.23. The dislocation pileup acts similarly to an internal crack with a length that scales with the grain size $d$, intensifying the stress in the surrounding grains. Replacing $a$ by $d$ in the modified Griffith equation (Eq. 7.13), the applied stress needed to cause fracture in adjacent grains is related to the grain size as

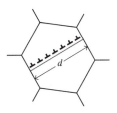

**FIGURE 7.23** Dislocation pileup within a grain.

$$\sigma_f = k_f d^{-1/2}, \qquad k_f \propto \sqrt{\frac{E\mathcal{G}_c}{\pi}}$$

These two relations for yielding and fracture are plotted in Fig. 7.24 against inverse root grain size (so grain size increases to the left), with the slopes being $k_Y$ and $k_f$, respectively. When $k_f > k_Y$, fracture will not occur until $\sigma = \sigma_Y$ for values of $d$ to the left of point $A$, because yielding and slip are a prerequisite for cleavage. In this region the yielding and fracture stresses are the same, and the failure appears brittle, because large-scale yielding will not have a chance to occur. To the right of point $A$, yielding takes place prior to fracture and the material appears ductile. The point $A$ therefore defines a critical grain size $d^*$ at which a "nil-ductility" transition from ductile (grains smaller than $d^*$) to brittle failure will take place.

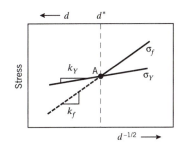

**FIGURE 7.24** Effect of grain size on yield and fracture stress.

As the temperature is lowered, the yield stress $\sigma_Y$ will increase as described in Chapter 6, and the fracture stress $\sigma_f$ will decrease (because atomic mobility, and thus $\mathcal{G}_c$, decrease). Therefore, point $A$ shifts to the right as temperature is lowered. The critical grain size for nil ductility now occurs at a smaller value; that is, the grains must be smaller to avoid embrittling the material. Equivalently, refining the grain size has the effect of lowering the ductile-brittle transition temperature. Hence, grain size refinement raises both the yield and fracture stress, lowers the ductile-brittle transition temperature, and promotes toughness as well. This is a singularly useful strengthening mechanism, because other techniques such as strain hardening and solid-solution hardening tend to achieve strengthening at the expense of toughness.

Factors other than temperature can also embrittle steel. Inclusions such as carbon and phosphorus act to immobilize slip systems that might otherwise relieve the stresses associated with dislocation pileups, and these inclusions can raise the yield stress and thus the ductile-brittle transition temperature markedly. Similar effects can be induced by damage from high-energy radiation, so embrittlement of nuclear reactor components is of great concern. Embrittlement is also facilitated by the presence of notches, because they generate triaxial stresses that constrain plastic flow. High strain rates promote brittleness, because the flow stress needed to accommodate the strain rate is higher. Improper welding can lead to brittleness, both by altering the steel's microstructure and by generating residual internal stresses.

## 7.4 FATIGUE

### 7.4.1 Phenomenology

We have mentioned the phenomenon of "fatigue" several times previously, as in the growth of cracks in the Comet aircraft that led to disaster when they became large enough to propagate catastrophically as predicted by the Griffith criterion. Fatigue, as understood by materials technologists, is a process in which damage accumulates due to the repetitive application of loads that may be well below the yield point. The process is dangerous because a single application of the load would not produce any ill effects, and a conventional stress analysis might lead to an assumption of safety that does not exist.

In one popular view of fatigue in metals, the fatigue process is thought to begin at an internal or surface flaw, where the stresses are concentrated, and consists initially of shear flow along slip planes. Over a number of cycles this slip generates intrusions and extrusions that begin to resemble a crack. A true crack running inward from an intrusion region may propagate initially along one of the original slip planes, but eventually it turns to propagate transversely to the principal normal stress, as seen in Fig. 7.25.

When the failure surface of a fatigued specimen is examined, a region of slow crack growth is usually evident in the form of a "clamshell" concentric around the location of the initial flaw. (See Fig. 7.26.) The clamshell region often contains concentric "beach marks" at which the crack was arrested for some number of cycles before resuming its growth. Eventually, the crack may become large enough to satisfy the energy or stress intensity criteria for rapid propagation, following the previous expressions for fracture mechanics. This final phase produces the rough surface typical of fast fracture. In postmortem examination of failed parts, it is often possible to correlate the beach marks with specific instances of overstress and to estimate the applied stress at failure from the size of the crack just before rapid propagation and the fracture toughness of the material.

The modern study of fatigue is generally dated from the work of A. Wöhler, a technologist in the German railroad system in the mid-nineteenth century. Wöhler was concerned by the failure of axles after various times in service, at loads considerably less than expected. A railroad car axle is essentially a round beam in four-point bending, which produces a compressive stress along the top surface and a tensile stress along the bottom (see Fig. 7.27). After the axle has rotated a half turn, the bottom becomes the top and vice versa, so the stress on a particular region of material at the surface varies sinusoidally from tension to compression and back again. This is now known as *fully reversed* fatigue loading.

### 7.4.2 *S-N* Curves

Well before a microstructural understanding of fatigue processes was developed, engineers had developed empirical means of quantifying the fatigue process and designing against it. Perhaps the most important concept is the *S-N diagram,* such as those

**FIGURE 7.25**
Intrusion-extrusion model of fatigue crack initiation.

**FIGURE 7.26**  Typical fatigue-failure surfaces. (From B. Chalmers, *Physical Metallurgy,* Wiley, 1959, p. 212. Reprinted by permission.)

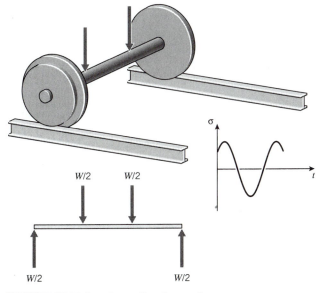

**FIGURE 7.27**  Fatigue in a railroad car axle.

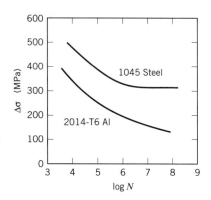

**FIGURE 7.28** *S-N* curves for aluminum and low-carbon steel. (From H. W. Hayden, W. G. Moffatt, and J. Wulff, *The Structure and Properties of Materials,* Vol. III, Wiley, 1965.)

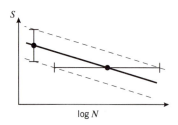

**FIGURE 7.29** Variability in fatigue lifetimes and fracture strengths.

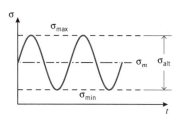

**FIGURE 7.30** Simultaneous mean and cyclic loading.

shown in Fig. 7.28,[16] in which a constant cyclic stress of amplitude $S$ is applied to a specimen and the number of loading cycles $N$ until the specimen fails is determined. Millions of cycles might be required to cause failure at lower loading levels, so the abscissa is usually plotted logarithmically.

In some materials, notably ferrous alloys, the *S-N* curve flattens out eventually, so that below a certain *endurance limit, $\sigma_e$,* failure does not occur no matter how long the loads are cycled. Obviously, the designer will size the structure to keep the stresses below $\sigma_e$ by a suitable safety factor if cyclic loads are to be withstood. For some other materials, such as aluminum, no endurance limit exists, and the designer must arrange for the planned lifetime of the structure to be less than the failure point on the *S-N* curve.

Statistical variability is troublesome in fatigue testing; it is necessary to measure the lifetimes of perhaps 20 specimens at each of 10 or so load levels to define the *S-N* curve with statistical confidence.[17] It is generally impossible to cycle the specimen at more than approximately 10 Hz (inertia in components of the testing machine and heating of the specimen often become problematic at higher speeds), and at that speed it takes 11.6 days to reach $10^7$ cycles of loading. Obtaining a full *S-N* curve is obviously a tedious and expensive procedure.

At first glance the scatter in measured lifetimes seems enormous, especially given the logarithmic scale of the abscissa. If the coefficient of variability in conventional tensile testing is usually only a few percent, why do the fatigue lifetimes vary over orders of magnitude? It must be remembered that in tensile testing we are measuring the variability in stress at a given number of *cycles* (one), while in fatigue we are measuring the variability in cycles at a given *stress.* Stated differently, in tensile testing we are generating vertical scatter bars, but in fatigue they are horizontal (see Fig. 7.29). Note that we must expect more variability in the lifetimes as the *S-N* curve becomes flatter, so materials that are less prone to fatigue damage require more specimens to provide a given confidence limit on lifetime.

### 7.4.3 Effect of Mean Load

Of course, not all actual loading applications involve fully reversed stress cycling. A more general sort of fatigue testing adds a *mean stress, $\sigma_m$,* on which a sinusoidal cycle is superimposed, as shown in Fig. 7.30. Such a cycle can be phrased in several ways, one common way being to state the alternating stress $\sigma_{alt}$ and the *stress ratio* $R = \sigma_{min}/\sigma_{max}$. For fully reversed loading, $R = -1$. A stress cycle of $R = 0.1$ is often used in aircraft component testing, and corresponds to a tension-tension cycle in which $\sigma_{min} = 0.1\sigma_{max}$.

A very substantial amount of testing is required to obtain an *S-N* curve for the simple case of fully reversed loading, and it will usually be impractical to determine whole families of curves for every combination of mean and alternating stress. There are a number of strategems for finessing this difficulty, one common one being the *Goodman diagram,* shown in Fig. 7.31. Here a graph is constructed with mean stress as the abscissa and alternating stress as the ordinate, and a straight "lifeline" is drawn from $\sigma_e$ on the $\sigma_{alt}$ axis to the ultimate tensile stress $\sigma_f$ on the $\sigma_m$ axis. Then for any given mean stress, the endurance limit—the value of alternating stress at which fatigue fracture never occurs—can be read directly as the ordinate of the lifeline at that value of $\sigma_m$. Alternatively, if the design application dictates a given ratio of $\sigma_e$

---

[16]H.W. Hayden, W.G. Moffatt, and J. Wulff, *The Structure and Properties of Materials,* Vol. III, John Wiley & Sons, 1965. Reprinted by permission.

[17]*A Guide for Fatigue Testing and the Statistical Analysis of Fatigue Data,* ASTM STP-91-A, 1963.

to $\sigma_{\text{alt}}$, a line is drawn from the origin with a slope equal to that ratio. Its intersection with the lifeline then gives the effective endurance limit for that combination of $\sigma_f$ and $\sigma_m$.

## 7.4.4 Miner's Law for Cumulative Damage

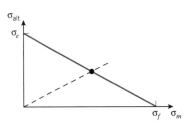

**FIGURE 7.31** The Goodman diagram.

When the cyclic load level varies during the fatigue process, a cumulative damage model along the lines of our earlier discussion of Zhurkov's theory can be hypothesized. To illustrate, take the lifetime to be $N_1$ cycles at a stress level $\sigma_1$ and $N_2$ at $\sigma_2$. If damage is assumed to accumulate at a constant rate during fatigue and a number of cycles $n_1$ is applied at stress $\sigma_1$, where $n_1 < N_1$ as shown in Fig. 7.32, then the fraction of lifetime consumed will be $n_1/N_1$. To determine how many additional cycles the specimen will survive at stress $\sigma_2$, an additional fraction of life will be available such that the sum of the two fractions equals one:

$$\frac{n_1}{N_1} + \frac{n_2}{N_2} = 1$$

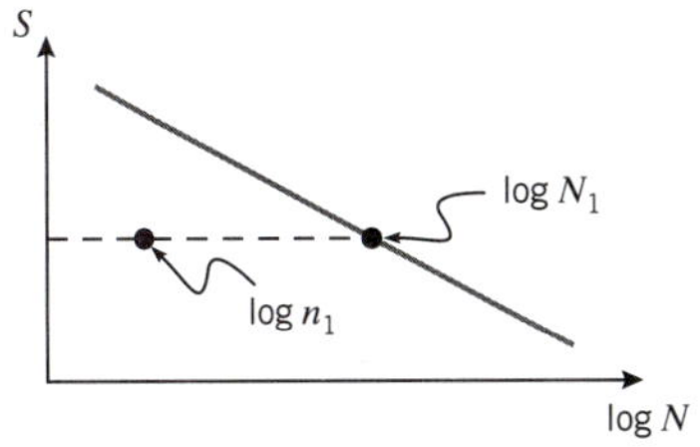

Note that absolute cycles and not log cycles are used here. Solving for the remaining cycles permissible at $\sigma_2$, we have

$$n_2 = N_2\left(1 - \frac{n_1}{N_1}\right)$$

The generalization of this approach is called *Miner's law*, and can be written

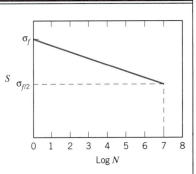

**FIGURE 7.32** The concept of fractional lifetime.

$$\boxed{\sum \frac{n_j}{N_j} = 1} \tag{7.19}$$

where $n_j$ is the number of cycles applied at a load corresponding to a lifetime of $N_j$.

## EXAMPLE 7.15

Consider a hypothetical material in which the *S-N* curve is linear from a value equal to the fracture stress $\sigma_f$ at one cycle ($\log N = 0$), falling to a value of $\sigma_f/2$ at $\log N = 7$, as shown in Fig. 7.33. This behavior can be described by the relation

$$\log N = 14\left(1 - \frac{S}{\sigma_f}\right)$$

The material has been subjected to $n_1 = 10^5$ load cycles at a level $S = 0.6\sigma_f$, and we wish to estimate how many cycles $n_2$ the material can now withstand if we raise the load to $S = 0.7\sigma_f$. From the *S-N* relationship, we know the lifetime at $S = 0.6\sigma_f =$ constant would be $N_1 = 3.98 \times 10^5$ and the lifetime at $S = 0.7\sigma_f =$ constant would be $N_2 = 1.58 \times 10^4$. Now we apply Eq. 7.19:

$$\frac{n_1}{N_1} + \frac{n_2}{N_2} = \frac{1 \times 10^5}{3.98 \times 10^5} + \frac{n_2}{1.58 \times 10^4} = 1$$

$$n_2 = 1.18 \times 10^4$$

**FIGURE 7.33** Linear *S-N* curve.

Miner's "law" should be viewed like many other materials "laws": a useful approximation, quite easy to apply, that *might* be accurate enough to use in design. But damage accumulation in fatigue is usually a complicated mixture of several different mechanisms, and the assumption of linear damage accumulation inherent in Miner's law should be viewed skeptically. If portions of the material's microstructure become unable to bear load as fatigue progresses, the stress must be carried by the surviving microstructural elements. The rate of damage accumulation in these elements then increases, so that the material suffers damage much more rapidly in the last portions of its fatigue lifetime. If, on the other hand, cyclic loads induce strengthening mechanisms such as molecular orientation or crack blunting, the rate of damage accumulation could drop during some part of the material's lifetime. Miner's law ignores such effects and often fails to capture the essential physics of the fatigue process.

### 7.4.5 Crack Growth Rates

Certainly in aircraft, but also in other structures as well, it is vital that engineers be able to predict the rate of crack growth during load cycling, so that the part in question can be replaced or repaired before the crack reaches a critical length. A great deal of experimental evidence supports the view that the crack growth rate can be correlated with the cyclic variation in the stress intensity factor:

$$\frac{da}{dN} = A \, \Delta K^m \tag{7.20}$$

where $da/dN$ is the fatigue crack growth rate per cycle, $\Delta K = K_{max} - K_{min}$ is the stress intensity factor range during the cycle, and $A$ and $m$ are parameters that depend on the material, environment, frequency, temperature, and stress ratio. This is sometimes known as the *Paris law* and leads to plots similar to that shown in Fig. 7.34.

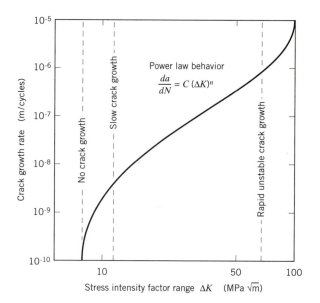

**FIGURE 7.34** The Paris law for fatigue crack growth rates in steel.

| **TABLE 7.4** Numerical parameters in the Paris equation | | |
|---|---|---|
| **Alloy** | $m$ | $A$ |
| Steel | 3 | $10^{-11}$ |
| Aluminum | 3 | $10^{-12}$ |
| Nickel | 3.3 | $4 \times 10^{-12}$ |
| Titanium | 5 | $10^{-11}$ |

The exponent $m$ is often near 4 for metallic systems, which might be rationalized as the damage accumulation being related to the volume $V_p$ of the plastic zone: Since the volume $V_p$ of the zone scales with $r_p^2$ and $r_p \propto K_I^2$, then $da/dn \propto \Delta K^4$. Some specific values of the constants $m$ and $A$ for various alloys are given in Table 7.4.

# GENERAL REFERENCES

**1.** Anderson, T. L., *Fracture Mechanics: Fundamentals and Applications,* CRC Press, Boca Raton, FL, 1991.

**2.** Barsom, J. M., ed., *Fracture Mechanics Retrospective,* American Society for Testing and Materials, Philadelphia, 1987.

**3.** Collins, J. A., *Failure of Materials in Mechanical Design,* Wiley, New York, 1981.

**4.** Courtney, T. H., *Mechanical Behavior of Materials,* McGraw-Hill, New York, 1990.

**5.** Gordon, J. E., *The New Science of Strong Materials, or Why You Don't Fall Through the Floor,* Princeton University Press, 1976.

**6.** Hertzberg, R. W., *Deformation and Fracture Mechanics of Engineering Materials,* Wiley, New York, 1976.

**7.** Knott, J. F., *Fundamentals of Fracture Mechanics,* John Wiley—Halsted Press, New York, 1973.

**8.** Mendenhall, W., R. L. Scheaffer, and D. D. Wackerly, *Mathematical Statistics with Applications,* Duxbury Press, Boston, 1986.

**9.** Strawley, J. E., and W. F. Brown, *Fracture Toughness Testing,* ASTM STP **381,** 133, 1965.

**10.** Tetelman, A. S., and A. J. McEvily, Jr., *Fracture of Structural Materials,* Wiley, New York, 1967.

# 7.5 PROBLEMS

**1.** Ten strength measurements have produced a mean tensile strength of $\overline{\sigma}_f = 100$ MPa, with 95 percent confidence limits of $\pm 8$ MPa. How many additional measurements would be necessary to reduce the confidence limits by half, assuming the mean and standard deviation of the measurements remain unchanged?

**2.** The 30 measurements of the tensile strength of graphite/epoxy composite listed in Example 7.2 were made at room temperature. Thirty additional tests conducted at 93°C gave the values (in kpsi):

| | | |
|---|---|---|
| 63.40 | 62.08 | 67.80 |
| 69.70 | 61.53 | 72.68 |
| 72.80 | 70.53 | 75.09 |
| 63.60 | 72.88 | 67.23 |
| 71.20 | 74.90 | 64.80 |
| 72.07 | 78.61 | 75.84 |
| 76.97 | 68.72 | 63.87 |
| 70.94 | 72.87 | 72.46 |
| 76.22 | 64.49 | 69.54 |
| 64.65 | 75.12 | 76.97 |

For these data:

(*a*) Determine the arithmetic mean, standard deviation, and coefficient of variation.

(*b*) Determine the 95 percent confidence limits on the mean strength.

(*c*) Determine whether the average strengths at 23°C and 93°C are statistically different.

(*d*) Determine the normal and Weibull B-allowable strengths.

(*e*) Plot the cumulative probability of failure $P_f$ versus the failure stress on normal probability paper.

(*f*) Do the data appear to be distributed normally, based on the $\chi^2$ test?

(*g*) Plot the cumulative probability of survival $P_s$ versus the failure stress on Weibull probability paper.

(*h*) Determine the Weibull parameters $\sigma_0$ and $m$.

(*i*) Estimate how the mean strength would change if the specimens were made 10 times smaller or 10 times larger.

**3.** Repeat Problem 2, but using data for −59°C:

| | | |
|---|---|---|
| 55.62 | 63.69 | 68.84 |
| 55.91 | 63.8 | 69.15 |
| 56.73 | 64.7 | 69.3 |
| 57.54 | 65.2 | 69.37 |
| 58.28 | 65.33 | 69.82 |
| 59.23 | 66.39 | 70.94 |
| 60.39 | 66.43 | 71.39 |
| 60.62 | 66.72 | 71.74 |
| 61.1 | 67.05 | 72.2 |
| 62.1 | 67.76 | 73.46 |

**4.** Using the kinetic data in Table 7.1, estimate the creep-rupture lifetime for aluminum at a stress of 6.4 kpsi and a temperature of 85°F.

**5.** Using the kinetic data in Table 7.1, estimate the lifetime of a silver chloride specimen subjected to a constant stress rate of $R_\sigma = 2$ MPa/s at 30°C.

**6.** Estimate the apparent activation energy and volume for fracture of a material whose creep-rupture lifetimes for various applied stresses and temperatures are as follows:

| $T$, °C | $\sigma$, MPa | $t_f$, s |
|---|---|---|
| 18 | 53.9 | 17400 |
| | 58.8 | 132 |
| | 63.7 | 1 |
| 50 | 49.0 | 27800 |
| | 53.9 | 341 |
| | 58.8 | 4.21 |
| 70 | 49.0 | 2671 |
| | 53.9 | 42.4 |
| | 58.8 | 0.67 |

**7.** Using a development analogous to that employed in Chapter 6 for the theoretical yield stress, show that the theoretical ultimate tensile strength is $\sigma_{th} \approx E/10$ (much larger than that observed experimentally). Assume a harmonic atomic force function $\sigma = \sigma_{th} \sin(2\pi x/\lambda)$, where $x$ is the displacement of an atom from its equilibrium position and $\lambda \approx a_0$ is the interatomic spacing, as graphed in Fig. P.7.7. The maximum stress $\sigma_{th}$ can then be found by using

$$E = \left(\frac{d\sigma}{d\epsilon}\right)_{x\to 0} \quad \text{and} \quad \epsilon = \frac{x}{a_0}$$

**8.** Using a safety factor of 2, find the safe operating pressure in a closed-end steel pressure vessel 1′ in diameter and 0.2″ in wall thickness.

**9.** A pressure vessel is constructed with a diameter of $d = 18''$ and a length of $L = 6'$. The vessel is to be capable of withstanding an internal pressure of $p = 1000$ psi, and the wall thickness is such as

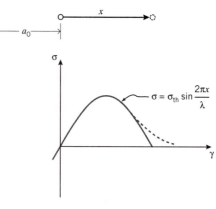

**FIGURE P7.7**

to keep the nominal hoop stress under 2500 psi. However, the vessel bursts at an internal pressure of only 500 psi, and a micrographic investigation reveals the fracture to have been initiated by an internal crack 0.1″ in length. Calculate the fracture toughness ($K_{Ic}$) of the material.

**10.** A highly cross-linked epoxy resin has a coefficient of linear thermal expansion $\alpha = 5 \times 10^{-5}$ K$^{-1}$, $G_{Ic} = 120$ J/m$^2$, $E = 3.2$ GPa, and $\nu = 0.35$. A thick layer of resin is cured and is firmly bonded to an aluminum part ($\alpha = 2.5 \times 10^{-5}$ K$^{-1}$) at 180°C. Calculate the minimum defect size needed to initiate cracking in the resin on cooling to 20°C. Take $\alpha$ in Eq. 7.17 to be $2/\pi$ for penny-shaped cracks of radius $a$ in a wide sheet.

**11.**

**(a)** A thick plate of aluminum alloy, 175 mm wide, contains a centrally located crack 75 mm in length. The plate experiences brittle fracture at an applied stress (uniaxial, transverse to the crack) of 110 MPa. Determine the fracture toughness of the material.

**(b)** What would the fracture stress be if the plate were wide enough to permit an assumption of infinite width?

**12.** In order to obtain valid plane-strain fracture toughness, the plastic zone size must be small with respect to the specimen thickness $B$, the crack length $a$, and the "ligament" width $W - a$. The established criterion is

$$(W - a), B, a \geq \left(\frac{K_{Ic}}{\sigma_Y}\right)^2$$

Rank the materials in the database in terms of the parameter given on the right-hand side of this expression.

**13.** When a 150 kN load is applied to a tensile specimen containing a 35 mm crack, the overall displacement between the specimen ends is 0.5 mm. When the crack has grown to 37 mm, the displacement for this same load is 0.505 mm. The specimen is 40 m thick. The fracture load of an identical specimen, but with a crack length of 36 mm, is 175 kN. Find the fracture toughness $K_{Ic}$ of the material.

**14.** A steel has an ultimate tensile strength of 110 kpsi and a fatigue endurance limit of 50 kpsi. The load is such that the alternating stress is 0.4 of the mean stress. Using the Goodman method with a safety factor of 1.5, find the magnitude of alternating stress that gives safe operation.

**15.** A titanium alloy has an ultimate tensile strength of 120 kpsi and a fatigue endurance limit of 60 kpsi. The alternating stress is 20 kpsi. Find the allowable mean stress, using a safety factor of 2.

**16.** A material has an $S$-$N$ curve that is linear from a value equal to the fracture stress $\sigma_f$ at one cycle ($\log N = 0$), falling to a value of $\sigma_f/3$ at $\log N = 7$, as shown in Fig. P.7.16. The material has been subjected to $n_1 = 3 \times 10^4$ load cycles at a level $S = 0.7\sigma_f$. Estimate how many cycles, $n_2$, the material can withstand if the stress amplitude is now raised to $S = 0.8\sigma_f$.

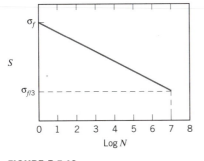

**FIGURE P.7.16**

**17.** A steel alloy has an $S$-$N$ curve that falls linearly from 240 kpsi at $10^4$ cycles to 135 kpsi at $10^6$ cycles, as shown in Fig. P.7.17. A specimen is loaded at 160 kpsi alternating stress for $10^5$ cycles, after which the alternating stress is raised to 180 kpsi. How many additional cycles at this higher stress would the specimen be expected to survive?

**18.** Consider a body, large enough to be considered infinite in lateral dimension, containing a central through-thickness crack initially of length $2a_0$ and subjected to a cyclic stress of amplitude $\Delta\sigma$. Using the Paris law (Eq. 7.20), show that the number of cycles $N_f$ needed

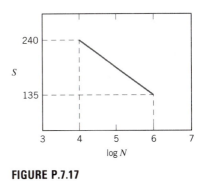

**FIGURE P.7.17**

for the crack to grow to a length $a_f$ is given by the relation

$$\ln\left(\frac{a_f}{a_0}\right) = A(\Delta\sigma)^2 \pi N_f$$

when $m = 2$, and for other values of $m$

$$\left| a_f^{1-m/2} - a_0^{1-m/2} \right| = \frac{2-m}{2m} A(\Delta\sigma)^m \pi^{m/2} N_f$$

**19.** Use the expression obtained in Problem 18 to compute the number of cycles a steel component can sustain before failure, where the initial crack is 0.1 mm and the critical crack length to cause fracture is 2.5 mm. The stress amplitude per cycle is 950 MPa. Take the crack to be that of a central crack in an infinite plate.

**20.** Use the expression developed in Problem 18 to investigate whether it is better to limit the size $a_0$ of initial flaws or to extend the size $a_f$ of the flaw at which fast fracture occurs. Limiting $a_0$ might be done with improved manufacturing or better inspection methods, and increasing $a_f$ could be done by selecting a material with greater fracture toughness. For the "baseline" case, take $m = 3.5$, $a_0 = 2$ mm, $a_f = 5$ mm. Compute the percentage increase in $N_f$ by letting ($a$) the initial flaw size to be reduced to $a_0 = 1$ mm, and ($b$) increasing the final flaw size to $N_f = 10$ mm.

# A PROPERTIES OF COMMON STRUCTURAL MATERIALS

Table A.1 lists typical numerical values for selected mechanical properties of a number of important structural materials. All are isotropic except for spruce, for which the strength values apply in the direction of the grain. It must be emphasized that these values are typical but that in many cases a very wide range of values can be exhibited for ostensibly identical materials. Some properties, such as density, do not vary much, and the tabulated values are quite reliable. Others, such as fracture toughness, can be varied over an extremely large range by the material's processing history as well as other factors. The cost values are especially prone to variation, depending on current economic conditions.

*Therefore, these values should be used only for preliminary design and analysis purposes.* Current vendor data along with actual laboratory testing should be used for final design, *especially when human safety issues are present.*

**TABLE A.1** Properties of structural materials

| Material | Type | Cost $/kg | Density Mg/m³ | Young's modulus, GPa | Shear modulus, GPa | Poisson's ratio | Yield stress, MPa | UTS, MPa | Fracture toughness, MPa-√m | Thermal expansion 10⁻⁶/K |
|---|---|---|---|---|---|---|---|---|---|---|
| 1 Alumina (Al$_2$O$_3$) | Ceramic | 18.10 | 3.75 | 310 | 125 | 0.26 | 4800 | 320 | 4.4 | 8.1 |
| 2 Aluminum alloy | Metal | 3.60 | 2.72 | 76 | 28 | 0.34 | 330 | 550 | 40.7 | 33.1 |
| 3 Beryllium alloy | Metal | 315.00 | 2.88 | 345 | 110 | 0.12 | 360 | 500 | 4.2 | 13.7 |
| 4 Bone (compact) | Natural | 1.90 | 1.95 | 14 | 3.5 | 0.43 | 100 | 100 | 4.9 | 20.0 |
| 5 CFRP laminate | Composite | 113.00 | 1.55 | 1.5 | 53 | 0.28 | 200 | 550 | 37.1 | 12.5 |
| 6 Cermets | Composite | 78.60 | 11.74 | 520 | 200 | 0.3 | 650 | 1200 | 12.9 | 5.8 |
| 7 Concrete | Ceramic | 0.05 | 2.45 | 48 | 20 | 0.2 | 25 | 3 | 0.7 | 11.0 |
| 8 Copper alloys | Metal | 2.25 | 8.33 | 138 | 50 | 0.35 | 510 | 720 | 93.5 | 18.4 |
| 9 Cork | Natural | 9.95 | 0.18 | 0.032 | 0.005 | 0.25 | 1.4 | 1.5 | 0.1 | 180.3 |
| 10 GFRP laminate | Composite | 3.90 | 1.75 | 26 | 10 | 0.28 | 125 | 530 | 29.7 | 18.5 |
| 11 Glass (soda) | Ceramic | 1.35 | 2.39 | 63 | 26 | 0.23 | 1500 | 75 | 0.7 | 8.8 |
| 12 Granite | Ceramic | 3.15 | 2.65 | 66 | 26 | 0.25 | 2500 | 2500 | 1.5 | 6.5 |
| 13 Ice (H$_2$O) | Ceramic | 0.23 | 0.92 | 9.1 | 3.6 | 0.28 | 88 | 6.5 | 0.1 | 55.1 |
| 14 Lead alloys | Metal | 1.20 | 11.10 | 16 | 5.5 | 0.45 | 33 | 42 | 39.6 | 28.8 |
| 15 Nickel alloys | Metal | 6.10 | 8.49 | 180 | 70 | 0.31 | 900 | 1200 | 93.5 | 13.0 |
| 16 Polyamide (nylon) | Polymer | 4.30 | 1.15 | 2 | 0.76 | 0.42 | 40 | 55 | 3.0 | 102.6 |
| 17 Polybutadiene elastomer | Polymer | 1.20 | 0.91 | 0.0016 | 0.0005 | 0.5 | 2.1 | 2.1 | 0.1 | 140.1 |
| 18 Polycarbonate | Polymer | 4.90 | 1.20 | 2.7 | 0.97 | 0.42 | 70 | 77 | 2.6 | 70.2 |
| 19 Polyethylene (HDPE) | Polymer | 1.00 | 0.95 | 1 | 0.31 | 0.42 | 25 | 33 | 3.5 | 225.4 |
| 20 Polypropylene | Polymer | 1.10 | 0.90 | 1.2 | 0.42 | 0.42 | 35 | 45 | 3.0 | 86.0 |
| 21 Polyurethane elastomer | Polymer | 4.00 | 1.17 | 0.025 | 0.0086 | 0.5 | 30 | 30 | 0.3 | 125.1 |
| 22 Polyvinyl chloride (rigid PVC) | Polymer | 1.50 | 1.37 | 1.5 | 0.6 | 0.42 | 53 | 60 | 0.5 | 75.1 |
| 23 Silicon | Ceramic | 2.35 | 2.32 | 110 | 44 | 0.24 | 3200 | 2500 | 1.5 | 6.0 |
| 24 Silicon carbide (SiC) | Ceramic | 36.15 | 2.85 | 430 | 190 | 0.15 | 9800 | 630 | 4.2 | 4.2 |
| 25 Spruce (parallel to grain) | Natural | 1.00 | 0.40 | 9 | 0.8 | 0.3 | 48 | 50 | 2.5 | 4.0 |
| 26 Steel, high carbon | Metal | 0.65 | 7.85 | 210 | 76 | 0.29 | 590 | 1200 | 49.5 | 13.5 |
| 27 Steel, stainless austenitic | Metal | 2.70 | 7.85 | 210 | 786 | 0.28 | 870 | 1200 | 49.5 | 16.6 |
| 28 Titanium alloys | Metal | 16.25 | 4.65 | 100 | 39 | 0.36 | 70 | 850 | 86.9 | 9.4 |
| 29 Tungsten carbide (WC) | Ceramic | 49.70 | 15.50 | 640 | 270 | 0.21 | 6800 | 450 | 3.7 | 5.8 |

E.R. Parker, *Materials Data Book*, McGraw-Hill, 1967.

J.A. Charles and F.A.A. Quinn, *Selection and Use of Engineering Materials*, Butterworths, London, 1989.

M.F. Ashby and D.R.H. Jones, *Engineering Materials*, Pergamon Press, Oxford, 1980.

*Materials Engineering*, December 1991 (Materials Selector Issue).

*Advanced Materials and Processes*, June 1990.

S.H. Crandall, N.C. Dahl, and T.J. Lardner, *An Introduction to the Mechanics of Solids*, McGraw-Hill, 1978.

F.P. Gerstle, "Composites," *Encyclopedia of Polymer Science and Engineering*, Wiley, 1991.

# B MECHANICAL PROPERTIES OF COMPOSITE MATERIALS

Table B.1 lists physical and mechanical property values for representative ply and core materials widely used in fiber-reinforced composite laminates. Ply properties are taken from F. P. Gerstle, "Composites," *Encyclopedia of Polymer Science and Engineering,* Wiley, New York, 1991, which should be consulted for data from a wider range of materials. See also G. Lubin, *Handbook of Composites,* Van Nostrand, New York, 1982.

**TABLE B.1** Properties of composite materials

|  | S-glass/ epoxy | Kevlar/ epoxy | HM Graphite/ epoxy | Pine | Rohacell 51 rigid foam |
|---|---|---|---|---|---|
| Elastic properties: | | | | | |
| $E_1$, GPa | 55 | 80 | 230 | 13.4 | 0.07 |
| $E_2$, GPa | 16 | 5.5 | 6.6 | 0.55 | 0.07 |
| $G_{12}$, GPa | 7.6 | 2.1 | 4.8 | 0.83 | 0.021 |
| $\nu_{12}$ | 0.26 | 0.31 | 0.25 | 0.30 | |
| Tensile strengths: | | | | | |
| $\sigma_1$, MPa | 1800 | 2000 | 1100 | 78 | 1.9 |
| $\sigma_2$, MPa | 40 | 20 | 21 | 2.1 | 1.9 |
| $\sigma_{12}$, MPa | 80 | 40 | 65 | 6.2 | 0.8 |
| Compressive strengths: | | | | | |
| $\sigma_1$, MPa | 690 | 280 | 620 | 33 | 0.9 |
| $\sigma_2$, MPa | 140 | 140 | 170 | 3.0 | 0.9 |
| Physical properties: | | | | | |
| $\alpha_1$, $10^{-6}/°C$ | 2.1 | −4.0 | −0.7 | | 33 |
| $\alpha_2$, $10^{-6}/°C$ | 6.3 | 60 | 28 | | 33 |
| Volume fraction | 0.7 | 0.54 | 0.7 | | |
| Thickness, mm | 0.15 | 0.13 | 0.13 | | |
| Density, Mg/m$^3$ | 2.0 | 1.38 | 1.63 | 0.55 | 0.05 |

# LAPLACE TRANSFORMS

## BASIC DEFINITION

$$\mathscr{L} f(t) = \overline{f}(s) = \int_0^\infty f(t) e^{-st}\, dt$$

## FUNDAMENTAL PROPERTIES

$$\mathscr{L}[c_1 f_1(t) + c_2 f_2(t)] = c_2 \overline{f}_1(s) c_1 \overline{f}_2(s)$$

$$\mathscr{L}\left[\frac{\partial f}{\partial t}\right] = s\overline{f}(s) - f(0^-)$$

## SOME USEFUL TRANSFORM PAIRS

| $f(t)$ | $\overline{f}(s)$ |
|---:|:---|
| $u(t)$ | $1/s$ |
| $t^n$ | $n!/s^{n+1}$ |
| $e^{-at}$ | $1/(s + a)$ |
| $\frac{1}{a}(1 - e^{-at})$ | $1/s(s + a)$ |
| $\frac{t}{a} - \frac{1}{a}(1 - e^{-at})$ | $1/s^2(s + a)$ |

Here $u(t)$ is the Heaviside or unit step function, defined as

$$u(t) = \begin{cases} 0, & x < 0 \\ 1, & x \geq 0 \end{cases}$$

## THE CONVOLUTION INTEGRAL

$$\mathscr{L}f \cdot \mathscr{L}g = \overline{f} \cdot \overline{g} = \mathscr{L}\left[\int_0^t f(t-\lambda)g(\lambda)\,d\lambda\right] = \mathscr{L}\left[\int_0^t f(\lambda)g(t-\lambda)\,d\lambda\right]$$

# MATRIX AND INDEX NOTATION

A *vector* can be described by listing its components along the $xyz$ Cartesian axes. For instance the displacement vector **u** can be denoted as $u_x$, $u_y$, $u_z$, using letter subscripts to indicate the individual components. The subscripts can employ numerical indices as well, with 1, 2, and 3 indicating the $x$, $y$, and $z$ directions; the displacement vector can therefore be written equivalently as $u_1$, $u_2$, $u_3$.

A common and useful shorthand is simply to write the displacement vector as $u_i$, where the $i$ subscript is an *index* that is assumed to range over 1,2,3 (or simply 1 and 2 if the problem is a two-dimensional one). This is called the *range convention* for index notation. Using the range convention, the vector equation $u_i = a$ implies three separate scalar equations:

$$u_1 = a$$
$$u_2 = a$$
$$u_3 = a$$

We will often find it convenient to denote a vector by listing its components in a vertical list enclosed in braces, and this form will help us keep track of matrix-vector multiplications a bit more easily. We therefore have the following equivalent forms of vector notation:

$$\mathbf{u} = u_i = \begin{Bmatrix} u_1 \\ u_2 \\ u_3 \end{Bmatrix} = \begin{Bmatrix} u_x \\ u_y \\ u_z \end{Bmatrix}$$

Second-rank quantities such as stress, strain, moment of inertia, and curvature can be denoted as $3 \times 3$ matrix arrays; for instance, the stress can be written using numerical indices as

$$[\sigma] = \begin{bmatrix} \sigma_{11} & \sigma_{12} & \sigma_{13} \\ \sigma_{21} & \sigma_{22} & \sigma_{23} \\ \sigma_{31} & \sigma_{32} & \sigma_{33} \end{bmatrix}$$

Here the first subscript index denotes the row and the second the column. The indices also have a physical meaning. For instance, $\sigma_{23}$ indicates the stress on the 2 face (the plane whose normal is in the 2, or $y$, direction) and acting in the 3, or $z$, direction. To help distinguish them, we'll use brackets for second-rank tensors and braces for vectors.

Using the range convention for index notation, the stress can also be written as $\sigma_{ij}$, where both the $i$ and the $j$ range from 1 to 3; this gives the nine components just listed explicitly. (Since the stress matrix is symmetric, $\sigma_{ij} = \sigma_{ji}$, only six of these nine components are independent.)

A subscript that is repeated in a given term is understood to imply summation over the range of the repeated subscript; this is the *summation convention* for index notation. For instance, to indicate the sum of the diagonal elements of the stress matrix we can write

$$\sigma_{kk} = \sum_{k=1}^{3} \sigma_{kk} = \sigma_{11} + \sigma_{22} + \sigma_{33}$$

The multiplication rule for matrices can be stated formally by taking $\mathbf{A} = (a_{ij})$ to be an $(M \times N)$ matrix and $\mathbf{B} = (b_{ij})$ to be an $(R \times P)$ matrix. The matrix product $\mathbf{AB}$ is defined only when $R = N$, and is the $(M \times P)$ matrix $\mathbf{C} = (c_{ij})$ given by

$$c_{ij} = \sum_{k=1}^{N} a_{ik}b_{kj} = a_{i1}b_{1j} + a_{i2}b_{2j} + \cdots + a_{iN}b_{Nk}$$

Using the summation convention, this can be written simply

$$c_{ij} = a_{ik}b_{kj}$$

where the summation is understood to be over the repeated index $k$. In the case of a $3 \times 3$ matrix multiplying a $3 \times 1$ column vector we have

$$\begin{bmatrix} a_{11} & a_{12} & a_{13} \\ a_{21} & a_{22} & a_{23} \\ a_{31} & a_{32} & a_{33} \end{bmatrix} \begin{Bmatrix} b_1 \\ b_2 \\ b_3 \end{Bmatrix} = \begin{Bmatrix} a_{11}b_1 & +a_{12}b_2 & +a_{13}b_3 \\ a_{21}b_1 & +a_{22}b_2 & +a_{23}b_3 \\ a_{31}b_1 & +a_{32}b_2 & +a_{33}b_3 \end{Bmatrix} = a_{ij}b_j$$

The *comma convention* uses a subscript comma to imply differentiation with respect to the variable following, so $f_{,2} = \partial f/\partial y$ and $u_{i,j} = \partial u_i/\partial x_j$. For instance, the expression $\sigma_{ij,j} = 0$ uses all of the three previously defined index conventions: range on $i$, sum on $j$, and differentiate:

$$\frac{\partial \sigma_{xx}}{\partial x} + \frac{\partial \sigma_{xy}}{\partial y} + \frac{\partial \sigma_{xz}}{\partial z} = 0$$

$$\frac{\partial \sigma_{yx}}{\partial x} + \frac{\partial \sigma_{yy}}{\partial y} + \frac{\partial \sigma_{yz}}{\partial z} = 0$$

$$\frac{\partial \sigma_{zx}}{\partial x} + \frac{\partial \sigma_{zy}}{\partial y} + \frac{\partial \sigma_{zz}}{\partial z} = 0$$

The *Kroenecker delta* is a useful entity defined as

$$\delta_{ij} = \begin{cases} 0, & i \neq j \\ 1, & i = j \end{cases}$$

This is the index form of the *unit matrix* **I**:

$$\delta_{ij} = \mathbf{I} = \begin{bmatrix} 1 & 0 & 0 \\ 0 & 1 & 0 \\ 0 & 0 & 1 \end{bmatrix}$$

So, for instance,

$$\sigma_{kk}\delta_{ij} = \begin{bmatrix} \sigma_{kk} & 0 & 0 \\ 0 & \sigma_{kk} & 0 \\ 0 & 0 & \sigma_{kk} \end{bmatrix}$$

where $\sigma_{kk} = \sigma_{11} + \sigma_{22} + \sigma_{33}$.

# E CRYSTALLOGRAPHIC NOTATION

Specific planes within crystal lattices can be specified using a notation called *Miller indices*.[1] These are simply the reciprocals of the intersections the plane makes with the axes along the unit cell, expressed in terms of the lattice spacing.

Referring to a cubic unit cell of lattice spacing $a$, for example, let the $xyz$ axes be established along the cell edges as shown in Fig. E.1. Now consider a plane parallel to the $xy$ plane and intersecting the $z$ axis at $a$. The intersection points along the three axes are therefore $\infty,\infty,a$. Dividing these values by the lattice spacing $a$ and then taking reciprocals (to avoid having to use $\infty$ in the indices), the Miller indices are (001). Similarly, the (112) plane is the one intersecting the $xyz$ axes at $a,a,a/2$.

A number of useful crystallographic parameters can be calculated directly from the Miller indices. For instance, the interplanar spacing $d$ between two parallel planes in a cubic lattice whose indices are $hkl$ can be shown to be

$$d_{hkl} = \frac{a}{\sqrt{h^2 + k^2 + l^2}}$$

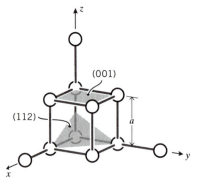

**FIGURE E.1** The (001) and (112) planes in a cubic lattice.

---

[1] For a more complete description of crystal systems and notation, see for instance W.D. Callister, *Materials Science and Engineering,* Wiley, 1985.

In the face-centered cubic (fcc) system, the (111) plane is close-packed, as shown in Fig E.2. The distance between one such plane and the one just above or below it is

$$d_{111} = \frac{a}{\sqrt{1^2 + 1^2 + 1^2}} = \frac{a}{\sqrt{3}}$$

Symmetry considerations dictate that a number of lattice planes will be equivalent. For instance, the four nonparallel (111), $(1\bar{1}1)$, $(\bar{1}11)$, and $(\bar{1}\bar{1}1)$ planes are all close-packed and are crystallographically equivalent. (The overhead bar indicates a negative value: the $(\bar{1}11)$ plane intersects the $xyz$ axes at $-a,a,a$.) These sets of equivalent planes are indicated with braces rather than parentheses; hence, the designation {111} refers to the whole set of such planes.

The *direction* of any line in a crystal system is indicated by drawing it so as to pass through the origin of the $xyz$ axes, and then giving the coordinates of any other point on the line, expressed in terms of the lattice spacing $a$ and reduced to a set of smallest integers. The reduced coordinates are enclosed in brackets rather than parentheses, to distinguish directions from planes. Thus a line passing along the close-packed direction in the $xy$ plane of the fcc lattice, as shown in Fig. E.3, will be denoted by $[\bar{1}10]$, even though in this figure the line has been displaced along the $x$ axis by a distance $a$. The line passing through 0,0,0 and $-a,a,0$ is crystallographically identical to this one.

As with sets of planes, symmetry will render sets of directions equivalent. As shown in Fig. E.3, the three directions $[\bar{1}10]$, $[\bar{1}01]$, and $[0\bar{1}1]$ are all along closest-packed directions on the (111) plane and are thus crystallographically equivalent. The family of all such directions is indicated by using angle brackets, as in $\langle\bar{1}10\rangle$.

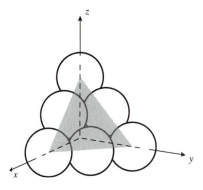

**FIGURE E.2** The close-packed (111) plane in the fcc lattice.

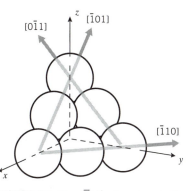

**FIGURE E.3** The $\langle\bar{1}10\rangle$ closest-packed directions on the (111) plane in the fcc system.

APPENDIX **F** **COMPUTER PROGRAM FOR 3-DIMENSIONAL STRESS TRANSFORMATIONS**

A Fortran program named strs3d has been included on the software diskette to reduce the tedium in exploring the three-dimensional stress manipulations discussed in Chapter 3. The code allows the user to transform the stress state to new axes described by Euler angles, to determine values for the characteristic stress equation, and to determine three-dimensional principal values.

Initially, the strs3d code allocates array memory, prompts the user to enter the stress matrix by rows, and then prompts the user for a choice of options:

1. Transform the stress state using Euler angles
2. Explore the characteristic stress equation or compute principal values
3. Restart to enter a new stress matrix
4. Quit

The Fortran code for this initial section is

```
c
c            s t r s 3 d
c
c    an instructional code for transformations and principal values
c    of three-dimensional stress states
c
      dimension sig(3,3),sigp(3),temp(3,3),signew(3,3),
     1   t(3,3),tt(3,3)
      common /inv/a(4)
      data   rad/57.29578/
c
c    read stress matrix
c
   10 write (6,15)
   15 format (5x,'enter stress matrix by rows'/)
      read (5,*) ((sig(i,j),j=1,3),i=1,3)
```

```
c
c       read branch option
c
   20 write (6,30)
   30 format (/5x,'enter problem option:',
      1            /7x,'1 - transformation',
      2            /7x,'2 - principal values',
      3            /7x,'3 - restart',
      4            /7x,'4 - stop'/)
      read (5,*) ibr
      go to (100,200,10,35),ibr
c
   35 stop 'program terminated'
```

If the "transformation" option is selected, the code branches to label 100 and prompts the user to enter Euler angles $\psi,\theta,\phi$, as explained in Section 3.3.3. As discussed in Sections 3.3.3 and 3.3.4, the stresses can be transformed from unprimed to primed axes by the matrix multiplication

$$[\sigma'] = \mathbf{A}[\sigma]\mathbf{A}^{\mathrm{T}}$$

where $\mathbf{A}$ is a matrix of direction cosines; $a_{ij}$ are the cosines of the angles between the $i$th *unprimed* axis and the $j$th *primed* axis. The components of the $\mathbf{A}$ matrix can be computed in terms of Euler angles as

$$\mathbf{A} = \begin{bmatrix} \cos\phi & \sin\phi & 0 \\ -\sin\phi & \cos\phi & 0 \\ 0 & 0 & 1 \end{bmatrix}\begin{bmatrix} 1 & 0 & 0 \\ 0 & \cos\theta & \sin\theta \\ 0 & -\sin\theta & \cos\theta \end{bmatrix}\begin{bmatrix} \cos\psi & \sin\psi & 0 \\ -\sin\psi & \cos\psi & 0 \\ 0 & 0 & 1 \end{bmatrix}$$

$$= \begin{bmatrix} \cos\psi\cos\phi - \sin\phi\cos\theta\sin\psi & \sin\psi\cos\phi + \sin\phi\cos\theta\cos\psi & \sin\theta\sin\phi \\ -\cos\psi\sin\phi - \sin\psi\cos\theta\cos\phi & -\sin\phi\sin\psi + \cos\phi\cos\theta\cos\psi & \sin\theta\cos\phi \\ \sin\theta\sin\psi & -\sin\theta\cos\phi & \cos\theta \end{bmatrix}$$

Using the Fortran variables t and tt for the transformation matrix $\mathbf{A}$ and its transpose $\mathbf{a}^{\mathrm{T}}$, the code computes the transformation matrices and carries out the matrix triple product for the transformation. The transformation matrix $\mathbf{A}$ and the transformed stress $[\sigma']$ are printed out, and then the code returns to the initial section for another option choice. The transformation portion of the code is as follows:

```
  100 continue
c
c***********************************************************************
c
c       compute transformed stresses
c
c***********************************************************************
c
      write (6,120)
  120 format (/5x,'enter euler angles psi, theta, phi (degrees):'/)
      read (5,*) psi, theta, phi
c
c       compute transformation matrix t(i,j)
c
      spsi=sin(psi/rad)
      cpsi=cos(psi/rad)
```

```
      stheta=sin(theta/rad)
      ctheta=cos(theta/rad)
      sphi=sin(phi/rad)
      cphi=cos(phi/rad)
c
      t(1,1)=cpsi*cphi-sphi*ctheta*spsi
      t(2,1)=-sphi*cpsi-spsi*ctheta*cphi
      t(3,1)=stheta*spsi
      t(1,2)=cphi*spsi+sphi*ctheta*cpsi
      t(2,2)=-sphi*spsi+cphi*ctheta*cpsi
      t(3,2)=-stheta*cphi
      t(1,3)=stheta*sphi
      t(2,3)=stheta*cphi
      t(3,3)=ctheta
c
c     inverse (transpose) transformation matrix tt(i,j)
c
      do 130 i=1,3
      do 130 j=1,3
  130 tt(i,j)=t(j,i)
c
c     compute transformed stresses signew = t*(sig*tt) = t*temp
c
      call mult (sig,tt,3,3,3,temp)
      call mult (t,temp,3,3,3,signew)
c
c     print results, then branch for new option
c
      write (6,140)
  140 format(//1x,'transformation matrix:',18x,'stress:'/)

      do 150 i=1,3
  150 write (6,160) (t(i,j),j=1,3),(signew(i,j),j=1,3)
  160 format (1x,3f10.4,10x,3f10.4)
c
      go to 20
```

The call to subroutine `mult` invokes a routine for matrix multiplication, listed at the end of the main Fortran program.

The three principle stresses are the roots of the characteristic equation

$$f(\sigma_p) = \sigma_p^3 - I_1\sigma_p^2 + I_2\sigma_p - I_3 = 0$$

where the coefficients are

$$I_1 = \sigma_x + \sigma_y + \sigma_z$$

$$I_2 = \sigma_x\sigma_y + \sigma_x\sigma_z + \sigma_y\sigma_z - \tau_{xy}^2 - \tau_{yz}^2 - \tau_{xz}^2$$

$$I_3 = \tfrac{1}{3}\sigma_{ij}\sigma_{jk}\sigma_{ki} = \sigma_x\sigma_y\sigma_z + 2\tau_{xy}\tau_{yz}\tau_{xz} - \sigma_x\tau_{xz}^2 - \sigma_y\tau_{xz}^2 - \sigma_z\tau_{xy}^2$$

If the "principal values" option is selected, the code branches to allow the user to explore the characteristic equation and, if desired, solve it for the three roots using a numerical method known as *Newton-Raphson iteration*. The user is given the option of computing $f(\sigma_p)$ for various values of $\sigma_p$; after a bit of searching and watching for sign changes, the user develops a sense for the approximate location of the roots.

While a number of numerical "recipes" are available to solve the cubic characteristic equation directly, strs3d is writen to use the Newton-Raphson iterative scheme
in order to sneak a little numerical analysis into the text. This method requires that
both the function whose roots are being sought, in our case $f(\sigma)$ and its first derivative $f'(\sigma) = df/d\sigma$ can be computed for arbitrary values of the independent variable
$\sigma$. A tangent line is drawn to the function at the value corresponding to the current
estimate $\sigma^i$ for the root, and this line is extended to the abscissa to obtain the next
estimate for the root. The value of the updated estimate is therefore

$$\sigma^{i+1} = \sigma^i - \Delta\sigma, \qquad \Delta\sigma = \frac{f(\sigma^i)}{f'(\sigma^i)}$$

where the superscripts $i$ and $i + 1$ indicate the current and the subsequent estimate.
The process is begun by the user providing an initial estimate for $\sigma^i$, probably based
on having explored the characteristic equation as discussed earlier. The interaction
continues until the increment $\Delta\sigma$ becomes sufficiently small or a maximum predeter-
mined number of iterations is completed.

```
      c
      c*********************************************************************
      c
      c      compute principal values and directions of principal planes
      c
      c*********************************************************************
      c
      c      compute invariants
      c
  200 a(1)=1.
      a(2)=-1.*(sig(1,1)+sig(2,2)+sig(3,3))
      a(3)= sig(1,1)*sig(2,2) + sig(1,1)*sig(3,3) + sig(2,2)*sig(3,3)
     1      -sig(1,2)*sig(1,2) - sig(2,3)*sig(2,3) - sig(1,3)*sig(1,3)
      a(4)=-1.*(sig(1,1)*(sig(2,2)*sig(3,3) - sig(3,2)*sig(2,3))
     1           -sig(1,2)*(sig(2,1)*sig(3,3) - sig(3,1)*sig(2,3))
     2           +sig(1,3)*(sig(2,1)*sig(3,2) - sig(3,1)*sig(2,2)))
      c
      c      compute principal stresses as roots of cubic equation
      c
      c
      c      interactive evaluation of characteristic equation
      c
  205 write (6,210)
  210 format (/5x,'enter solution option:',
     1         /7x,'1 - evaluate characteristic equation',
     2         /7x,'2 - newton-raphson iteration',
     3         /7x,'3 - compute direction cosines',
     4         /7x,'4 - return for new problem option'/)
      read (5,*) ibr
      go to (215,220,245,20),ibr
      c
      c
  215 write (6,216)
  216 format (' Enter sigma (999 to stop)'/)
      read (5,*) sigg
      if (sigg.eq.999) go to 205
      ysg=ysig(sigg)
```

```
                              write (6,217)  sigg,ysg
                   217 format (' y(',g12.4,') = ',g12.4/)
                              go to 215
          c
          c        Newton-Raphson iteration to refine roots
          c
             220 kmax=10
             222 write (6,223)
             223 format (' Enter initial guess (999 to stop)'/)
                      read (5,*) sigpi
                      if (sigpi.eq.999) go to 205
                        do 230 k=1,kmax
                          ysg=ypsig(sigpi)
                          if (ysg.ne.0.) go to 226
                            write (6,225)
             225            format (' Zero slope - try another guess')
                            go to 222
             226          delta=ysig(sigpi)/ysg
                          sigpi=sigpi-delta
                           if (sigpi.ne.0) go to 228
                            if (delta.lt.0.01) go to 240
                            go to 230
             228          if (abs(delta/sigpi).lt.0.01) go to 240
             230       continue
                      write (6,*) ' No convergence for root'
                      go to 222
             240 write (6,241) sigpi,k
             241 format (' root =',g12.4,' (',i3,' iterations)')
                      go to 222
```

The functions ysig and ypsig to compute the stress function and its derivative are listed after the main program code.

Once all three principal stresses have been determined, the code can compute and print out the direction cosines of the three corresponding principal planes. In principle, the direction cosines could be computed by substituting each principal stress successively into $(\boldsymbol{\sigma} - \sigma_p\mathbf{I})\hat{\mathbf{n}} = \mathbf{0}$ (Eq. 3.40) and solving for the corresponding orientation vector $\hat{\mathbf{n}}$. (Only two of these equations are independent, so the condition $\hat{\mathbf{n}} \cdot \hat{\mathbf{n}} = 1$ must be used as well.) However, the code uses a simpler procedure given by Ugural.[1] If we denote the cofactors of the coefficient matrix of the foregoing equation as

$$a = \begin{vmatrix} (\sigma_y - \sigma_p) & \tau_{yz} \\ \tau_{yz} & (\sigma_z - \sigma_p) \end{vmatrix}$$

$$b = -\begin{vmatrix} \tau_{xy} & \tau_{yz} \\ \tau_{xz} & (\sigma_z - \sigma_p) \end{vmatrix}$$

$$c = \begin{vmatrix} \tau_{xy} & (\sigma_y - \sigma_p) \\ \tau_{xz} & \tau_{yz} \end{vmatrix}$$

[1] A.C. Ugural and S.K. Fenster, *Advanced Strength and Applied Elasticity,* Elsevier, New York, 1987.

and further define a parameter $k$ as

$$k = \frac{1}{\sqrt{a^2 + b^2 + c^2}}$$

the direction cosines $\hat{\mathbf{n}} = (l, m, n)$ corresponding to a particular choice of one of the principal stresses in these calculations are

$$l = ak, \qquad m = bk, \qquad n = ck$$

Inspection will show that this gives the required relation $l^2 + m^2 + n^2 = 1$.

The following section of the strs3d code prompts the user to enter the values of the three principal stresses and for each computes and prints the corresponding orientation vector $\hat{\mathbf{n}}$:

```
c
c
c       compute direction cosines and print results
c
  245 write (6,*) ' Enter 3 principal stresses'
      read (5,*) sigp
      write (6,250)
  250 format (//5x,'stress',12x,'l',9x,'m',9x,'n'/)
c
      do 270 i=1,3
c
          a2=(sig(2,2)-sigp(i))*(sig(3,3)-sigp(i))-(sig(2,3)*sig(2,3))
          b= sig(1,2)*(sig(3,3)-sigp(i))-(sig(1,3)*sig(2,3))
          c=(sig(1,2)*sig(2,3))-sig(1,3)*(sig(2,2)-sigp(i))
c
          akk=sqrt (a2*a2 + b*b + c*c)
          ak=0.
          if (abs(akk).gt. 0.001) ak=1./akk
c
          al=a2*ak
          am=b*ak
          an=c*ak
c
          write (6,260) sigp(i),al,am,an
  260     format (1x,f10.4,5x,3f10.4)
  270 continue
c
      go to 20
      end
c
c
      function ysig(sig)
      common /inv/ a(4)
      ysig=a(1)*sig**3+a(2)*sig*sig+a(3)*sig+a(4)
      return
      end
c
c
      function ypsig(sig)
      common /inv/ a(4)
```

```
            ypsig=3.*a(1)*sig*sig+2.*a(2)*sig+a(3)
            return
            end
c
c

            subroutine mult (a,b,l,m,n,c)
c
c           returns the matrix product c = a * b
c
            dimension a(l,m),b(m,n),c(l,n)
c
            do 20 i=1,l
            do 20 j=1,n
               cc=0.
               do 10 k=1,m
                   cc=cc+a(i,k)*b(k,j)
    10       continue
    20 c(i,j)=cc
            return
            end
```

The following is an example of str3d in use, in this case a simple example of pure shear ($\tau_{xy} = 1$, all other stresses zero). As was shown in Example 3.7, the principal stresses for this stress state (that is, the roots of the characteristic equation) are $\sigma_{p1} = 1$, $\sigma_{p2} = 0$, $\sigma_{p3} = -1$, and occur on planes 45° away from the $xy$ axes.

```
strs3d
     enter stress matrix by rows

0,1,0
1,0,0
0,0,0

     enter problem option:
       1 - transformation
       2 - principal values
       3 - restart
       4 - stop

1

     enter euler angles psi, theta, phi (degrees):
45,0,0


transformation matrix:                        stress:

     0.7071    0.7071    0.0000        1.0000    0.0000    0.0000
    -0.7071    0.7071    0.0000        0.0000   -1.0000    0.0000
     0.0000    0.0000    1.0000        0.0000    0.0000    0.0000

     enter problem option:
       1 - transformation
       2 - principal values
       3 - restart
       4 - stop
```

```
      enter solution option:
          1 - evaluate characteristic equation
          2 - newton-raphson iteration
          3 - compute direction cosines
          4 - return for new problem option

1
 enter sigma (999 to stop):
-5
 y(   -5.00000) =     -120.000
 enter sigma (999 to stop):
-2
 y(   -2.00000) =     -6.00000
 enter sigma (999 to stop):
-.5
 y(   -0.500000) =     0.375000
 enter sigma (999 to stop):
.5
 y(   0.500000) =     -0.375000
 enter sigma (999 to stop):
2
 y(    2.00000) =      6.00000
 enter sigma (999 to stop):
999

      enter solution option:
          1 - evaluate characteristic equation
          2 - newton-raphson iteration
          3 - compute direction cosines
          4 - return for new problem option

2
 enter initial guess (999 to stop):
-2
      root = -1.00     after  5 iterations
 enter initial guess (999 to stop):
.5
      root = -1.00     after  2 iterations
 enter initial guess (999 to stop):
-.1
      root =    0.     after  3 iterations
 enter initial guess (999 to stop):
.8
      root =  1.00     after  4 iterations
 enter initial guess (999 to stop):
999

      enter solution option:
          1 - evaluate characteristic equation
          2 - newton-raphson iteration
          3 - compute direction cosines
          4 - return for new problem option

3
```

```
enter 3 principal stresses:
-1,0,1

        stress          l          m          n

        -1.0000        0.7071     0.7071     0.0000
         0.0000        0.0000     0.0000     0.0000
         1.0000        0.7071    -0.7071     0.0000

        enter problem option:
          1 - transformation
          2 - principal values
          3 - restart
          4 - stop

        4
```

# ANALYSIS OF LAMINATED PLATES

*APPENDIX* G

## THE plate CODE

The laminate analysis procedure described in Section 4.7 has been coded in Fortran to make a small code named plate. This program is not as elaborate as others available commercially, but it does suffice for many situations and is useful for following how the calculations are carried out.

The code is user-prompting; that is, the user simply types in data requested by the computer. The computer begins by asking for the properties of the various ply types to be used in making up the laminate, followed by the stacking sequence from the bottom up. The code then carries out the calculations leading to the formation of the matrices in Eq. 4.57. The computer then asks for the applied tractions and bending moments (per unit of plate width), after which it calculates the resulting extensional strains and curvatures.

```
c       plate - a prompt-driven routine for laminated plate calculations
        dimension S(3,3),Sbar(3,3),Dbar(3,3),E(6,7),kwa(6),
     *           A(3,3),Ainv(3,3),R(3,3),Rinv(3,3),Et(6,6),
     *           temp1(3,3),temp2(3,3),temp3(3,3),
     *           eps0(3),xkappa(3),sigbar(3),sig(3),vtemp1(3),
     *           vtemp2(3),E1(10),E2(10),Gnu12(10),G12(10),thk(10),
     *           z(21),mat(20),theta(20),Dsave(3,3,20),Tsave(3,3,20)
       data R/1.,3*0.,1.,3*0.,2./,Rinv/1.,3*0.,1.,3*0.,.5/,
     *      E/42*0./

c----------------------------------------------------------------------
c       input material set selections

        i=1
10      write(6,20) i
20      format (' assign properties for lamina type ',i2,'...'/)

           write(6,*) 'enter modulus in fiber direction...'
           write(6,*) '   (enter -1 to stop): '
           read (5,*) E1(i)
           if (E1(i) .lt. 0.) go to 30
```

```fortran
      write(6,*) 'enter modulus in transverse direction: '
      read (5,*) E2(i)
      write(6,*) 'enter principal Poisson ratio: '
      read (5,*) Gnu12(i)

c     check for isotropy
      check=abs((E1(i)-E2(i))/E1(i))
      if (check.lt.0.001) then
         G12(i)=E1(i)/(2.*(1.+Gnu12(i)))
      else
         write(6,*) 'enter shear modulus: '
         read (5,*) G12(i)
      end if

      write(6,*) 'enter ply thickness: '
      read (5,*) thk(i)
      i=i+1

      go to 10

c-------------------------------------------------------------------
c     define layup

30    iply=1
      z(1)=0.

      write(6,*) 'define layup sequence, starting at bottom...'
      write(6,*) '   (use negative material set number to stop)'

40    write (6,50) iply
50    format (/' enter material set number for ply number',i3,': ')
      read (5,*) m
      if (m.lt.0) go to 60
         mat(iply)=m
         write(6,*) 'enter ply angle: '
         read (5,*) theta(iply)
         z(iply+1)=z(iply)+thk(m)
         iply=iply+1

      go to 40

c     compute boundary coordinates  (measured from centerline)

60    thick=z(iply)
      N = iply-1
      z0 = thick/2.
      np1=N+1

      do 70 i=1,np1
         z(i)=z(i)-z0
70    continue

c-------------------------------------------------------------------
c-------------------------------------------------------------------
c     loop over plies, form stiffness matrix

      do 110 iply=1,N
```

```
          m=mat(iply)
c         form lamina compliance in 1-2 directions (Eq. 3.54)


          S(1,1) = 1./E1(m)
          S(2,1) = -Gnu12(m) / E1(m)
          S(3,1) = 0.

          S(1,2) = S(2,1)
          S(2,2) = 1./E2(m)
          S(3,2) = 0.

          S(1,3) = 0.
          S(2,3) = 0.
          S(3,3) = 1./G12(m)


c---------------------------------------------------------
c         transform to x-y axes
c         obtain transformation matrix A (Eq. 3.27)



          thet = theta(iply) * 3.14159/180.

          sc = sin(thet)*cos(thet)
          s2 = (sin(thet))**2
          c2 = (cos(thet))**2

          A(1,1) = c2
          A(2,1) = s2
          A(3,1) = -1.*sc

          A(1,2) = s2
          A(2,2) = c2
          A(3,2) = sc

          A(1,3) = 2.*sc
          A(2,3) = -2.*sc
          A(3,3) = c2 - s2

c         inverse transformation matrix

          Ainv(1,1) = c2
          Ainv(2,1) = s2
          Ainv(3,1) = sc

          Ainv(1,2) = s2
          Ainv(2,2) = c2
          Ainv(3,2) = -1.*sc

          Ainv(1,3) = -2.*sc
          Ainv(2,3) = 2.*sc
          Ainv(3,3) = c2 - s2

c         transformed compliance matrix Sbar (Eq. 3.55)
c              [Sbar] = [R][T]-1[R]-1[S][T]
```

```
          call matmul (3,3,3,3,3,3,R,Ainv,temp1)
          call matmul (3,3,3,3,3,3,temp1,Rinv,temp2)
          call matmul (3,3,3,3,3,3,temp2,S,temp3)
          call matmul (3,3,3,3,3,3,temp3,A,Sbar)


c-----------------------------------------------------------------
c          invert Sbar (transformed compliance matrix) to obtain
c              Dbar (transformed stiffness matrix)
c          start by setting Dbar = Sbar, then call inversion routine

          do 80 i=1,3
          do 80 j=1,3
             Dbar(i,j)=Sbar(i,j)
80        continue

          call matinv(isol,idsol,3,3,Dbar,3,kwa,det)

c        save Dbar and Tinv matrices

          do 90 i=1,3
          do 90 j=1,3
             Dsave(i,j,iply)=Dbar(i,j)
             Tsave(i,j,iply)=Tinv(i,j)
90        continue

c        add to laminate stiffness matrix
c        Eqs. 4.52, 4.53, 4.56

          ip1=iply+1
          z1=        (z(ip1)   -z(iply)   )
          z2=     0.5*(z(ip1)**2-z(iply)**2)
          z3=(1./3.)*(z(ip1)**3-z(iply)**3)
          do 100 i=1,3
          do 100 j=1,3
             E(i,j)    = E(i,j) +      Dbar(i,j)*z1
             xx        =              Dbar(i,j)*z2
             E(i+3,j)  = E(i+3,j) +   xx
             E(i,j+3)  = E(i,j+3) +   xx
             E(i+3,j+3)= E(i+3,j+3) + Dbar(i,j)*z3
100       continue

c        end loop over plies; stiffness matrix now formed
110       continue


c-----------------------------------------------------------------
c-----------------------------------------------------------------
c     output stiffness matrix

      write(6,120)
120   format(/' laminate stiffness matrix:',/)
      do 140 i=1,6
          write(6,130) (e(i,j),j=1,6)
130       format (4x,3e12.4,2x,3d12.4)
          if (i.eq.3) write(6,*)
140   continue
```

```
c----------------------------------------------------------
c       obtain and print laminate compliance matrix

c       do 300 i=1,6
c       do 300 j=1,6
c          Et(i,j)=E(i,j)
c300    continue

c       call matinv(isol,idsol,6,6,Et,6,kwa,det)

c       write(6,310)
c310    format(/' laminate compliance matrix:',/)
c       do 320 i=1,6
c          write(6,130) (Et(i,j),j=1,6)
c          if (i.eq.3) write(6,*)
c320    continue

c----------------------------------------------------------
c       obtain traction-moment vector

        write(6,*)
        write(6,*) 'input tractions and moments...'
        write(6,*)
        write(6,*) '    Nx: '
        read (5,*) e(1,7)
        write(6,*) '    Ny: '
        read (5,*) e(2,7)
        write(6,*) '   Nxy: '
        read (5,*) e(3,7)
        write(6,*) '    Mx: '
        read (5,*) e(4,7)
        write(6,*) '    My: '
        read (5,*) e(5,7)
        write(6,*) '   Mxy: '
        read (5,*) e(6,7)

c----------------------------------------------------------
c       solve resulting system; print strains and rotations

        call matinv(isol,idsol,6,7,e,6,kwa,det)
        write(6,150) (e(i,7),i=1,6)
150     format(/' midplane strains:',//3x,'eps-xx =',e12.4,
     *     /3x,'eps-yy =',e12.4,/3x,'eps-xy =',e12.4,
     *     //' rotations:',//3x,'kappa-xx =',e12.4,
     *     /3x,'kappa-yy= ',e12.4,/3x,'kappa-xy =',e12.4//)

c----------------------------------------------------------
c       compute ply stresses

        write(6,160)
160     format (/' stresses:',/2x,'ply',5x,'sigma-1',
     *           5x,'sigma-2',4x,'sigma-12'/)

        do 210 iply=1,N

           do 180 i=1,3
              eps0(i)=e(i,7)
```

```
                                xkappa(i)=e(i+3,7)
                                do 170 j=1,3
                                    Dbar(i,j)=Dsave(i,j,iply)
                                    Tinv(i,j)=Tsave(i,j,iply)
        170                     continue
        180             continue

                        call matmul (3,3,3,3,3,1,Qbar,eps0,vtemp1)
                        call matmul (3,3,3,3,3,1,Qbar,xkappa,vtemp2)

                        zctr=(z(iply)+z(iply+1))/2.
                        do 190 i=1,3
                            sigbar(i) = vtemp1(i) + zctr*vtemp2(i)
        190             continue

                        call matmul (3,3,3,3,3,1,Tinv,sigbar,sig)
                        write(6,200) iply,sig
        200             format (3x,i2,3e12.4)

        210     continue
                stop
                end
```

In this code, `matmul` and `matinv` are library routines for matrix multiplication and inversion.

## NUMERICAL EXAMPLE

Following is a log of a typical user interaction with the `plate` code, in this case for a $0°$–$90°$–$0°$ layup of graphite-epoxy with a unit traction applied in the *x*-direction.

```
plate

assign properties for lamina type 1...

enter modulus in fiber direction...
   (enter -1 to stop): 230e9
enter modulus in transverse direction: 6.6e9
enter principal Poisson ratio: .25
enter shear modulus: 4.8e9
enter ply thickness: .13e-3

assign properties for lamina type 2...

enter modulus in fiber direction...
   (enter -1 to stop): -1
define layup sequence, starting at bottom...
   (use negative material set number to stop)

enter material set number for ply number 1: 1
enter ply angle: 0

enter material set number for ply number 2: 1
```

```
enter ply angle: 90

enter material set number for ply number 3: 1
enter ply angle: 0

enter material set number for ply number 4: -1


laminate stiffness matrix:

      .6077E+08    .6447E+06    .7648E+00    .0000D+00    .0000D+00    .0000D+00
      .6447E+06    .3167E+08    .3611E+02    .0000D+00    .0000D+00    .0000D+00
      .7648E+00    .3611E+02    .1872E+07    .0000D+00    .0000D+00    .0000D+00

     -.2888E-05    .4390E-06    .0000E+00    .1098D+01    .8171D-02    .1077D-08
      .4390E-06    .2780E-05    .0000E+00    .8171D-02    .7366D-01    .5086D-07
      .0000E+00    .0000E+00    .3068E-05    .1077D-08    .5086D-07    .2373D-01


input tractions and moments...

   Nx: 1
   Ny: 0
  Nxy: 0
   Mx: 0
   My: 0
  Mxy: 0


midplane strains:

   eps-xx =    .1646E-07
   eps-yy =   -.3350E-09
   eps-xy =   -.2614E-15


rotations:

   kappa-xx =    .4410E-13
   kappa-yy=   -.9036E-13
   kappa-xy =    .2255E-18


stresses:
   ply       sigma-1      sigma-2      sigma-12

    1     .3792E+04    .2499E+02   -.1255E-05
    2    -.4998E+02    .1083E+03    .1981E-03
    3     .3792E+04    .2499E+02   -.1255E-05
Stop - Program terminated.
```

The x-direction stiffness $E_x$ of the laminate can be obtained by dividing the applied stress in that direction, $\sigma_x$, by the midplane strain $\epsilon_x^0$ computed by the code. Since the stress is related to the traction by $N_x = t\sigma_x$, where $t = 3 \times 0.13 \times 10^{-3}$ is the laminate thickness, we can write

$$E_x = \frac{\sigma_x}{\epsilon_x^0} = \frac{[1/(3 \times 0.13 \times 10^{-3})]}{0.1646 \times 10^{-7}} = 1.55 \times 10^9 \text{ Pa}$$

We can check this result by examining the values in the laminate stiffness matrix, where we see that $E_{1,1} = 0.6077 \times 10^8$. Since all tractions and bending moments other than $N_x$ were taken to be zero, matrix multiplication of Eq. 4.56 using these values would give

$$E_{1,1}\epsilon_x^0 = N_x = \sigma_x t$$

$$E_x = \frac{\sigma_x}{\epsilon_x^0} = \frac{E_{1,1}}{t} = \frac{0.6077 \times 10^8 \text{ N/m}}{3 \times 0.13 \times 10^{-3} \text{ m}} = 1.55 \times 10^9 \text{ Pa}$$

as before. Similar manipulations could provide the other constants ($E_y$, $G_{xy}$, $\nu_{xy}$, $\nu_{yx}$) as desired.

## TEMPERATURE EFFECTS

There are a number of improvements one might consider for the plate code described in the foregoing section: it could be extended to include interlaminar shear stresses between plies; it could incorporate a database of commercially available prepreg and core materials; or the user interface could be made "friendlier" and graphically oriented. Many such features are available in commercial codes, or could be added by the user, and will not be discussed further here. However, thermal expansion effects are so important in application that a laminate code almost *must* have this feature to be usable, and the general approach will be outlined here.

In general, an increase in temperature $\Delta T$ causes a thermal expansion given by the well-known relation $\epsilon_T = \alpha \Delta T$, where $\epsilon_T$ is the thermally induced strain and $\alpha$ is the coefficient of linear thermal expansion. This thermal strain is obtained without any need to apply stress, so that when Hooke's law is used to compute the stress from the strain, the thermal component is substracted first: $\sigma = E(\epsilon - \alpha \Delta T)$. The thermal expansion causes normal strain only, so shearing components of strain are unaffected. Equation 3.50 can thus be extended as

$$\boldsymbol{\sigma} = \mathbf{D}(\boldsymbol{\epsilon} - \boldsymbol{\epsilon}_T)$$

where the thermal strain vector in the 1-2 coordinate frame is

$$\boldsymbol{\epsilon}_T = \begin{Bmatrix} \alpha_1 \\ \alpha_2 \\ 0 \end{Bmatrix} \Delta T$$

Here $\alpha_1$ and $\alpha_2$ are the anisotropic thermal expansion coefficients in the fiber and transverse directions. Transforming to common $x$-$y$ axes, this relation becomes

$$\begin{Bmatrix} \sigma_x \\ \sigma_y \\ \tau_{xy} \end{Bmatrix} = \begin{bmatrix} \overline{D}_{11} & \overline{D}_{12} & \overline{D}_{13} \\ \overline{D}_{12} & \overline{D}_{22} & \overline{D}_{23} \\ \overline{D}_{13} & \overline{D}_{23} & \overline{D}_{33} \end{bmatrix} \left( \begin{Bmatrix} \epsilon_x \\ \epsilon_y \\ \gamma_{xy} \end{Bmatrix} - \begin{Bmatrix} \alpha_x \\ \alpha_y \\ \alpha_{xy} \end{Bmatrix} \Delta T \right) \qquad (G.1)$$

The subscripts on the $\overline{\mathbf{D}}$ elements refer to row and column positions within the stiffness matrix rather than coordinate directions; the overbar serves as a reminder that these elements refer to $x$-$y$ axes. The thermal expansion vector on the right-hand side, $\alpha = (\alpha_x, \alpha_y, \alpha_{xy})$, is essentially a strain vector, so it can be obtained from $(\alpha_1, \alpha_2, 0)$ as in Eq. 3.31:

$$\left\{ \begin{array}{c} \alpha_x \\ \alpha_y \\ \alpha_{xy} \end{array} \right\} = \mathbf{R}\mathbf{A}^{-1}\mathbf{R}^{-1} \left\{ \begin{array}{c} \alpha_1 \\ \alpha_2 \\ 0 \end{array} \right\}$$

Note that in the common $x$-$y$ direction, thermal expansion induces both normal and shearing strains.

The previous temperature-independent development can now be repeated, modified only by carrying along the thermal expansion terms. As before, the strain vector for any position $\mathbf{z}$ from the midplane is given in terms of the midplane strain $\boldsymbol{\epsilon}^0$ and curvature $\boldsymbol{\kappa}$ by

$$\boldsymbol{\epsilon} = \boldsymbol{\epsilon}^0 + \mathbf{z}\boldsymbol{\kappa}$$

The corresponding stress is then

$$\boldsymbol{\sigma} = \overline{\mathbf{D}}(\boldsymbol{\epsilon}^0 + \mathbf{z}\boldsymbol{\kappa} - \boldsymbol{\alpha}\Delta T)$$

Balancing the stresses against the applied tractions and moments as before, we obtain

$$\mathbf{N} = \int \boldsymbol{\sigma} \, d\mathbf{z} = \mathcal{A}\boldsymbol{\epsilon}^0 + \mathcal{B}\boldsymbol{\kappa} - \int \overline{\mathbf{D}}\boldsymbol{\alpha}\Delta T \, d\mathbf{z}$$

$$\mathbf{M} = \int \boldsymbol{\sigma}\mathbf{z} \, d\mathbf{z} = \mathcal{B}\boldsymbol{\epsilon}^0 + \mathcal{D}\boldsymbol{\kappa} - \int \overline{\mathbf{D}}\boldsymbol{\alpha}\Delta T\mathbf{z} \, d\mathbf{z}$$

This result is identical to that of Eqs. 4.51 and 4.54, other than the addition of the integrals representing the "thermal loads." This permits temperature-dependent problems to be handled by an "equivalent mechanical formulation"; the overall governing equations can be written as

$$\left\{ \begin{array}{c} \overline{\mathbf{N}} \\ \overline{\mathbf{M}} \end{array} \right\} = \left[ \begin{array}{cc} \mathcal{A} & \mathcal{B} \\ \mathcal{B} & \mathcal{D} \end{array} \right] \left\{ \begin{array}{c} \boldsymbol{\epsilon}^0 \\ \boldsymbol{\kappa} \end{array} \right\}, \quad \text{or} \quad \left\{ \begin{array}{c} \boldsymbol{\epsilon}^0 \\ \boldsymbol{\kappa} \end{array} \right\} = \left[ \begin{array}{cc} \mathcal{A} & \mathcal{B} \\ \mathcal{B} & \mathcal{D} \end{array} \right]^{-1} \left\{ \begin{array}{c} \overline{\mathbf{N}} \\ \overline{\mathbf{M}} \end{array} \right\} \qquad \text{(G.2)}$$

where the "equivalent thermal loads" are given as

$$\overline{\mathbf{N}} = \mathbf{N} + \int \overline{\mathbf{D}}\boldsymbol{\alpha}\Delta T \, d\mathbf{z}$$

$$\overline{\mathbf{M}} = \mathbf{M} + \int \overline{\mathbf{D}}\boldsymbol{\alpha}\Delta T\mathbf{z} \, d\mathbf{z}$$

The extension of the `plate` code to accommodate thermal effects thus consists of modifying the $6 \times 1$ loading vector by adding the two $3 \times 1$ vector integrals in the foregoing expression. This thermal loading vector can probably be handled most easily within the loop over the laminate plies, because the individual $\overline{\mathbf{D}}_i$ arrays are already available there.

# H STATISTICAL TABLES

## TABLE 1 NORMAL DISTRIBUTION

The probability density function $f(X)$ for the standard normal distribution has the form

$$f(X) = \frac{1}{\sqrt{2\pi}} e^{-X^2/2}$$

where $X$ is the difference between the random variable's value and the mean, expressed as a multiple of the standard deviation.

To find $f(X)$ for a given $X$, from the table, first locate the row containing the $X$ value to one decimal place. The value of $f(X)$ is in the column headed by the number in the second decimal place. For example, for $X = 1.25$, the value of $f(X)$ will be found in the column headed by ".05" on the row with "1.2" at the left; the value of $f(X)$ is found to be .1826. For negative $X$ values, ignore the minus sign.

| X | .00 | .01 | .02 | .03 | .04 | .05 | .06 | .07 | .08 | .09 |
|---|-----|-----|-----|-----|-----|-----|-----|-----|-----|-----|
| .0 | .3989 | .3989 | .3989 | .3988 | .3986 | .3984 | .3982 | .3980 | .3977 | .3973 |
| .1 | .3970 | .3965 | .3961 | .3956 | .3951 | .3945 | .3939 | .3932 | .3925 | .3918 |
| .2 | .3910 | .3902 | .3894 | .3885 | .3876 | .3867 | .3857 | .3847 | .3836 | .3825 |
| .3 | .3814 | .3802 | .3790 | .3778 | .3765 | .3752 | .3739 | .3725 | .3712 | .3697 |
| .4 | .3683 | .3668 | .3653 | .3637 | .3621 | .3605 | .3589 | .3572 | .3555 | .3538 |
| .5 | .3521 | .3503 | .3485 | .3467 | .3448 | .3429 | .3410 | .3391 | .3372 | .3352 |
| .6 | .3332 | .3312 | .3292 | .3271 | .3251 | .3230 | .3209 | .3187 | .3166 | .3144 |
| .7 | .3123 | .3101 | .3079 | .3056 | .3034 | .3011 | .2989 | .2966 | .2943 | .2920 |
| .8 | .2897 | .2874 | .2850 | .2827 | .2803 | .2780 | .2756 | .2732 | .2709 | .2685 |
| .9 | .2661 | .2637 | .2613 | .2589 | .2565 | .2541 | .2516 | .2492 | .2468 | .2444 |
| 1.0 | .2420 | .2396 | .2371 | .2347 | .2323 | .2299 | .2275 | .2251 | .2227 | .2203 |
| 1.1 | .2179 | .2155 | .2131 | .2107 | .2083 | .2059 | .2036 | .2012 | .1989 | .1965 |
| 1.2 | .1942 | .1919 | .1895 | .1872 | .1849 | .1826 | .1804 | .1781 | .1758 | .1736 |
| 1.3 | .1714 | .1691 | .1669 | .1647 | .1626 | .1604 | .1582 | .1561 | .1539 | .1518 |
| 1.4 | .1497 | .1476 | .1456 | .1435 | .1415 | .1394 | .1374 | .1354 | .1334 | .1315 |
| 1.5 | .1295 | .1276 | .1257 | .1238 | .1219 | .1200 | .1182 | .1163 | .1145 | .1127 |
| 1.6 | .1109 | .1092 | .1074 | .1057 | .1040 | .1023 | .1006 | .0989 | .0973 | .0957 |
| 1.7 | .0940 | .0925 | .0909 | .0893 | .0878 | .0863 | .0848 | .0833 | .0818 | .0804 |
| 1.8 | .0790 | .0775 | .0761 | .0748 | .0734 | .0721 | .0707 | .0694 | .0681 | .0669 |
| 1.9 | .0656 | .0644 | .0632 | .0620 | .0608 | .0596 | .0584 | .0573 | .0562 | .0551 |
| 2.0 | .0540 | .0529 | .0519 | .0508 | .0498 | .0488 | .0478 | .0468 | .0459 | .0449 |
| 2.1 | .0440 | .0431 | .0422 | .0413 | .0404 | .0396 | .0387 | .0379 | .0371 | .0363 |
| 2.2 | .0355 | .0347 | .0339 | .0332 | .0325 | .0317 | .0310 | .0303 | .0297 | .0290 |
| 2.3 | .0283 | .0277 | .0270 | .0264 | .0258 | .0252 | .0246 | .0241 | .0235 | .0229 |
| 2.4 | .0224 | .0219 | .0213 | .0208 | .0203 | .0198 | .0194 | .0189 | .0184 | .0180 |
| 2.5 | .0175 | .0171 | .0167 | .0163 | .0158 | .0154 | .0151 | .0147 | .0143 | .0139 |
| 2.6 | .0136 | .0132 | .0129 | .0126 | .0122 | .0119 | .0116 | .0113 | .0110 | .0107 |
| 2.7 | .0104 | .0101 | .0099 | .0096 | .0093 | .0091 | .0088 | .0086 | .0084 | .0081 |
| 2.8 | .0079 | .0077 | .0075 | .0073 | .0071 | .0069 | .0067 | .0065 | .0063 | .0061 |
| 2.9 | .0060 | .0058 | .0056 | .0055 | .0053 | .0051 | .0050 | .0048 | .0047 | .0046 |
| 3.0 | .0044 | .0043 | .0042 | .0040 | .0039 | .0038 | .0037 | .0036 | .0035 | .0034 |
| 3.1 | .0033 | .0032 | .0031 | .0030 | .0029 | .0028 | .0027 | .0026 | .0025 | .0025 |
| 3.2 | .0024 | .0023 | .0022 | .0022 | .0021 | .0020 | .0020 | .0019 | .0018 | .0018 |
| 3.3 | .0017 | .0017 | .0016 | .0016 | .0015 | .0015 | .0014 | .0014 | .0013 | .0013 |
| 3.4 | .0012 | .0012 | .0012 | .0011 | .0011 | .0010 | .0010 | .0010 | .0009 | .0009 |
| 3.5 | .0009 | .0008 | .0008 | .0008 | .0008 | .0007 | .0007 | .0007 | .0007 | .0006 |
| 3.6 | .0006 | .0006 | .0006 | .0005 | .0005 | .0005 | .0005 | .0005 | .0005 | .0004 |
| 3.7 | .0004 | .0004 | .0004 | .0004 | .0004 | .0004 | .0003 | .0003 | .0003 | .0003 |
| 3.8 | .0003 | .0003 | .0003 | .0003 | .0003 | .0002 | .0002 | .0002 | .0002 | .0002 |
| 3.9 | .0002 | .0002 | .0002 | .0002 | .0002 | .0002 | .0002 | .0002 | .0001 | .0001 |

# TABLE 2  CUMULATIVE NORMAL DISTRIBUTION

The cumulative distribution function $F(X)$ for the standard normal distribution has the form

$$F(X) = \int_{-\infty}^{X} \frac{1}{\sqrt{2\pi}} e^{-t^2/2} \, dt$$

$F(X)$ is the probability that the value of the normally distributed random variable is less than or equal to $X$ standard deviations above the mean. For negative $X$ values, subtract $F(|X|)$ from 1.

| X | .00 | .01 | .02 | .03 | .04 | .05 | .06 | .07 | .08 | .09 |
|---|-----|-----|-----|-----|-----|-----|-----|-----|-----|-----|
| .0 | .5000 | .5040 | .5080 | .5120 | .5160 | .5199 | .5239 | .5279 | .5319 | .5359 |
| .1 | .5398 | .5438 | .5478 | .5517 | .5557 | .5596 | .5636 | .5675 | .5714 | .5753 |
| .2 | .5793 | .5832 | .5871 | .5910 | .5948 | .5987 | .6026 | .6064 | .6103 | .6141 |
| .3 | .6179 | .6217 | .6255 | .6293 | .6331 | .6368 | .6406 | .6443 | .6480 | .6517 |
| .4 | .6554 | .6591 | .6628 | .6664 | .6700 | .6736 | .6772 | .6808 | .6844 | .6879 |
| .5 | .6915 | .6950 | .6985 | .7019 | .7054 | .7088 | .7123 | .7157 | .7190 | .7224 |
| .6 | .7257 | .7291 | .7324 | .7357 | .7389 | .7422 | .7454 | .7486 | .7517 | .7549 |
| .7 | .7580 | .7611 | .7642 | .7673 | .7704 | .7734 | .7764 | .7794 | .7823 | .7852 |
| .8 | .7881 | .7910 | .7939 | .7967 | .7995 | .8023 | .8051 | .8078 | .8106 | .8133 |
| .9 | .8159 | .8186 | .8212 | .8238 | .8264 | .8289 | .8315 | .8340 | .8365 | .8389 |
| 1.0 | .8413 | .8438 | .8461 | .8485 | .8508 | .8531 | .8554 | .8577 | .8599 | .8621 |
| 1.1 | .8643 | .8665 | .8686 | .8708 | .8729 | .8749 | .8770 | .8790 | .8810 | .8830 |
| 1.2 | .8849 | .8869 | .8888 | .8907 | .8925 | .8944 | .8962 | .8980 | .8997 | .9015 |
| 1.3 | .9032 | .9049 | .9066 | .9082 | .9099 | .9115 | .9131 | .9147 | .9162 | .9177 |
| 1.4 | .9192 | .9207 | .9222 | .9236 | .9251 | .9265 | .9279 | .9292 | .9306 | .9319 |
| 1.5 | .9332 | .9345 | .9357 | .9370 | .9382 | .9394 | .9406 | .9418 | .9429 | .9441 |
| 1.6 | .9452 | .9463 | .9474 | .9484 | .9495 | .9505 | .9515 | .9525 | .9535 | .9545 |
| 1.7 | .9554 | .9564 | .9573 | .9582 | .9591 | .9599 | .9608 | .9616 | .9625 | .9633 |
| 1.8 | .9641 | .9649 | .9656 | .9664 | .9671 | .9678 | .9686 | .9693 | .9699 | .9706 |
| 1.9 | .9713 | .9719 | .9726 | .9732 | .9738 | .9744 | .9750 | .9756 | .9761 | .9767 |
| 2.0 | .9772 | .9778 | .9783 | .9788 | .9793 | .9798 | .9803 | .9808 | .9812 | .9817 |
| 2.1 | .9821 | .9826 | .9830 | .9834 | .9838 | .9842 | .9846 | .9850 | .9854 | .9857 |
| 2.2 | .9861 | .9864 | .9868 | .9871 | .9875 | .9878 | .9881 | .9884 | .9887 | .9890 |
| 2.3 | .9893 | .9896 | .9898 | .9901 | .9904 | .9906 | .9909 | .9911 | .9913 | .9916 |
| 2.4 | .9918 | .9920 | .9922 | .9925 | .9927 | .9929 | .9931 | .9932 | .9934 | .9936 |
| 2.5 | .9938 | .9940 | .9941 | .9943 | .9945 | .9946 | .9948 | .9949 | .9951 | .9952 |
| 2.6 | .9953 | .9955 | .9956 | .9957 | .9959 | .9960 | .9961 | .9962 | .9963 | .9964 |
| 2.7 | .9965 | .9966 | .9967 | .9968 | .9969 | .9970 | .9971 | .9972 | .9973 | .9974 |
| 2.8 | .9974 | .9975 | .9976 | .9977 | .9977 | .9978 | .9979 | .9979 | .9980 | .9981 |
| 2.9 | .9981 | .9982 | .9982 | .9983 | .9984 | .9984 | .9985 | .9985 | .9986 | .9986 |
| 3.0 | .9987 | .9987 | .9987 | .9988 | .9988 | .9989 | .9989 | .9989 | .9990 | .9990 |
| 3.1 | .9990 | .9991 | .9991 | .9991 | .9992 | .9992 | .9992 | .9992 | .9993 | .9993 |
| 3.2 | .9993 | .9993 | .9994 | .9994 | .9994 | .9994 | .9994 | .9995 | .9995 | .9995 |
| 3.3 | .9995 | .9995 | .9995 | .9996 | .9996 | .9996 | .9996 | .9996 | .9996 | .9997 |
| 3.4 | .9997 | .9997 | .9997 | .9997 | .9997 | .9997 | .9997 | .9997 | .9997 | .9998 |

In the following table the third row indicates the probability that the value of the normally distributed random variable will differ from the mean by more than $X$.

| X | 1.282 | 1.645 | 1.960 | 2.326 | 2.576 | 3.090 | 3.291 | 3.891 | 4.417 |
|---|-------|-------|-------|-------|-------|-------|-------|-------|-------|
| $F(X)$ | .90 | .95 | .975 | .99 | .995 | .999 | .9995 | .99995 | .999995 |
| $2[1 - F(X)]$ | .20 | .10 | .05 | .02 | .01 | .002 | .001 | .0001 | .00001 |

# TABLE 3 CHI-SQUARE TABLE

Maximum $\chi^2$ values for $\alpha =$

| Degrees of Freedom | .995 | .990 | .975 | .950 | .900 | .750 | .500 | .250 | .100 | .050 | .025 | .010 | .005 |
|---|---|---|---|---|---|---|---|---|---|---|---|---|---|
| 1 | .0000393 | .000157 | .000982 | .00393 | .0158 | .102 | .455 | 1.32 | 2.71 | 3.84 | 5.02 | 6.63 | 7.88 |
| 2 | .0100 | .0201 | .0506 | .103 | .211 | .575 | 1.39 | 2.77 | 4.61 | 5.99 | 7.38 | 9.21 | 10.6 |
| 3 | .0717 | .115 | .216 | .352 | .584 | 1.21 | 2.37 | 4.11 | 6.25 | 7.81 | 9.35 | 11.3 | 12.8 |
| 4 | .207 | .297 | .484 | .711 | 1.06 | 1.92 | 3.36 | 5.39 | 7.78 | 9.49 | 11.1 | 13.3 | 14.9 |
| 5 | .412 | .554 | .831 | 1.15 | 1.61 | 2.67 | 4.35 | 6.63 | 9.24 | 11.1 | 12.8 | 15.1 | 16.7 |
| 6 | .676 | .872 | 1.24 | 1.64 | 2.20 | 3.45 | 5.35 | 7.84 | 10.6 | 12.6 | 14.4 | 16.8 | 18.5 |
| 7 | .989 | 1.24 | 1.69 | 2.17 | 2.83 | 4.25 | 6.35 | 9.04 | 12.0 | 14.1 | 16.0 | 18.5 | 20.3 |
| 8 | 1.34 | 1.65 | 2.18 | 2.73 | 3.49 | 5.07 | 7.34 | 10.2 | 13.4 | 15.5 | 17.5 | 20.1 | 22.0 |
| 9 | 1.73 | 2.09 | 2.70 | 3.33 | 4.17 | 5.90 | 8.34 | 11.4 | 14.7 | 16.9 | 19.0 | 21.7 | 23.6 |
| 10 | 2.16 | 2.56 | 3.25 | 3.94 | 4.87 | 6.74 | 9.34 | 12.5 | 16.0 | 18.3 | 20.5 | 23.2 | 25.2 |
| 11 | 2.60 | 3.05 | 3.82 | 4.57 | 5.58 | 7.58 | 10.3 | 13.7 | 17.3 | 19.7 | 21.9 | 24.7 | 26.8 |
| 12 | 3.07 | 3.57 | 4.40 | 5.23 | 6.30 | 8.44 | 11.3 | 14.8 | 18.5 | 21.0 | 23.3 | 26.2 | 28.3 |
| 13 | 3.57 | 4.11 | 5.01 | 5.89 | 7.04 | 9.30 | 12.3 | 16.0 | 19.8 | 22.4 | 24.7 | 27.7 | 29.8 |
| 14 | 4.07 | 4.66 | 5.63 | 6.57 | 7.79 | 10.2 | 13.3 | 17.1 | 21.1 | 23.7 | 26.1 | 29.1 | 31.3 |
| 15 | 4.60 | 5.23 | 6.26 | 7.26 | 8.55 | 11.0 | 14.3 | 18.2 | 22.3 | 25.0 | 27.5 | 30.6 | 32.8 |
| 16 | 5.14 | 5.81 | 6.91 | 7.96 | 9.31 | 11.9 | 15.3 | 19.4 | 23.5 | 26.3 | 28.8 | 32.0 | 34.3 |
| 17 | 5.70 | 6.41 | 7.56 | 8.67 | 10.1 | 12.8 | 16.3 | 20.5 | 24.8 | 27.6 | 30.2 | 33.4 | 35.7 |
| 18 | 6.26 | 7.01 | 8.23 | 9.39 | 10.9 | 13.7 | 17.3 | 21.6 | 26.0 | 28.9 | 31.5 | 34.8 | 37.2 |
| 19 | 6.84 | 7.63 | 8.91 | 10.1 | 11.7 | 14.6 | 18.3 | 22.7 | 27.2 | 30.1 | 32.9 | 36.2 | 38.6 |
| 20 | 7.43 | 8.26 | 9.59 | 10.9 | 12.4 | 15.5 | 19.3 | 23.8 | 28.4 | 31.4 | 34.2 | 37.6 | 40.0 |
| 21 | 8.03 | 8.90 | 10.3 | 11.6 | 13.2 | 16.3 | 20.3 | 24.9 | 29.6 | 32.7 | 35.5 | 38.9 | 41.4 |
| 22 | 8.64 | 9.54 | 11.0 | 12.3 | 14.0 | 17.2 | 21.3 | 26.0 | 30.8 | 33.9 | 36.8 | 40.3 | 42.8 |
| 23 | 9.26 | 10.2 | 11.7 | 13.1 | 14.8 | 18.1 | 22.3 | 27.1 | 32.0 | 35.2 | 38.1 | 41.6 | 44.2 |
| 24 | 9.89 | 10.9 | 12.4 | 13.8 | 15.7 | 19.0 | 23.3 | 28.2 | 33.2 | 36.4 | 39.4 | 43.0 | 45.6 |
| 25 | 10.5 | 11.5 | 13.1 | 14.6 | 16.5 | 19.9 | 24.3 | 29.3 | 34.4 | 37.7 | 40.6 | 44.3 | 46.9 |
| 26 | 11.2 | 12.2 | 13.8 | 15.4 | 17.3 | 20.8 | 25.3 | 30.4 | 35.6 | 38.9 | 41.9 | 45.6 | 48.3 |
| 27 | 11.8 | 12.9 | 14.6 | 16.2 | 18.1 | 21.7 | 26.3 | 31.5 | 36.7 | 40.1 | 43.2 | 47.0 | 49.6 |
| 28 | 12.5 | 13.6 | 15.3 | 16.9 | 18.9 | 22.7 | 27.3 | 32.6 | 37.9 | 41.3 | 44.5 | 48.3 | 51.0 |
| 29 | 13.1 | 14.3 | 16.0 | 17.7 | 19.8 | 23.6 | 28.3 | 33.7 | 39.1 | 42.6 | 45.7 | 49.6 | 52.3 |
| 30 | 13.8 | 15.0 | 16.8 | 18.5 | 20.6 | 24.5 | 29.3 | 34.8 | 40.3 | 43.8 | 47.0 | 50.9 | 53.7 |

# TABLE 4  *t*-DISTRIBUTION

| Degrees of Freedom | Percentile | | | | | | |
|---|---|---|---|---|---|---|---|
| | 50 | 80 | 90 | 95 | 98 | 99 | 99.9 |
| 1 | 1.000 | 3.078 | 6.314 | 12.706 | 31.821 | 63.657 | 636.610 |
| 2 | .816 | 1.886 | 2.920 | 4.303 | 6.965 | 9.925 | 31.598 |
| 3 | .765 | 1.638 | 2.353 | 3.182 | 4.541 | 5.841 | 12.941 |
| 4 | .741 | 1.533 | 2.132 | 2.776 | 3.747 | 4.604 | 8.610 |
| 5 | .727 | 1.476 | 2.015 | 2.571 | 3.365 | 4.032 | 6.859 |
| 6 | .718 | 1.440 | 1.943 | 2.447 | 3.143 | 3.707 | 5.959 |
| 7 | .711 | 1.415 | 1.895 | 2.365 | 2.998 | 3.499 | 5.405 |
| 8 | .706 | 1.397 | 1.860 | 2.306 | 2.896 | 3.355 | 5.041 |
| 9 | .703 | 1.383 | 1.833 | 2.262 | 2.821 | 3.250 | 4.781 |
| 10 | .700 | 1.372 | 1.812 | 2.228 | 2.764 | 3.169 | 4.587 |
| 11 | .697 | 1.363 | 1.796 | 2.201 | 2.718 | 3.106 | 4.437 |
| 12 | .695 | 1.356 | 1.782 | 2.179 | 2.681 | 3.055 | 4.318 |
| 13 | .694 | 1.350 | 1.771 | 2.160 | 2.650 | 3.012 | 4.221 |
| 14 | .692 | 1.345 | 1.761 | 2.145 | 2.624 | 2.977 | 4.140 |
| 15 | .691 | 1.341 | 1.753 | 2.131 | 2.602 | 2.947 | 4.073 |
| 16 | .690 | 1.337 | 1.746 | 2.120 | 2.583 | 2.921 | 4.015 |
| 17 | .689 | 1.333 | 1.740 | 2.110 | 2.567 | 2.898 | 3.965 |
| 18 | .688 | 1.330 | 1.734 | 2.101 | 2.552 | 2.878 | 3.922 |
| 19 | .688 | 1.328 | 1.729 | 2.093 | 2.539 | 2.861 | 3.883 |
| 20 | .687 | 1.325 | 1.725 | 2.086 | 2.528 | 2.845 | 3.850 |
| 21 | .686 | 1.323 | 1.721 | 2.080 | 2.518 | 2.831 | 3.819 |
| 22 | .686 | 1.321 | 1.717 | 2.074 | 2.508 | 2.819 | 3.792 |
| 23 | .685 | 1.319 | 1.714 | 2.069 | 2.500 | 2.807 | 3.767 |
| 24 | .685 | 1.318 | 1.711 | 2.064 | 2.492 | 2.797 | 3.745 |
| 25 | .684 | 1.316 | 1.708 | 2.060 | 2.485 | 2.787 | 3.725 |
| 26 | .684 | 1.315 | 1.706 | 2.056 | 2.479 | 2.779 | 3.707 |
| 27 | .684 | 1.314 | 1.703 | 2.052 | 2.473 | 2.771 | 3.690 |
| 28 | .683 | 1.313 | 1.701 | 2.048 | 2.467 | 2.763 | 3.674 |
| 29 | .683 | 1.311 | 1.699 | 2.045 | 2.462 | 2.756 | 3.659 |
| 30 | .683 | 1.310 | 1.697 | 2.042 | 2.457 | 2.750 | 3.646 |
| 40 | .681 | 1.303 | 1.684 | 2.021 | 2.423 | 2.704 | 3.551 |
| 60 | .679 | 1.296 | 1.671 | 2.000 | 2.390 | 2.660 | 3.460 |
| 120 | .677 | 1.289 | 1.658 | 1.980 | 2.358 | 2.617 | 3.373 |
| ∞ | .674 | 1.282 | 1.645 | 1.960 | 2.326 | 2.576 | 3.291 |

# PROBABILITY PAPER

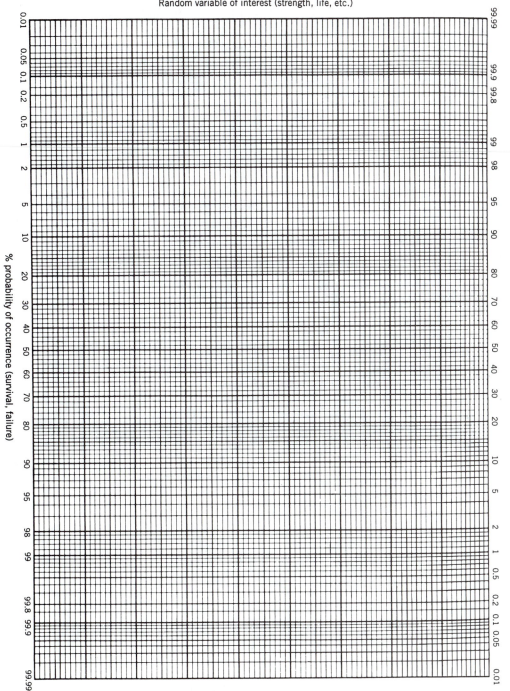

Random variable of interest (strength, life, etc.)

% probability of occurrence (survival, failure)

% probability of occurrence (survival, failure)

Random variable of interest (strength, life, etc.)

# INDEX